THE WORK ENVIRONMENT

VOLUME TWO

Healthcare, Laboratories, and Biosafety

Doan J. Hansen
EDITOR

Library of Congress Cataloging-in-Publication Data

(Revised for vol. 2)

The Work environment.

Includes bibliographical references and index.
Contents: v. 1. Occupational health fundamentals.--
v. 2. Health care, laboratories, and biosafety.
1. Industrial hygiene. I. Hansen, Doan J.
RC963.W67 1991 613.6′2 91-8213
ISBN 0-87371-303-6 (v. 1)
ISBN 0-87371-392-3 (v. 2)

Direct all inquiries to CRC Press, Inc., 2000 Corporate Blvd., N.W., Boca Raton, Florida, 33431.

PRINTED IN THE UNITED STATES OF AMERICA
1 2 3 4 5 6 7 8 9 0

Printed on acid-free paper

Introduction

Early humans, as hunters and gatherers, were by necessity opportunists. The same must have been true with respect to early scientific thinking and the practice of medicine. Chance observations were used to advantage; for example, juices obtained from crushing certain berries of other organisms could be used as dyes or to provide medical comfort, etc. It can be said that the first laboratories were the everyday environments in which primitive man lived. Finding a meal, or not becoming the meal of some contemporary predator, were among the hazards that were faced.

As time passed, science became formalized. Because of the limitations of the available technology, science was defined in a manner consistent with everyday observations of things. For example, Artistotle (384–322 BC), in *Meterologica*, defined everything on earth as being made from the four elements: *air, fire, earth,* and *water*. There were also four different properties (*hot, cold, moist,* and *dry*) which, combined with the ''elements,'' were thought to constitute everything.

Much early science was based on religion, superstition, and desire, not on natural law. Medieval alchemists tried in vain to turn lead into gold. Scientists, chemists, alchemists, philosophers, and physicians often attributed the phenomena they observed to the whim of gods or demons. Beginning around the year 1400, alchemists (or anyone else suspected of practicing ''magic'') was an enemy of the Church because of the potential conflicts.

Physicians of the time felt that illness was caused by imbalances of the four body humors: *blood, black bile, yellow bile,* and *phlegm*. They seldom considered that disease could be caused by external agents. In general, prayer, pilgrimages, and superstitions were employed along with alleged medical remedies, like bleeding, to cure illness. The Black Death (bubonic plague), which started in 1347, became a pandemic which easily exceeded the ability of science to control it. Within a year or two the plague killed perhaps one quarter of the population at that time. The likelihood of contracting the Black Death was known to physicians and scientists of the day; however, they had no valid methods with which to protect themselves (let alone their patients).

In 1676 Anton van Leeuwenhoek was the first to use a microscope to discover the ''animalcules'' that existed in stagnant water. In 1773, Otto Frederick Miller observed *bacilli* (from the Latin ''little rods'') and *spirilla* (named for their spiral shape). Among his other work Louis Pasteur, in the 1860s, used the microscope and discovered that bacteria had infected the silkworms and their food, mulberry, adversely affecting the French silk industry. From his research, Pasteur developed the germ theory of disease: disease is not caused by the gods, but by some hazardous agent, whether a virus, bacteria, chemical radiation, or a physical stressor.

Today we are usually aware at least in part of what the cause of a disease is even when science has not found a solution or cure. As in the past, science and society today are rife with puzzles which may outlive those working on them.

Some scientists, healthcare practitioners, and others, may contract and needlessly suffer from the ailments they treat or research. Simply stated, anyone who provides medical care, or who works as a scientist or technician in a laboratory, faces occupational hazards which may be subtle but are as real and dangerous to them as black lung disease is to a coal miner.

Scientific research in today's world is unique, complex, and extraordinary. In medicine, in healthcare, and in the social sciences, any facet of a specialty addresses its own unusual issues. There are many, many potential hazards (such as HIV, hepatitis, and tuberculosis) that these professions have in common. Also in common are the problems of accidental chemical or biological spills, and medical waste disposal; there are regulations, government guidelines, and control measures with which to comply.

This volume of *The Work Environment: Healthcare, Laboratories and Biosafety* discusses occupational hazards which may be encountered by the very diverse populations who work in laboratories, or who treat or work with patients, clients, inmates, or long-term care populations in any capacity. These chapters, along with an extensive glossary, provide contemporary guidance on topical issues.

Doan J. Hansen, PhD, MPH, CIH, is an industrial hygienist engaged in Hazard Assessment and Chemical Emergency Response Planning at Brookhaven National Laboratory, Upton, NY. He is also the acting director of the Occupational Safety and Health Program at New York University, a clinical assistant professor in Allied Health Sciences for the State University of New York at Stony Brook, and a consultant. He has an MPH in Environmental and Industrial Health and a PhD in Industrial Health, both from The University of Michigan, and a BA from Albion College. He is a certified industrial hygienist whose professional activities include participation on the AIHA Committee on Emergency Response Planning, the AIHA Committee on Exposure Assessment, including chairing its Subcommittee on Acute Exposure Assessment; and is a member of the DOE Subcommittee on Consequence Assessment and Protective Actions (SCAPA). Dr. Hansen also edited *The Work Environment, Volume One: Occupational Health Fundamentals,* and is the series editor and originator.

Contributors

Arthur F. Brings, CSP, is director of Environmental Health and Safety at Cold Spring Harbor Laboratory, Cold Spring Harbor, NY. He previously held positions in occupational health and safety at New York University Medical Center and at the State University of New York at Stony Brook. He holds a BA in biology from Long Island University and is a certified safety professional. He is an active member of the American Society of Safety Engineers and has served as president of the Long Island Chapter.

Thomas J. Cuthel, BS, has received his BS in chemistry from Manhattan College (1980) and a Certificate in Environmental/Occupational Health and Safety from the State University of New York at Stony Brook (1992). He has worked for the Department of Environmental Health and Safety at SUNY Stony Brook since 1984, his most recent position as chemical safety laboratory manager. Previously, he worked for the New York Institute of Technology in the Life Sciences Department.

Marc A. Gomez, MPH, CIH, CSP, is currently director of Safety and Occupational Health for the North Broward Hospital District, a multi-hospital system in Broward County, FL. Prior to this position, he had been an industrial hygiene and safety engineer with the IBM Corporation in Boca Raton, FL. He received his BS in biology from Tulane University in 1979 and his MPH from The University of Michigan in 1981. He is currently a member of the AIHA Biohazards Committee. He is also a past president of the Florida Local Section, AIHA, and is currently a director of the organization. His credentials include being a certified industrial hygienist, certified safety professional, and a certified healthcare risk manager. He also has worked as a consultant for a number of organizations in indoor air quality and ergonomics.

Audra S. Gomez, BA, RDH, is currently a dental hygienist practicing in South Florida. She lectures to various dental organizations on subjects related to periodontics. She has also developed a comprehensive OSHA Compliance Manual for Dental Offices and is working as a consultant in this area. She received her BA from Nova University in 1992 and her AS in dental hygiene from Palm Beach Community College in 1981.

Neil Hawkins, ScD, CIH, studied nuclear engineering and health physics at Georgia Tech and later earned a Masters in industrial hygiene and a Doctorate in environmental health sciences from Harvard University, School of Public

Health. Neil joined The Dow Chemical Company in 1988 in the Health and Environmental Sciences Department. Currently, he is environmental specialist in chemicals and metals environmental affairs, where he supports CHEMAWARE Enhanced Environmental and Product Stewardship and provides environmental issues management for the Chemical Intermediates Business. This includes products such as ethylene oxide, propylene oxide, glycols, alkanolamines, ethyleneamines, acetone and phenol. Additionally, Neil is vice chair of Dow's Indoor Air Issues Management Team. He has published many risk assessment papers, and also has edited a book on occupational exposure assessment strategies. He has given numerous technical presentations, serves on the Editorial Board of the American Industrial Hygiene Association Journal, and is certified in the comprehensive practice of industrial hygiene.

Fredrick T. Horn, BS, AAS, RN, received his BS degree in biology and chemistry from the State University of New York at Stony Brook in 1975 and his AAS in marine biology environmental sciences from Suffolk Community College in 1971. He has held corporate industrial hygiene positions with the Long Island Lighting Company and with Grumman Corporation including overseas assignments. He has been a public health officer in the USAF stationed at the Suffolk County Airport at Westhampton Beach, Long Island, since 1979, and has been responsible for the application of industrial hygiene to USAF personnel and public health. He is presently employed as a project engineer industrial hygienist at Brookhaven National Laboratory in Upton, NY.

Stan Lengerich, MS, graduated from Ball State University with a bachelor's degree in environmental science and from The University of Michigan with a master's degree in industrial health. He spent four years working for The Dow Company and Merrell Dow Pharmaceuticals and as industrial hygienist. He is currently employed as an industrial hygienist at Eli Lilly and Company. In both graduate school and industry, Mr. Lengerich has been involved with numerous projects involving the research, design, and maintenance of laboratory ventilation systems.

John Marchese, MS, BS, is an assistant director of Environmental Health and Safety for the State University of New York at Stony Brook. His responsibility is for the Safety Program for the University Hospital. John has been at Stony Brook for 12 years. He received a BS in biology, an MS in environmental engineering and, most recently, a certificate in waste management, all from SUNY Stony Brook. John has lectured at many seminars including the Third International Conference on Nosocomial Infection in 1990. In addition, he is an adjunct professor for the Continuing Education Department at SUNY Stony Brook in the Occupational Safety and Health Certificate Program.

George B. Marshall, MS, CSP, PE, has been director of the Department of Environmental Health and Safety at the State University of New York at Stony Brook since 1975. Before working for the University at Stony Brook he had a thirteen-year career in Loss Prevention with major insurance carriers in the New York metropolitan area. Mr. Marshall has lectured both in the United States and in the People's Republic of China on a broad spectrum of safety issues. Programs developed under Mr. Marshall's direction have captured the Award of Honor from the National Safety Council's Campus Safety Association.

Contents

PART I

Laboratory and Clinical Environments

Part I, **Laboratory and Clinical Environments,** contains the following chapters:

Chapter 1, **The Laboratory Safety Standard,** discusses 29 CFR 1910.1450, "Occupational Exposure to Hazardous Chemicals in Laboratories." This standard is the laboratory version of Worker Right to Know. The chapter breaks down the Laboratory Safety Standard, as it is more commonly known, into component parts (including the Chemical Hygiene Plan), explains them, and makes compliance with the standard a less daunting prospect. The standard itself is included as Appendix A to this volume.

Chapter 2, **Working With Biohazards,** examines the use of biohazardous agents in the laboratory. This chapter includes biosafety principles and explanations of concepts such as pathogenic, nonpathogenic, virulence, and infectivity. Measures for controlling exposure to biohazards are presented, for example, through reduction in the generation of infectious aerosols, by understanding the different classes of biosafety cabinets, and by application of the principles of the four biosafety levels, among other topics.

Chapter 3, **Research Laboratory Ventilation Systems,** addresses the importance of ventilation in providing control of hazardous agents in laboratories. This chapter describes the typical laboratory (lab hood) system for contaminant control. It then identifies common problems associated with lab hoods and sashes, with balancing and maintaining constant, adequate airflows, and identifies possible solutions to those problems.

CHAPTER 1

The Laboratory Safety Standard

George B. Marshall

INTRODUCTION

The Occupational Safety and Health Administration (OSHA) has promulgated a safety standard for the *Occupational Exposure to Hazardous Chemicals in Laboratories* (29 CFR Section 1910.1450), commonly referred to as the Laboratory Safety Standard. Employers need to have been in compliance with this standard by January 31, 1991. The entire text of this standard may be found in Appendix A to this volume.

The purpose of this Laboratory Safety Standard is to provide a higher level of employee protection from exposure to toxic substances than the General Industrial Standards (29 CFR 1910) previously have done, while allowing for more flexible compliance methods. Because this is a performance standard, it allows employers to tailor an individualized Chemical Hygiene Plan to the particular circumstances and practices of their laboratories. Though this standard may parallel the Hazard Communication Standard (29 CFR 1910.1200) in some ways, it does not replace it. (Refer to *The Work Environment,* Volume I, Chapter 5, for details of the Hazard Communication Standard.)

0-87371-392-3/93/$0.00 + $.50

The necessity to develop a safety standard for laboratories stemmed from a feeling that the needs of the laboratory environment were not adequately addressed by existing safety and health regulations. Though many of the provisions of the Laboratory Safety Standard are similar to the Hazard Communication Standard, the emphasis of the former is directed more toward having standard operating procedures, adequate control measures, and emergency response plans. This chapter introduces and explains the Laboratory Safety Standard (29 CFR 1910.1450). Central to this standard is the Chemical Hygiene Plan which will be discussed in detail later.

COVERAGE — WHO IS AFFECTED BY THE LABORATORY SAFETY STANDARD?

People employed in laboratories in the private sector are covered under this standard. However, many laboratories are found in educational institutions, a significant number of which are in the public sector. Laboratory workers in these public institutions are also covered if they are under a state plan. All laboratories, whether under this standard or not, should follow its provisions to provide a safe work environment.

When considering employees as *laboratory workers,* do not forget to include custodial, maintenance, and any other workers who might be required to enter the laboratory as part of their regular job duties. These employees are covered by this standard too, and must receive the appropriate training.

This standard does not apply to the use of hazardous chemicals that do not meet the definition of *laboratory* use, nor does it apply to the laboratory use of hazardous chemicals that do not present a potential for employee exposure, such as reagent test strips and certain commercially prepared test kits. If the laboratory use of chemicals does not fall within the coverage of the Laboratory Safety Standard, employees will still be protected by the Hazard Communication Standard.

DEFINITIONS

Some of the more pertinent definitions from this standard may be helpful. While a laboratory is defined as a workplace where relatively small quantities of hazardous chemicals are used on a non-production basis,[1] a more detailed definition of the laboratory use of hazardous chemicals can be found in List 1. When referring to laboratory scale, the standard means work where the containers used for reactions, transfers, and other handling of substances are designed to be easily and safely manipulated by one person.[2]

Regarding hazardous chemicals, this standard means a chemical for which there is statistically significant evidence, based on at least one study conducted

List 1.
Laboratory Use of Hazardous Chemicals

1. Chemical manipulations are carried out on a *laboratory scale*
2. Multiple chemical procedures or chemicals are used
3. Procedures are not part of a production process
4. *Protective laboratory practices and equipment* are available and in common use to minimize the potential for employee exposure

in accordance with established scientific principles, that acute or chronic health effects may occur in exposed employees.[3] The term health hazard is broad and includes chemicals which are carcinogens, toxic or highly toxic agents, reproductive toxins, irritants, corrosives, sensitizers, hepatotoxins, nephrotoxins, neurotoxins, agents which act on the hematopoietic systems, and agents which damage the lungs, skin, eyes, or mucous membranes.[4] Substances that are listed in the latest printed edition of the National Institute for Occupational Safety and Health's *Registry of Toxic Effects of Chemical Substances* (RTECS),[5] the International Agency for Research in Cancer's (IARC) *Monographs,*[6] the National Toxicology Program's (NTP) *Annual Report on Carcinogens,*[7] the American Conference of Governmental Industrial Hygienist's (ACGIH) *Threshold Limit Values for Chemical Substances and Physical Agents in the Work Environment,*[8] or OSHA's 29 CFR part 1910: Subpart Z Toxic and Hazardous Substances[9] list, or have yielded evidence of acute or chronic health hazards in human, animal, or other biological testing, should be considered as toxic chemicals.

An action level is a designated concentration of a chemical, usually one half of the permissible exposure limit (PEL) of the substance specified in 29 CFR 1910 subpart Z, which initiates required activities, such as exposure monitoring and medical surveillance.

CHEMICAL HYGIENE PLAN

The Chemical Hygiene Plan is at the heart of the Laboratory Safety Standard. The Chemical Hygiene Plan is to be a written program detailing procedures and equipment capable of protecting employees from health hazards presented by hazardous chemicals in the laboratory and maintaining exposures below the permissible exposure limits. It needs to meet the requirements of paragraph (e) of the standard, all of which are encompassed by the outline in List 2. This plan is to be reviewed and updated on an annual basis.

STANDARD OPERATING PROCEDURES

Standard operating procedures (SOPs) need to be developed for the safe use and handling of hazardous chemicals. It is important to have clearly written and

List 2.
Chemical Hygiene Plan (as outlined in paragraph (e) of the Standard)

1. Standard operating procedures when working with hazardous chemicals
2. Control measures to reduce exposure
3. Engineering controls and other protective equipment
4. Training and information
5. Procedures for prior approval of operations
6. Medical exams and consultations
7. Chemical hygiene officer
8. Additional protections for use of particularly hazardous substances
9. Emergency procedures
10. Recordkeeping

easily understood procedures for each analytical protocol or major function in a laboratory to be sure that manipulations are consistent and that safety is a consideration in each step of a process. List 3 is a listing of SOPs items that should be included when developing the written Chemical Hygiene Plan. The listing (List 3) is a guide and should be supplemented depending on the needs of the individual laboratory. Each standard operating procedure listed in List 3 is explained more fully below.

List 3.
Standard Operating Procedures

1. General safety procedures
2. Housekeeping
3. Glassware
4. Electrical equipment
5. Compressed gas tanks
6. Flammable chemicals
7. Corrosive chemicals
8. Reactive chemicals
9. Toxic chemicals

General Standard Operating Procedures

List 4 contains general standard operating procedures that should be developed for laboratories. Smoking should be prohibited in laboratories. A burning cigarette is an ignition source for combustible materials and flammable solvents often found in laboratories.

There is a concern for hand-to-mouth exposure from both bacteria and certain toxic substances. By prohibiting smoking, one source of this exposure is eliminated. In addition, eating and drinking also must be prohibited in laboratories. These are poor laboratory practices and another common source of hand-to-mouth contamination. Food should not be permitted in laboratory refrigerators. Biological specimens containing a variety of pathogens and toxic chemical products may contaminate the food stored there. Chemical and biological material should only be stored in properly labeled laboratory refrigerators. One last item

List 4.
Standard Operating Procedures: General

1. Smoking
2. Eating or drinking
3. Food storage
4. Personal protective equipment
5. Personal hygiene
6. Pipetting
7. Labeling
8. MSDSs

to help prevent this exposure is to prohibit the application of cosmetics in laboratories.

The use of proper Personal Protective Equipment (PPE) needs to be mandatory whenever it is required by policy, procedure, standard, or Material Safety Data Sheets (MSDS). Such items as gloves need to be worn whenever handling chemicals that can cause harm to the skin or can be absorbed through the skin. When there is a danger of chemical splashes, approved eye protection (goggles, safety glasses with side shields, or full face shields) specific to the biological or chemical product being used needs to be worn. In regard to such splashes, contact lenses should not be worn in the laboratory. Contact lenses, especially the soft ones, will absorb certain solvent vapors, offer no protection during a splash or spill, may concentrate caustic agents against the cornea, and may prevent tears from washing contaminants from the eye. Protective clothing should also be worn whenever there is the potential for splashes. Laboratory coats with high percentages of acetate or other highly combustible materials should be avoided. Laboratory coats should be long-sleeved, worn in the laboratory at all times, and removed when leaving the laboratory. Shoes are also a form of PPE. They should be comfortable, rubber-soled, and cover the entire foot. Shoes with open toes, open heels, or open weave fabric should not be permitted in laboratories as they afford little protection in case of a spill.

Personal hygiene is an important factor in laboratory safety, and long hair should be secured back and off the shoulders in such a manner as to prevent it from coming in contact with contaminated materials or surfaces and to prevent shedding of organisms into the work area. It is also important to keep hair out of moving machinery and instrumentation. Hands should be washed frequently during the day, especially before leaving the laboratory, before and after handling chemicals or specimens, and before eating, smoking, or using a restroom. Mouth pipetting should not be permitted in any modern laboratory. Mechanical pipetting aids are available for every task.

Other items that need to be addressed in the standard operating procedures section of a Chemical Hygiene Plan include the labeling of all chemical containers in accordance with the Hazard Communication Standard (29 CFR 1910.1200). In addition, MSDSs need to be available in each laboratory (or group of laboratories) for the products used within those areas.

Housekeeping Standard Operating Procedures

The legal and common sense rules of good housekeeping must be meticulously followed in a laboratory environment. Housekeeping standard operating procedures, List 5, are discussed below.

List 5.
Standard Operating Procedures: Housekeeping

1. Exit obstructions
2. Trash
3. Needles and syringes
4. Decorations

Exits and aisles must not be obstructed in any way. No equipment, chairs, supplies, or trash should be permitted in exit routes. Laboratory doors should be kept closed to discourage unauthorized entry, but exit doors must never be blocked, bolted, or obstructed in any way that might prevent egress. Trash should not be allowed to accumulate in any area, and should be disposed of daily. Flammable solvents and solvent soaked rags need special handling. These items need to be stored, and disposed of, in self-closing metal containers that are approved by the local fire jurisdiction. Needles and syringes also need special handling for disposal. They should be placed in impervious containers that are conspicuously labeled to ensure safe disposal. Universal precautions for dealing with sharps are discussed in Chapter 4 of this volume.

Festive occasions in the laboratory need special attention. Holidays, birthdays, and awarding of grants may be occasions for celebration, but decorations should be limited to designs on glass outside the laboratory work areas. Hanging decorations and candles should be prohibited as well as decorations on lights and instruments.

Glassware Standard Operating Procedures

Glassware is common to most laboratories, and it needs to be handled in an appropriate manner (standard operating procedures for glassware are listed in List 6). Heated glass and glass containers are very dangerous and should be handled with special heat-resistant gloves. Pipettes should not be left sticking out of bottles, flasks, or beakers since they could easily cause a serious puncture wound. Laboratory workers should be instructed not to attempt forceable removal of stoppers that become stuck on glass tubing. If stoppers are stuck, they should be carefully cut off. Glassware that has been exposed to infectious substances or materials must be decontaminated before reuse or disposal. Chemical bottles need to be completely emptied, rinsed with an appropriate solvent, and have the label crossed out before disposal in impervious containers. Laboratory waste disposal is further discussed in Chapters 7 and 8 of this volume. Chipped or broken glassware should not be used. Glass and sharp objects need to be disposed of in properly labeled impervious containers to prevent accidental cuts and punc-

List 6.
Standard Operating Procedures:
Glassware

1. Handling
2. Disposal

tures. Disposal of broken glass along with paper and trash is unsafe and is a hazard to the janitorial/custodial staff.

Electrical Equipment Standard Operating Procedures

A wide variety of electrical equipment will be found in any modern laboratory, from low to high voltage, and care should be taken when using any energized equipment. All instruments should be grounded, including household-type appliances such as microwave ovens. Standard operating procedures for electrical equipment are given in List 7.

List 7.
Standard Operating Procedures:
Electrical Equipment

1. Grounding
2. Shocks
3. Repairs

It should be a standard operating procedure to immediately report all electrical shocks, including small tingles, to the laboratory supervisor. Small shocks often precede major shocks, and even a small tingle may indicate a potential problem, such as improper grounding of equipment. Laboratory workers should not attempt to use an instrument that is causing shocks. They should shut off the current to the instrument and/or unplug it, and know the location of the circuit breakers for each laboratory area. Extension cords should be avoided in laboratories. If used, they should be the properly grounded, three-wire type. The only multiple adapters that should be permitted are those equipped with an overload protecting fuse. Whenever possible, circuits should be equipped with Ground-Fault Circuit Interrupter (GFCI) protection. Never attempt to make repairs to any instrument while it is energized. An exception to this is when adjustments have to be made which require the instrument to be on. In this case, a worker's hands should be dry and free of all jewelry, and extreme caution should be used. Repairs or additions to the electrical system of the laboratory (switches, outlets, circuit boxes) should only be done by a competent electrician.

Compressed Gas Tank Standard Operating Procedures

Standard operating procedures for compressed gas tanks are shown in List 8. Some definitions will help to introduce this section. A compressed gas is defined

List 8.
Standard Operating Procedures:
Compressed Gas Tanks

1. Definitions
2. Storage and handling
3. Regulators
4. Transport
5. Special precautions

here as any material with a gauge pressure exceeding 25 lb/psi at 70°F, or any liquid flammable material having a vapor pressure exceeding 40 psi absolute at 100°F. A flammable gas is defined as one for which the U.S. Department of Transportation (DOT) requires their red flammable gas label. This type of gas would have a flash point less than 100°F. A toxic gas is defined as one for which the DOT requires the white poison gas label. Toxic chemicals will produce acute or chronic health effects in those who have been improperly exposed. All laboratory workers should be familiar with the particular characteristics of the gas, i.e., flammability, reactivity, and toxicity.

All compressed gases received, used, or stored, in the laboratory need to be labeled according to the DOT regulations (49 CFR 100-199). Damaged cylinders should neither be accepted nor used. In addition, each cylinder should be marked by label or tag with the name of its contents. Compressed gas cylinders should be stored and used in the upright position, be securely fastened to prevent their falling or being knocked over, and have their valves protected with a standard cap whenever not in use. Compressed gas cylinders are under considerable pressure and should not be exposed to temperature extremes. They should neither be stored in the vicinity of combustibles nor exposed to excessive dampness or to corrosive chemicals or fumes. Except for dry ice and cryogenic materials, it is not recommended to transfer gases from one vessel to another, and the refilling of compressed gas cylinders should be prohibited in laboratories.

A compressed gas cylinder should not be used without a regulator. After attaching the regulator, and before the cylinder is opened, the adjusting screw of the regulator should be released. Regulator damage can result if the gas is permitted to enter suddenly. Before using a cylinder, the valve should be slowly opened to clear dust or dirt. It is important that the opening is not pointed toward anyone during this procedure, and no one should stand in front of the regulator gauge glass when opening the valve. If a leak develops between a cylinder and regulator, the valve should be closed before attempting to tighten the union nut. Laboratory personnel should not be permitted to make any repairs or alterations to cylinders or accessories.

When transporting cylinders, whether empty or full, the valves should be closed and the protective cap should always be in place. A suitable hand truck should be used for transport and the cylinder firmly secured to it with straps or chains. It is important not to drag, slide, or roll cylinders when moving them.

Some special procedures and precautions need to be taken with laboratory use of certain compressed gases. Flow experiments with flammable gases should never be left unattended, and the use of an explosimeter or combustible gas alarm is highly recommended. Consideration should also be given to the need for electrically bonding and/or ground cylinders containing flammables. All connections in an experiment employing flammable or toxic gases should be leak tested before use. Leak testing may be accomplished by something as simple as a soap solution that will indicate leaks by bubbles, or more sophisticated instrumentation, such as an explosimeter in the case of flammables.

Some common laboratory gases have their own specific cautions. For example, safety plugs in the valves of chlorine cylinders fuse at 157°F. Care must be exercised to ensure that they are not exposed to steam, hot water, etc., which could produce this temperature. Chlorine leaks may be located using a cloth wet with aqua ammonia which will produce white fumes (ammonium chloride) in the presence of chlorine. In order to perform this testing, the appropriate respiratory protection must be worn. Oil, grease, or other flammable material should not be permitted to come in contact with the valves, regulators, gauges or any fittings of an oxygen cylinder because oil and grease in the presence of oxygen under pressure may ignite violently. Such cylinders must never be handled with oily hands or gloves.

Flammable Chemicals

Quantities of one gallon or more of flammables should be stored in a safety container approved by the local fire authority. If a reagent must be stored in glass for purity, the glass container should be placed in a rubber bottle carrier or other carrying device to lessen the danger of breakage when being moved. Only small quantities (i.e., one day working amounts) of flammables should be stored on open shelves. Bulk storage (i.e., more than one day's supply or more than 10 gal) should be stored in either a flammable safety cabinet or flammable storage room that meets the requirements and approval of the local fire authority. Fume hoods should not be used for bulk storage of flammables.

Flammables need to be used and stored in well-ventilated areas and kept away from fire, sparks, and reactive chemicals. Ether or any other flammable liquid should not be stored in a closed area, such as a refrigerator, unless the refrigerator is rated as explosion proof. Flammables should also be stored separately from oxidizers since these two chemicals can react violently with each other. (Refer to *The Work Environment,* Volume I, Chapter 13, for details on fire protection.)

Corrosive Chemicals

Acids and bases should be stored separately and kept in separate containers to facilitate handling. Organic acids (e.g., acetic acid or acetic anhydride) should be stored separately from strong oxidizing agents (e.g., sulfuric, nitric, or perchloric acid) to prevent corrosion of storage cabinets due to fume interaction. Caustic and corrosive materials should be stored near the floor to minimize the danger of bottles falling from high shelves. Corrosives should also be kept separate from flammable and combustible liquids.

When handling highly corrosive materials, appropriate PPE, such as aprons, gloves, and eye protection, needs to be worn. If moderate to large quantities of acids or alkalis are being used, employ a shield or barrier, or work in a sink so that spills or breaks can be controlled. Great care needs to be exercised when diluting acids. Always add the acid to the water and allow the acid to slowly run down the side of the mixing vessel. Mix by rotating slowly and avoid overheating. When transporting corrosive chemicals, even for short distances, safety bottle carriers should be used for containers over one quart in size.

Reactive Chemicals

Reactive chemicals need to be kept away from heat, sparks, and fire, so it is important to avoid smoking, cutting, welding, or creating friction near them. Incompatible chemicals should be kept separated (e.g., keep oxidizers away from flammables). Reactive chemicals need to be protected from sudden shock, and storage containers should be checked for leaks. Work with reactive chemicals should be done in well-ventilated areas, and PPE, such as safety glasses or chemical splash goggles, plus a face shield, proper gloves, and a long-sleeved laboratory coat should be worn. When conducting experiments that may result in polymerization, caution must be exercised because of the energy generated as a result of the reaction.

When using peroxides and other explosive chemicals, working quantities should be limited to the minimum amount required, and work should be performed in a way that minimizes vaporization. Because these chemicals are quite flammable, open flames and other sources of ignition should not be permitted near them. Peroxides should be stored at the lowest possible temperature to minimize their rate of decomposition. Peroxides can react with metals, so only ceramic or wood spatulas should be used when manipulating these chemicals in the laboratory. Pure peroxides should not be disposed of directly; these need to be diluted with water and a reducing agent (such as ferrous sulfate or sodium bisulfite) before disposal as chemical waste.

Toxic Chemicals

Laboratory workers must avoid skin contact with toxic chemicals by wearing the appropriate PPE, such as gloves, long-sleeved laboratory coats, or aprons. Wearing of open toe, open heel, or open weave fabric shoes should be avoided in laboratories. Procedures that generate aerosols of toxic substances should be performed in a fume hood or other suitable containment device. Good hygiene dictates that employees wash hands and arms immediately after working with toxic substances.

List 9.
Control Measures to Reduce Exposure

1. Engineering controls
2. Administrative controls
3. Personal protective equipment

CONTROL MEASURES TO REDUCE EXPOSURE

In order to reduce employee exposure to toxic substances, appropriate control measures need to be implemented (List 9). The determination of the exact type or form of control measure is often based upon the chemical manufacturer's recommendations found on the MSDS of that substance. Substitution of a less hazardous chemical is often the most desirable option. After product substitution, the implementation of other engineering controls, such as use of a fume hood, is the primary method of controlling hazards. If this is not appropriate or adequate, there should be an alteration in administrative controls, which involves charging the policies and procedures for performing a particular laboratory task. A simple change in work practices or procedures should reduce employee exposure. For example, the last method which should be used for reducing exposure is the issuance of PPE such as gloves, goggles, lab coats, or respirators. Personal protective equipment should not be substituted for engineering or administrative controls.

The selection of control measures will be largely based upon the information presented on the MSDS of the chemical product. In most laboratories, fume hoods will be the primary method used to control employee exposures. If this is not sufficient or appropriate, a change in procedure will need to follow. If necessary, and after all other control measure possibilities have been ruled out, a respirator may be issued to an employee after that employee has undergone specific respirator training, fit testing, and pulmonary function testing. Respirators may also be used in emergencies, to supplement engineering or administrative controls, while engineering controls are being installed, or when implementing administrative controls. Appropriate gloves should also be worn when an employee handles corrosives, solvents, irritants, carcinogens, teratogens, or other toxic substances. Goggles, or a combination of safety glasses with sideshields plus a faceshield, also should be worn whenever there is the possibility of a chemical splash. If respirators are issued to laboratory workers, they need to be inspected before each use and inspected and cleaned after each use. Gloves should be checked for tears and holes before use. All other forms of personal protective equipment should be inspected before use to ensure user safety. (Refer to 29 CFR 1910.134 and *The Work Environment,* Volume I, Chapter 14, for details on respiratory protection.)

The laboratory supervisor should be notified whenever there is a malfunction in any piece of safety equipment. When using any form of PPE, be sure that it is the proper type. For example, if the MSDS states that polypropylene gloves are required, then that is the only type to use. It is also important to remember that goggles considered adequate eye protection for biological splashes may not be appropriate for chemical splashes.

Permissible exposure limits (PEL) have been established for OSHA-regulated substances. This value is generally expressed in parts per million (ppm) or milligrams per cubic meter (mg/m^3). ACGIH has also established threshold limit values (TLV). Either or both of these values will be found on the MSDS, in the RTECS issued by the National Institute for Occupational Safety and Health (NIOSH), or in publications of the ACGIH. (See *The Work Environment,* Volume I, Chapter 4, for a complete discussion of occupational health standards and guidelines.)

CHEMICAL FUME HOODS, BIOSAFETY CABINETS, AND PROTECTIVE EQUIPMENT

In providing safe work environments, the emphasis should be on engineering controls to reduce or eliminate exposures. In laboratories, the most common form of engineering control is the hood and/or biosafety cabinet. All laboratory chemical fume hoods should be inspected at least annually. Laminar flow hoods and biosafety cabinets should be inspected by a qualified technician on at least an annual basis. All hoods should be rechecked after being moved or having major repair work performed. Safety practices for using chemical fume hoods are outlined in List 10.

TRAINING AND INFORMATION

Information and training are basic to all safety programs. Laboratory employees are to receive training on the contents of this standard. A written copy of this standard and its Chemical Hygiene Plan should be available in each laboratory during all work shifts.

This information should be kept in conjunction with a MSDS binder containing MSDSs for all chemicals used in each laboratory. The MSDSs are there to ensure that employees are aware of the hazards of chemicals in their workplace, the safe handling of toxic substances, and so they can make informed decisions as to their potential exposure. As with the Hazard Communication Standard, the information and training is to be provided at the time of an employee's initial assignment to the laboratory, if new chemical agents are introduced, or if exposures change. The frequency of refresher training is determined by the employer.

List 10.
Safety Practices For Using Chemical Fume Hoods

1. Hoods should be considered as safety devices that can contain and exhaust toxic, offensive, or flammable materials when the design of an experiment fails and vapors or dusts escape from the apparatus being used.
2. Hoods should **not** be regarded as a means for chemical disposal.
3. Hoods should be evaluated before use to ensure adequate face velocities (typically an average of 100 linear feet/min), and the absence of excessive turbulence; monitoring should be performed at regular intervals.
4. Except when adjustments of apparatus within the hood are being made, the hood should be kept closed, i.e., vertical sashes down and horizontal sashes closed; keeping the face opening as small as possible improves the overall performance of the hood.
5. The airflow pattern, and thus the performance of a hood, depends on such factors as placement of equipment in the hood, room drafts from open doors or windows, people walking by, or even the presence of the user in front of the hood.
6. Hoods are **not** intended for chemical storage; materials in hoods should be kept to a minimum, making sure that these materials are not blocking vents or disrupting air flow.
7. Solid objects and materials should not be permitted to enter exhaust ducts as they can lodge in the ducts or fans and adversely affect their operation.
8. A contingency plan for emergencies should be prepared in the event of ventilation failure, power failure, fire or explosion.
9. If it is certain that adequate general laboratory ventilation will be maintained when the fume hoods are not running, hoods should be turned off to conserve energy; if any doubt exists, or if toxic substances are being stored in the hood, the hood should be left on.

The acute signs and symptoms associated with exposures to hazardous chemicals used in the laboratory may vary and need to be discussed during training. Headaches, nausea, and dizziness can occur from chemical inhalation. Further inhalation could lead to more serious respiratory distress. Skin allergies and contact dermatitis can occur from absorbing chemical products through the skin. Nausea and vomiting can result from accidental chemical ingestion. Finally, chemicals which enter the body via an injection or a wound can also produce a variety of adverse symptoms. Certain chemical exposures can lead to chronic conditions, such as lung disease, liver disease, kidney disease, and cancer. In order for employees to know these potential hazards, it is important for them to read product labels and MSDSs in accordance with the Hazard Communication Standard. Training and information should include subjects such as the PEL of chemicals, the proper selection and use of PPE, and how to recognize hazardous or unsafe conditions. The training and information provided to employees is an important safety effort by employers, and that task should be well documented.

PRIOR APPROVAL PROCEDURES

No experimentation that can be expected to present hazards to employees should be performed unless the protocol has been specifically approved by the supervisor of the laboratory. Experimentation that would need such approval might include unattended experiments, after-hours work, or experimentation that has a high degree of hazard. It is suggested that any work of these types have written procedures, signed by the laboratory supervisor, for conducting them in a safe manner.

MEDICAL EXAMS AND CONSULTATIONS

Initial exposure monitoring must be performed if there is any reason to believe that exposure levels for a regulated substance may exceed the action level, or the PEL where there is no action level. Periodic monitoring will be required if initial results demonstrate elevated exposure levels. Monitoring frequency is determined by OSHA standards and guidelines for specific substances. Affected employees must be notified of the monitoring results within 15 days after receipt by the employer. Substances that require exposure monitoring to be performed under present OSHA standards are shown in List 11.

List 11.
Substances Requiring Exposure Monitoring (OSHA)

- Acrylonitrile
- Asbestos
- Benzene
- Coke oven emissions
- Cotton dust
- 1,2-Dibromo-3-chloropropane
- Ethylene oxide
- Formaldehyde
- Inorganic arsenic
- Lead
- Vinyl chloride

Employees who feel an overexposure to a toxic substance has taken place, or who have had an accident during work, should seek medical attention from either their personal physician or a hospital emergency room. Such incidents should also be reported to the employer. All medical evaluations and consultations are to be provided by, or under the direct supervision of, a licensed physician and without loss of pay, or cost to the employee. Specific information must be provided to the examining physician by both the employee and the employer. This is to include the identity of the hazardous chemicals to which the employee has been exposed in addition to the conditions under which the exposure occurred. It is also to include the symptoms of the employee and any available monitoring results.

A written opinion from the physician is to be obtained by the employee and provided to the employer. It must state recommendations for further medical follow-up. It should also include any medical conditions that were revealed during

List 12.
Select Carcinogen: Any Substance That Meets One of the Following Criteria

1. It is regulated by OSHA as a carcinogen
2. It is listed under the category, "known to be carcinogens," in the *Annual Report on Carcinogens,* published by the National Toxicology Program (NTP) (latest edition)
3. It is listed under Group 1 ("carcinogenic to humans") by the International Agency for Research on Cancer *Monographs* (IARC) (latest editions)
4. It is listed in either Group 2A or 2B by IARC or under the category, "reasonably anticipated to be carcinogens" by NTP, and causes statistically significant tumor incidence in experimental animals in accordance with the following criteria:
   a) After inhalation exposure of 6–7 hours per day, 5 days per week, for a significant portion of a lifetime to dosages of less than 10 mg/m^3
   b) After repeated skin application of less than 300 (mg/kg of body weight) per week
   c) After oral dosages of less than 50 mg/kg of body weight per day

the course of the examination that may result in increased occupational risk due to hazardous chemical exposure in the workplace. A final statement should be made that the employee has been informed of the results of the examination/consultation and if there is a need for further examinations or treatments. This written opinion is not to reveal any findings or conditions unrelated to the occupational exposure in question.

CHEMICAL HYGIENE OFFICER

The employer is responsible for overall chemical safety. A Chemical Hygiene Officer should be designated to develop and implement a Chemical Hygiene Plan. The Chemical Hygiene Officer should be qualified, by training or experience, to provide technical guidance in the development and implementation of the provisions of the Chemical Hygiene Plan.

CARCINOGENS

Designated areas need to be established and appropriately posted in those laboratories using "select carcinogens", reproductive toxins, or substances which have a high degree of acute toxicity. List 12 presents criteria for select carcinogens as defined by 29 CFR 1910.1450. (Refer to *The Work Environment,* Volume I, Chapter 5, on the Hazard Communication Standard for further guidance in defining the scope of health hazards and determining which chemicals are to be considered hazardous.)

Regulated areas are areas in which the PEL for specific substances is met or exceeded. Another value, known as the action level, is generally equivalent to one half of the PEL of a substance. Monitoring may have to be performed to determine actual exposure levels.

All work requiring the use of carcinogens or suspected carcinogens is to take place in a properly functioning hood, closed system, or other device of equal protection. A *closed system* would include a glove box or other system which physically encloses an operation or procedure, is constructed and maintained to provide a physical separation between the employee and the substance, is designed to prevent the escape of vapors into the laboratory, and allows manipulation of chemicals to be conducted in the enclosure by use of remote controls or gloves which are physically attached and sealed to the enclosure. If this is not achievable, then the appropriate respirator must be employed.

While working in designated areas, the appropriate gloves must be worn as well as an impervious garment to protect the employee's clothing. The appropriate protective faceshield and/or goggles must also be worn in conjunction with a respirator if, in fact, it is deemed necessary to wear one.

Entry and exit is restricted in regulated areas, and PPE may be required within such areas. Upon leaving a regulated area, employees should remove any contaminated articles of clothing and place them into properly labeled containers for disposal or laundering. Disposable items, such as gloves, must be discarded appropriately.

Hand washing should be performed as soon as all work has been finished. No personal hygiene practices, such as combing of hair and application of make-up, are to take place within designated or regulated areas. Smoking, eating, and drinking should also be forbidden in these areas.

Waste chemicals should be poured off into properly labeled containers or bottles. It is important that incompatible chemicals not be mixed together at any time. Contaminated objects must be disposed of in similarly labeled containers.

EMERGENCY PROCEDURES

All laboratories need to have emergency equipment. Fire extinguishers for chemical fires should be available as well as emergency showers and eye washes. Some areas should have acid and caustic spill kits to be used on those types of spills.

Although no one is required to use one, the proper way to use an extinguisher is to twist the pin, pull it out, then squeeze the trigger. For carbon dioxide (CO_2) or dry chemical extinguishers, stand approximately 6 to 8 feet from the fire; for water extinguishers, stand approximately 30 feet from the fire. Aim and direct the extinguishing media in a sweeping motion at an imaginary line just above the base of the burning item. Slowly move closer to the fire. Fire blankets may be used to smother a fire.

Emergency showers and eye wash stations should be available in or near all laboratories. Emergency showers are to be used when an employee catches fire or experiences a chemical splash. Standing under the shower, an employee who experiences a chemical splash should pull the ring or chain down with considerable force, or depress the lever, whichever is appropriate. Any contaminated clothing should be removed.

Eye washes are used for chemical splashes in the eye. The eye(s) should be rinsed for a minimum of 15 minutes and should not be rubbed. Medical attention should be sought for all chemical splashes in the eye. All accidents should be reported to the laboratory supervisor and documented.

Spill kits containing deactivating or neutralizing agents should be available to laboratories. These deactivating or neutralizing materials should be poured or sprinkled onto small chemical spills working from the outer edge in toward the center. Large spills should be taken care of by trained, certified emergency responders; however, the spill should be blocked off until emergency response personnel arrive. In instances where there is a highly toxic chemical spill, such as osmium, employees should evacuate the area immediately, close the doors, post a sign stating "DO NOT ENTER" and the nature of the spill.

Small spill cleanup should be done wearing the appropriate PPE after the area has been evacuated of any unnecessary personnel. An absorbent barrier to contain the spill should be created before trying to deactivate or neutralize the chemical (the MSDS should be consulted for this information). Again, cleanup operations should work from the outer edge in toward the center of the spill. All cleanup material should be disposed of as chemical waste in properly labeled containers.

For large radioactive, biohazard, toxic, or hazardous chemical or gas releases, evacuate the area. *Do not try to re-enter the area until emergency personnel arrive.* If the spill is not considered to be hazardous, try to contain the spill until emergency personnel arrive. Each laboratory should have the telephone number of the appropriate emergency response group posted. More specific information on responding to laboratory chemical spills or releases may be found in Chapter 7 of this volume.

RECORDKEEPING

Records of training, exposure monitoring, and medical consultation/examination are to be kept in accordance with 29 CFR 1910.1200 (Hazard Communication Standard). (Refer to *The Work Environment,* Volume I, Chapter 5, for details of the Hazard Communication Standard.)

SUMMARY

Because The Laboratory Safety Standard is a performance standard, every laboratory is given a fair degree of flexibility in its implementation. This can be a boon or a bane, as the efficacy of the program may have to be demonstrated to a compliance officer. Chemical Hygiene Plans have no set length or depth of detail required, but it will be important that they cover the topic areas noted

List 13.
Suggested Headings for Laboratory Safety Plans

Foreword
A. General principles
B. Responsibilities
C. The laboratory facility
D. Components of the Chemical Hygiene Plan
E. General procedures for working with chemicals
F. Safety recommendations
G. Material Safety Data Sheets

in 29 CFR 1910.1450. Laboratories will also need to be sure that they are actually implementing what has been written into their plan.

The Laboratory Safety Standard, contained in Appendix A to this volume, itself has two appendices. Appendix A to 1910.1450 provides some useful guidance when instituting laboratory safety plans, the main topic headings of which can be found in List 13. These serve only as guidelines, but are valuable in pointing the direction to be taken when coming into compliance with the Laboratory Safety Standard. Specific compliance efforts will be dependent on the size and complexity of the laboratory itself.

Appendix B to 1910.1450 provides a list of non-mandatory references presented in four sections (see List 14). All are valuable publications to have in a laboratory's reference library. One of the most useful references is the National Research Council's *Prudent Practices for Handling Hazardous Chemicals in Laboratories* and its companion piece, *Prudent Practices for Disposal of Chemicals from Laboratories*. Both are published by the National Academy Press, in Washington, D.C.

List 14.
29 CFR 1910.1450 Reference Sections

(a) Materials for the development of the Chemical Hygiene Plan
(b) Hazardous substances information
(c) Information on ventilation
(d) Information on availability of referenced material

REFERENCES

1. Code of Federal Regulations, *Occupational Exposure to Hazardous Chemicals in Laboratories,* OSHA, 29 CFR part 1910.1450(b). U.S. Government Printing Office, Washington, D.C.
2. OSHA, 29 CFR 1910.1450(b).
3. OSHA, 29 CFR 1910.1450(b).
4. OSHA, 29 CFR 1910.1450(b).

5. *Registry of Toxic Effects of Chemical Substances (RTECS),* U.S. Department of Health and Human Services, Public Health Service, Centers for Disease Control, National Institute for Occupational Safety and Health. U.S. Government Printing Office, Washington, D.C.
6. International Agency for Research in Cancer's (IARC) *Monographs* on the Evaluation of the Carcinogenic Risk of Chemicals to Man. World Health Organization Publications Center, Albany, NY.
7. *Annual Report on Carcinogens,* National Toxicology Program (NTP). U.S. Department of Health and Human Services, Public Health Service. U.S. Government Printing Office, Washington, D.C.
8. American Conference of Governmental Industrial Hygienists (ACGIH), Threshold Limit Values for Chemical Substances and Physical Agents in the Workroom Environment with Intended Changes, Cincinnati, OH.
9. OSHA, 29 CFR 1910.1000, subpart Z.

CHAPTER 2

Working With Biohazards

Arthur F. Brings

INTRODUCTION

The word biohazard is a contraction of the words "biological" and "hazardous" and is used to describe occupational hazards resulting from exposure to a biohazardous agent. According to the American Industrial Hygiene Association's Biohazards Committee, a biohazardous agent is one that is biological in nature, capable of self-replication, and has the capacity to produce deleterious effects upon other biological organisms, particularly humans.[1] Biohazardous agents include microorganisms such as bacteria, viruses, fungi, rickettsiae, and chlamydiae. In addition, blood and blood products, tissues and cell cultures, animal products, and any other biological material which can harbor the above microorganisms are also considered biohazardous agents.

Biohazards can be found in many industries and have been recognized as occupational hazards for over one hundred years. In the eighteenth century anthrax occurred in workers handling carcasses, skins, hides or hair from animals that had been infected with the bacteria anthrax bacillus. The first case of anthrax

0-87371-392-3/93/$0.00 + $.50

recorded in England appeared in woolsorters in 1847, soon after the introduction of alpaca and mohair, and became known as *woolsorters' disease.*[2] Psittacosis or ornithosis in man, which was identified during a pandemic in 1930,[3] is caused by a microorganism found in birds such as parrots, parakeets, pigeons, chickens, and ducks. While the disease is usually associated with people handling infected birds, it also can be spread to other occupations such as maintenance workers who clean pigeon droppings from buildings. This transmission of animal infections to man is known as zoonosis. For the most part, zoonoses in industry have been eliminated by controlling disease in animals. However, one should not lose sight of the fact that biohazards can exist in many industries and therefore cannot be discounted as a possible cause of occupational illness. Problems such as the incidence of Lyme disease in individuals working in wooded or grassy areas still exist. Lyme disease was first recognized in 1975 in Old Lyme, Connecticut. It is caused by a bacterium (a spirochete) and is transmitted in the northeast United States primarily by the deer tick. The disease is known to exist in over 43 states but predominates in the Northeast.[4]

Over the past 15 years a lot of attention has been focused on potential biohazards in the healthcare industry and in a relatively new industry called biotechnology. In these areas it is difficult to eliminate biohazards because much of the work is centered around patients harboring biohazardous agents and/or handling biohazardous materials. In the healthcare industry today, a widely recognized threat is the potential for infection with bloodborne pathogens including the human immunodeficiency virus (HIV) and the hepatitis B virus (HBV).[5] The potential threat of HIV and HBV as occupational hazards has had such an impact on healthcare workers, as well as workers in other occupations, that it is discussed in detail in Chapter 4. In addition to HIV and HBV, the healthcare worker is faced with handling a wide variety of materials containing potential biohazards which could result in an occupationally acquired infection. Much of the handling of these potentially biohazardous materials occurs in the laboratory, placing the laboratory worker in particular at risk.

Biotechnology includes any technology that uses living organisms, or parts of organisms, to make or modify products, to improve plants or animals, or to develop microorganisms for specific uses.[6] Biotechnology today consists of recombinant DNA technology, monoclonal antibody technology, and bioprocess technology, all of which require the laboratory manipulation of various microorganisms. When the technology was first developed there were many concerns over the safety of the process based on the immense power of the technique.[7] Today, however, most experts agree that the risks of biotechnology are no greater than the risks of the individual components used.[8] The biotechnology laboratory worker has no greater risk of contracting a laboratory-acquired infection than does a healthcare laboratory worker.

In this chapter the laboratory use of biohazardous agents is discussed, with an emphasis on the causes of laboratory-acquired infections and the principles of biosafety. This approach allows for the recognition of the hazards, evaluation of the risks, and the implementation of appropriate control measures based on a scheme of four biosafety levels corresponding to four classes of biohazardous agents.[9,10]

LABORATORY-ACQUIRED INFECTIONS

Any time there is a congregation of people, such as in the workplace, there is the potential for the spread of disease such as the common cold. However, only diseases or illnesses caused by a condition within the workplace are considered to be occupationally related. For example, an outbreak of conjunctivitis in a laboratory spread by the sharing of safety glasses that have not been sanitized is an example of an occupationally related illness, but it is not considered a laboratory-acquired infection. If a scientist is working with a microorganism, (which is not present in the community in which he lives) and becomes ill with a disease caused by that organism, then there is little doubt that he has become infected in the course of his work. If another individual, who does not handle that microorganism (but who works in the same room or another part of the building or is merely a visitor) becomes ill and the organism is recovered from his or her body, this also qualifies as a case of laboratory-acquired infection.[11]

Laboratory-acquired infections have been contracted throughout the history of microbiology with published reports of laboratory-associated cases of typhoid, cholera, glanders, brucellosis, and tetanus from as early as the turn of the century.[12] The first laboratory-acquired infections were all caused by a group of agents which were considered pathogenic microorganisms (discussed below).

Pathogenic, Nonpathogenic, and Risk

Traditionally microorganisms were separated into two categories: *pathogenic,* referring to organisms which are capable of causing disease, and *nonpathogenic,* referring to organisms which do not cause, or are not capable of causing, disease.[13,14] This distinction is not always valid since there have been infections caused by so called nonpathogens. In these infections, the nonpathogen's ability to cause disease was enhanced by the route of infection, the susceptibility of the individual, the dose, or any combination of the above. Thus, it follows that the concept of having two categories of microorganisms, pathogenic and nonpathogenic, is no longer tenable.[15] All microorganisms should be considered pathogenic. The degree of pathogenicity of a microorganism, that is, its comparative ability to cause disease, is known as virulence.[16]

When assessing the risk of working with a particular microorganism in the laboratory, the virulence of the microorganism, infectivity, and the consequences of an infection must be considered along with the operations and procedures to be performed. A useful guide for evaluating the risks associated with the use of various biohazardous agents in the laboratory is the *Classification of Etiologic*

Agents on the Basis of Hazard which is contained in the *NIH Guidelines for Research Involving Recombinant DNA Molecules.*[17]

Studies have shown that most laboratory-acquired infections occur in research laboratories and it is the trained scientific staff, not the support staff, that have the highest rate of infection.[18] It should be noted that in contrast to the documented occurrence of laboratory-acquired infections in laboratory personnel, laboratories where infectious agents are handled have not been shown to represent a threat to the community.[19] A 1976 study which reviewed 3921 cases of laboratory-acquired infections showed that less than 20% of all reported cases were associated with a known accident.[20] While the cause of the remaining 80% is unknown, exposure to infectious aerosols is considered to be the cause.

INFECTIOUS AEROSOLS

Aerosols, which are suspended liquid droplets in air, can be generated by many common laboratory techniques and can present a significant hazard to laboratory personnel. The potential for generating an aerosol exists any time there is an air to liquid interface. Aerosols are produced when liquids are subjected to bubbling, splashing, frothing, high frequency vibration, and centrifugal force.[21,22]

Aerosol production is only one of the hazards associated with many common laboratory procedures; other potential technique-related hazards exist. Pipettes in some form have probably been used as a basic tool by scientists as long as there have been laboratories.[23] Pipettes and pipetting have many potential hazards including the aspiration of liquids and aerosols. Mouth pipetting should be prohibited in all laboratories in order to prevent the accidental inhalation of aerosols and/or the ingestion of liquids. The cotton plug at the end of a pipette is not a filter and will not prevent the inhalation or ingestion of biohazardous agents; its only purpose is to prevent dust from collecting in the pipette. Mouth pipetting should also be prohibited to eliminate the potential for hand-to-mouth contamination which can result from a contaminated finger being used to seal the end of a pipette. Forceful ejection of the last drop in a pipette, and the mixing of liquids by drawing them up and down vigorously with a pipette, can produce aerosols. In addition, it always seems that there is a drop hanging from the end of the pipette that falls and hits the floor or work surface (this too produces an aerosol).

The platinum loop is used extensively in microbiology to inoculate cultures and many times is sterilized during use by heating with an open flame. Insertion of the loop in a flame, or immersion of a hot loop into a liquid, can cause sputtering creating aerosols. The loops tend to be made of very long and thin platinum wire which causes them to vibrate upon rapid movement in air or when they are streaked on dry agar, in either case generating aerosols.

Culture dishes and plates, whether they contain liquids or agar, are potential sources of aerosols if dropped and the contents are expelled. Other hazards include the release of airborne materials upon opening a dry plate, when opening a plate that has a film between the lid and the rim, and the dripping of the condensation or moisture which can develop on the lid of the plate.

Plugs and caps can present a hazard upon opening or removal. The container may already contain aerosols which can be released when the plug is removed or the cap is opened. In addition, there may be a difference in pressure inside the container which, when opened, can cause the generation of aerosols. Plugs and caps can become contaminated during handling resulting in the spread of the contaminant to work surfaces, equipment, and individuals.

Needles and syringes present the obvious hazard of potential for being wounded by a contaminated needle, but there are also some less obvious hazards that cannot be overlooked. Like platinum loops, small gauge needles have a tendency to vibrate when being removed from a rubber septum or upon withdrawal from an injection site, causing liquids to spew into the air. Liquids can escape from either end of a syringe and aerosols are generated upon the forceful discharge of liquids and during the expulsion of air bubbles from needle and syringe.

Centrifugation is an effective method for the generation of aerosols, especially when liquid droplets hit the wall of the centrifuge. Liquids can be released during centrifugation procedures by the disruption of caps, tubes, cups, and rotors. In addition, overflow can occur when angled tubes are overfilled, allowing liquids to hit the wall of the centrifuge and thereby generate aerosols.

Hazardous aerosols are created by most laboratory operations which involve the blending, mixing, stirring, grinding, or disruption of materials containing biohazardous agents. Even the use of a mortar and pestle can be a hazardous operation. Tissue grinders, ultrasonic cell disintegrators, magnetic mixers, stirrers, sonic cleaning devices, and shakers are other devices that can produce hazardous aerosols.

CONTROLLING INFECTIOUS AEROSOLS

The hazards associated with the techniques described above can be minimized if appropriate precautions are taken to prevent the generation of aerosols, or if the aerosols generated are confined to prevent their release into the workplace and/or the environment. Release of aerosols can be prevented through the proper use of available technology and through the use of proper work practices. For example, the use of mechanical pipetting devices is essential to prevent the aspiration and/or ingestion of infectious materials. Many mechanical pipetting devices are on the market today and it should not be difficult to find a device to meet the individual needs of the worker. Along with the mechanical pipetting device, a mark-to-mark pipette should be used so that the last drop of liquid does not have to be forcefully ejected to obtain an accurate measurement. Liquids should be discharged down the side walls of vessels to prevent them from splashing. When working with a pipette, an absorbent material such as "bench coat" or a towel that has been soaked in a disinfectant can be used to minimize aerosol generation by absorbing drops that may fall off the end of the pipette.

Inoculating loops are available with short stems and small loops which are made of soft material. These loops should be used to minimize the production of aerosols from vibration resulting from moving the loop through the air or streaking dry agar. The use of specially designed incinerators instead of open flames can also reduce the release of aerosols created by the sputtering that occurs when loops are sterilized by heating. Loops should be allowed to cool sufficiently before immersing them in liquids.

To prevent needles and syringes from coming apart or leaking at their connections, specially designed locking syringes should be used. When expelling the air bubbles from a needle and syringe, use an absorbent material at the end of the needle to absorb any liquids or aerosols generated or perform the procedure in a biological safety cabinet. Use an absorbent material to absorb any liquids or aerosols generated when removing a needle from a rubber septum or an injection site. Used needles and syringes should be placed in puncture-resistant containers for disposal. Sheaths should not be replaced on needles before disposal, unless self-sheathing needles are used, to reduce the risk of being wounded by the needle. Do not use needle destruction devices which clip the needle before disposal because such practices can result in the production of aerosols.

When using centrifuges, all tubes should first be inspected for cracks and potential leak sites. All tubes should also be balanced and not overfilled. If a tube breaks during centrifugation, allow the aerosols to settle before opening the lid and disinfect all spills immediately.

Confining the Agent: Primary Containment

There are many laboratory procedures in which it is difficult to prevent the generation of aerosols. These procedures include the use of blenders, mixers, sonicators, and other cell disruption equipment. Under these circumstances it is necessary to confine the aerosol of the biohazardous agent through the use of appropriate safety equipment which provides a *primary barrier*. Biological safety cabinets are among the most effective, as well as the most commonly used primary containment devices used for biohazardous agents, although any closed vessel can be used to provide primary containment.[24] In certain cases where it is impractical to work in a biological safety cabinet, personal protective equipment can be used to form the primary barrier.

There are three classes of biological safety cabinets: Class I, Class II (which has four types), and Class III, all of which provide protection for the laboratory worker and the environment. The classes of biological safety cabinets differ in

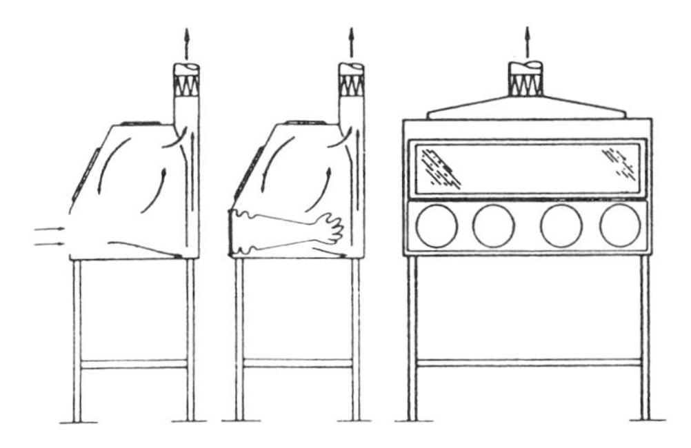

Figure 1. Class I cabinet.

their ability to handle hazardous chemicals including radioisotopes, and in their ability to prevent cross-contamination of biological materials being used within.

Chemical fume hoods should not be used to work with biohazardous agents because, while they may protect the laboratory worker from the biohazardous agent, they do not provide any protection for the environment. Also, horizontal clean benches, which are used in some laboratories for the preparation of sterile products, should not be used to work with biohazardous agents since they do not provide any protection for the laboratory worker or the environment. In fact, aerosols generated while handling biohazardous agents on a horizontal clean bench are blown directly into the face of the laboratory worker at the bench.

Class I biological safety cabinets, an example of which is shown in Figure 1, have an open front with a minimum face velocity of 75 ft/min which provides protection for the laboratory worker. To protect the environment from the biohazardous agents, all of the air is exhausted to the outside through a high efficiency particulate air (HEPA) filter. Class I cabinets can be used with most chemicals and radioisotopes, but do not offer any contamination protection for the materials used inside. The cabinet can be opened with an open front, with arm holes, or arm holes with full length protective gloves attached.

Class II biological safety cabinets, examples of which are shown in Figure 2, are divided into Class II, type A and Class II, type B. Class II, type A has a minimum face velocity of 75 ft/min and Class II, type B has a minimum face velocity of 100 ft/min. In addition, Class II, type B biological safety cabinets are further divided into three groups: B1, B2, and B3.

Class II, type A biological safety cabinets differ from Class I in that 70% of the air which passes through the HEPA filters is recirculated within the cabinet. The recirculated air moves down on the work surface in a laminar flow providing a clean work space and minimizing cross-contamination of the biological materials being used in the cabinet. The remaining 30% of filtered air is discharged into the laboratory or outdoors. Because this design recirculates approximately 70% of the total cabinet air, it should not be used with flammable solvents,

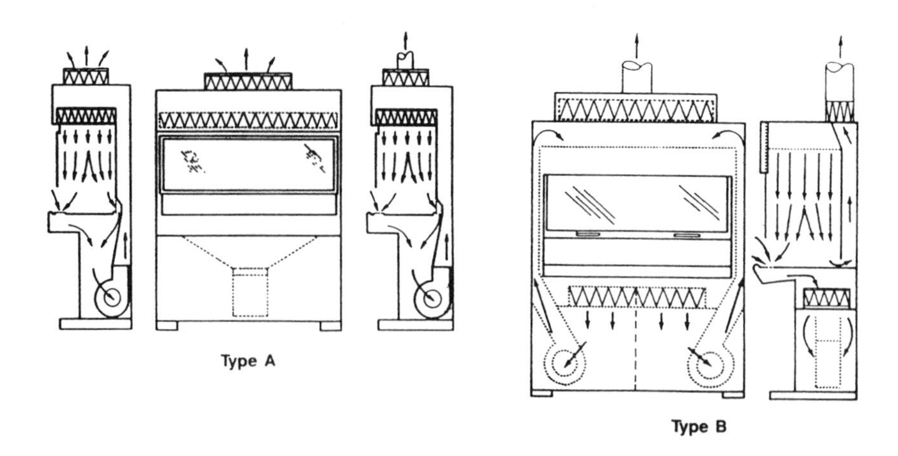

Figure 2. Class II cabinets.

volatile or toxic chemicals, or volatile radioactive materials.[25] Class II, type A cabinets provide protection for the laboratory worker, the environment, and the research materials.

Class II, type B1 biological safety cabinets recirculate 30% of the air which passes through the HEPA filter within the cabinet and exhausts the remaining 70% outdoors. Because 70% of the air is exhausted outdoors the cabinet can be used with limited quantities of volatile or toxic chemicals and trace amounts of radioactive materials.[26] Class II, type B1 cabinets provide protection for the laboratory worker, the environment, and the research materials.

Class II, type B2 biological safety cabinets exhaust 100% of the air outdoors through a HEPA filter. Since all of the air is exhausted outdoors, these cabinets can be used with volatile and toxic chemicals and volatile radioactive materials.[27] Class II, type B2 cabinets provide protection for the laboratory worker, the environment, and the research materials.

Class II, type B3 biological safety cabinets are similar to the Class II, type A cabinets except the type B3 cabinets have an inward face velocity of 100 ft/min. Because this design recirculates approximately 70% of the total cabinet air, it should not be used with flammable solvents, volatile or toxic chemicals, or volatile radioactive materials. Class II, type B3 cabinets provide protection for the laboratory worker, the environment, and the research materials.

Class III biological safety cabinets, an example of which is shown in Figure 3, are gas-tight, negative pressure containment systems which physically separate the biohazardous agent from the laboratory worker by the use of arm-length rubber gloves. These cabinets provide the highest degree of protection for the laboratory worker.[28] All of the air from the cabinet is exhausted through a HEPA filter and/or incinerated. The cabinets provide protection for the laboratory worker and the environment, but not the research materials.

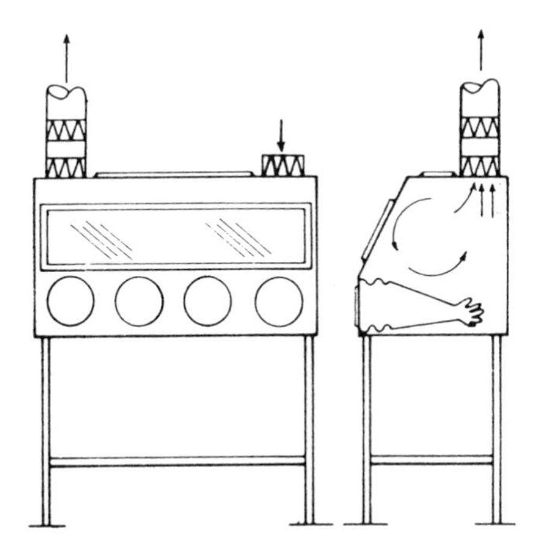

Figure 3. Class III cabinet.

Confining the Agent: Secondary Containment

In addition to primary containment of the biohazardous agent, secondary containment can be accomplished through facility design. There are three facility designs that provide increasing levels of containment: the basic laboratory, the containment laboratory, and the maximum containment laboratory.[29] The three facility designs and their uses are summarized below.

The basic laboratory is a laboratory of conventional design, which does not have any special engineering features and is suitable for use with biohazardous materials of low hazard. The basic laboratory would be used for work at Biosafety Levels 1 and 2 (which are described in the next section). In general, work in the basic laboratory is conducted on the open bench, with the exception of operations that have the potential to generate aerosols, which are conducted in a biological safety cabinet. The basic design features include a facility that can be easily cleaned with bench tops that are impervious to liquids. Each laboratory must have a sink for hand washing, and an autoclave for decontaminating infectious waste if the laboratory will be conducting work at Biosafety Level 2.

The containment laboratory differs from the basic laboratory in that it has engineering features, such as special ventilation, to contain biohazardous agents and access is controlled to limit entrance to the facility only to authorized personnel. The containment laboratory is suitable for work with biohazardous agents which require up to Biosafety Level 3. The facilities of the containment laboratory are similar to the basic laboratory with additional features (such as double doors that separate the laboratory from the rest of the building). Interior surfaces are sealed and water resistant. Work is usually conducted in biological safety cabinets and there must be directional air flow which provides for an influx of air from areas surrounding the laboratory.

The maximum containment laboratory is designed for work with extremely hazardous biological agents and is suitable for work with biohazardous agents up to the Biosafety Level 4. The maximum containment laboratory is usually in a separate building although it can be located within another building, as long as it is isolated from all building services. In addition to the features of the containment laboratory, the maximum containment laboratory is designed to provide primary and secondary containment through the use of a sealed facility with negative pressure and directional air flow. Access is limited through the use of airlocks and personnel are required to shower and change before leaving. All work in this facility is performed in a biological safety cabinet and all liquid effluents, waste materials, equipment, and air must be decontaminated before leaving the laboratory.

BIOSAFETY LEVELS

The concept of biosafety levels was first developed in 1969 with the publication of *Classification of Etiologic Agents on the Basis of Hazard.*[30] Individual etiologic agents are classified by degree of hazard and have been assigned to five classes (Class 1–4, with 4 being the most hazardous; Class 5 agents are excluded from the U.S. by law). Minimal conditions have been defined for the use of these agents[30] This same concept was used by the Centers for Disease Control and the National Institutes of Health as the basis for the development of the biosafety levels contained in the book *Biosafety in the Microbiological and Biomedical Laboratories.*[31] There are four biosafety levels described, each of which corresponds to the revised version of the *Classification of Etiologic Agents on the Basis of Hazard* which is contained in the *NIH Guidelines for Research Involving Recombinant DNA Molecules.*[32]

The four biosafety levels consist of combinations of laboratory practices and techniques, safety equipment, and facility design, appropriate for the operations performed and the risk posed by the biohazardous agent and for the laboratory function or activity.[33] The four biosafety levels are summarized in Table 1 and discussed below.

Biosafety Level 1

Biosafety Level 1 practices, safety equipment, and facilities are appropriate for undergraduate and secondary educational training and teaching laboratories and for other laboratory facilities where work is done with biological agents of no known or minimal potential hazard to laboratory personnel and the environment. Class 1 agents are handled at this biosafety level. Work at Biosafety Level 1 is generally conducted on the open bench in a basic laboratory which is not separated from the general traffic patterns in the building.

Table 1. Summary of Recommended Biosafety Levels for Infectious Agents

Biosafety Level	Practices and Techniques	Safety Equipment	Facilities
1	Standard microbiological practices	None: primary containment provided by adherence to standard laboratory practices during open bench operations.	Basic
2	Level 1 practices plus: Laboratory coats; decontamination of all infectious wastes; limited access; protective gloves and biohazard warning signs as indicated.	Partial containment equipment (i.e., Class I or II Biological Safety Cabinets) used to conduct mechanical manipulative procedures that have high aerosol potential that may increase the risk of exposure to personnel.	Basic
3	Level 2 practices plus: Special laboratory clothing; controlled access.	Partial containment equipment used for all manipulations of infectious material.	Containment
4	Level 3 practices plus: Entrance through change room where street clothing is removed and laboratory clothing is put on; shower on exit; all wastes are decontaminated on exit from the facility.	Maximum containment equipment (i.e., Class III biological safety cabinet or partial containment equipment in combination with full-body, air-supplied, positive-pressure personnel suit) used for all procedures and activities.	Maximum containment

The following practices, safety equipment, and facilities apply to agents assigned to Biosafety Level 1.

Standard Practices

1. Access to the laboratory may be restricted when experiments are in progress.
2. Work surfaces are decontaminated daily and after spills.
3. All contaminated wastes are decontaminated before disposal.
4. Mechanical pipetting devices are used; mouth pipetting is prohibited.
5. Smoking is prohibited and no food or drink is permitted in the laboratory.
6. Persons wash their hands frequently and before leaving the laboratory.
7. All procedures are performed to minimize the creation of aerosols.
8. Laboratory coats should be worn.

Safety Equipment

Special safety equipment is generally not required.

Facilities

1. The laboratory is designed to be easily cleaned.
2. Bench tops are impervious to liquids.
3. Each laboratory contains a sink for hand washing.

Biosafety Level 2

Biosafety Level 2 practices, equipment, and facilities are applicable to clinical, diagnostic, teaching, and other facilities for work with biohazardous agents of moderate potential hazard to personnel and the environment including Class 2 agents. Biosafety Level 2 is similar to Level 1 except that laboratory personnel have specific training in work with biohazardous agents, access to the laboratory is restricted, and procedures which have the potential to generate aerosols are conducted in a biological safety cabinet.

The following practices, safety equipment, and facilities apply to agents assigned to Biosafety Level 2.

Standard Practices

1. Access to the laboratory *is* restricted when experiments are in progress.
2. Work surfaces are decontaminated daily and after spills.
3. All contaminated wastes are decontaminated before disposal.
4. Mechanical pipetting devices are used; mouth pipetting is prohibited.
5. Smoking is prohibited and no food or drink is permitted in the laboratory.
6. Persons wash their hands frequently and before leaving the laboratory.
7. All procedures are performed to minimize the creation of aerosols.

8. Laboratory coats should be worn and must be removed when leaving the laboratory.
9. Gloves are worn to avoid skin contamination.
10. The use of needles and syringes is restricted.
11. Spills or overt exposures are reported to the laboratory supervisor.

Safety Equipment

1. Biological safety cabinets are used for all procedures which have the potential for generating aerosols.
2. High concentrations or large volumes of biohazardous agents are used in biological safety cabinets.

Facilities

1. The laboratory is designed to be easily cleaned.
2. Bench tops are impervious to liquids.
3. Each laboratory contains a sink for hand washing.
4. An autoclave is available for decontamination of contaminated wastes.

Biosafety Level 3

Biosafety Level 3 practices, safety equipment, and facilities are applicable to clinical, diagnostic, teaching, research, or production facilities in which work is done with indigenous or exotic agents where the potential for infection by aerosols is real and the disease may have serious or lethal consequences including Class 3 agents. All manipulations of biohazardous agents are performed in biological safety cabinets in a containment laboratory which has special engineering and design features.

The following practices, safety equipment, and facilities apply to agents assigned to Biosafety Level 3.

Standard Practices

1. Access to the laboratory *is* restricted and doors are closed when experiments are in progress.
2. Work surfaces are decontaminated daily and after spills.
3. All contaminated wastes are decontaminated before disposal.
4. Mechanical pipetting devices are used; mouth pipetting is prohibited.
5. Smoking is prohibited and no food or drink is permitted in the laboratory.
6. Persons wash their hands frequently and before leaving the laboratory.
7. All procedures are performed in a biological safety cabinet.
8. Laboratory coats must be worn and removed when leaving the laboratory.
9. Gloves are worn to avoid skin contamination.
10. The use needles and syringes is restricted.
11. Spills or overt exposures are reported to the laboratory supervisor.

12. Animals or plants not related to the work are not permitted in the laboratory.
13. Vacuum lines are protected with HEPA filters and liquid traps.
14. Baseline serum samples for all at-risk personnel should be collected and stored.

Safety Equipment

Biological safety cabinets are used for all procedures which involve the manipulation of biohazardous agents.

Facilities

1. The laboratory is separated from the rest of the building; the laboratory has a double door entry system.
2. The surfaces of walls, floors, and ceilings are water resistant and easy to clean.
3. Bench tops are impervious to liquids.
4. Each laboratory contains a sink for hand washing which is operated by foot or elbow and is located near the door.
5. An autoclave is available within the laboratory for decontamination of contaminated wastes.
6. A ducted exhaust ventilation system is provided which creates directional air flow into the laboratory; there is no recirculation of exhaust air.
7. All biological safety cabinets are exhausted outdoors.

Biosafety Level 4

Biosafety Level 4 practices, safety equipment and facilities are required for work with dangerous and exotic agents which pose a high individual risk of life-threatening disease including Class 4 agents. All work at Biosafety Level 4 is performed in a maximum containment laboratory in Class III biological safety cabinets or in Class I or Class II biological safety cabinets wearing a one-piece positive-pressure suit ventilated by a life support system. The facility is either in a separate building or in a controlled area within a building, completely isolated from all other areas of the building and building services. The maximum containment laboratory has special engineering and design features to prevent the release of biohazardous agents to the environment.

The following practices, safety equipment, and facilities apply to agents assigned to Biosafety Level 4.

Standard Practices

1. Access to the laboratory *is* restricted to those persons whose presence is required for program or support purposes.
2. Work surfaces are decontaminated daily and immediately after spills.

3. All contaminated wastes are decontaminated before disposal.
4. Mechanical pipetting devices are used; mouth pipetting is prohibited.
5. Smoking is prohibited and no food or drink is permitted in the laboratory.
6. Persons can enter the leave the laboratory only through a clothing change and shower room.
7. All procedures are performed in a biological safety cabinet.
8. Complete laboratory clothing, including undergarments, pants and shirts or jumpsuits, shoes and gloves, is provided and used by all personnel entering the facility. When leaving the laboratory and before proceeding to the shower area all personnel must remove their laboratory clothing and store it in a hamper or locker.
9. Supplies and materials needed in the facility are brought in by way of a double-doored autoclave, fumigation chamber, or air lock which is decontaminated between use.
10. The use of needles and syringes is restricted.
11. Spills or overt exposures are reported to the laboratory supervisor.
12. Animals or plants not related to the work are not permitted in the laboratory.
13. A medical surveillance program is required.

Safety Equipment

All procedures are performed in Class III biological safety cabinets or in Class I or Class II biological safety cabinets wearing a one-piece positive-pressure suit ventilated by a life support system.

Facilities

1. The facility is either in a separate building or in a controlled area within a building, which is completely isolated from all areas of the building and building services.
2. The interior surfaces of walls, floors and ceilings form a sealed internal shell which is water resistant and easy to clean; all penetrations are sealed and all discharges are decontaminated.
3. Bench tops have seamless surfaces which are impervious to liquids.
4. Each laboratory contains a sink for hand washing which is operated by foot or elbow and is located near the door.
5. Access doors to the laboratory are self-closing and lockable.
6. A double-door autoclave is provided to decontaminate materials being passed out of the laboratory.
7. A pass-through dunk tank or fumigation chamber is provided to decontaminate materials that cannot be autoclaved.
8. All liquid waste is decontaminated before being released from the laboratory.
9. An individual supply and exhaust air ventilation system is provided that maintains pressure differentials and directional air flow; the exhaust air from the laboratory is HEPA filtered and discharged outdoors.
10. All biological safety cabinets are exhausted outdoors.

SELECTING THE APPROPRIATE BIOSAFETY LEVEL

For any laboratory, the biosafety levels discussed above are only a summary and represent conditions under which classes of biohazardous agents ordinarily can be handled. Selection of an appropriate biosafety level for work with a particular agent is a subjective process which depends on a number of factors and should only be made by qualified individuals. Some of the most important factors that need to be considered when selecting an appropriate biosafety level are:

1. The pathogenicity or virulence of the biohazardous agent in terms of its ability to cause disease. Agents that are known to cause serious epidemic disease in healthy adults pose a much greater risk than agents that are not usually associated with any disease.
2. Biological stability of the biohazardous agent or its ability to survive in the environment. The more stable an agent is and the greater its resistance to disinfection, the greater the hazard.
3. The usual route of infection. An agent that is known to cause infection by airborne exposure should be considered more hazardous than one that is not.
4. The nature and function of the laboratory. Laboratories that conduct procedures and manipulations which have the potential to generate aerosols present a greater risk than those that do not.
5. The quantity and concentration of the biohazardous agent can increase the risk of infection.
6. The availability of an effective vaccine or therapeutic measures. Biohazardous agents which cause life-threatening diseases or diseases for which there are no known therapeutic measures are significantly more hazardous than those which are not.

Once an appropriate biosafety level is selected, biosafety can be maintained by adherence to good laboratory technique, use of the appropriate containment equipment, and by working in a laboratory designed for the biosafety level needed.

SUMMARY

The primary responsibility for the safe operation of any laboratory lies with the laboratory director whose judgment is critical in assessing the risks and helping to determine the appropriate biosafety level to be used for a particular agent. It is essential that the laboratory director establish policies and procedures whereby all laboratory personnel are advised of the potential hazards and receive the training necessary to handle biohazardous agents safely. Most (if not all) laboratory-acquired infections can be prevented by (1) assessing the risks,

(2) making laboratory workers aware of the potential hazards, (3) training laboratory workers in safe practices and techniques, (4) using the necessary safety and containment equipment, and (5) working in a properly designed facility.

REFERENCES

1. *Biohazards Reference Manual,* Akron, OH: American Industrial Hygiene Association (1985), p.7.
2. Hunter, D. *The Diseases of Occupations,* 6th ed., London: Hodder and Stoughton (1978), pp. 701–704.
3. Hunter, D. *The Diseases of Occupations,* 6th ed., London: Hodder and Stoughton (1978), p. 697.
4. Zaki, M. H. and D. C. Graham. *Lyme Disease,* Suffolk County Department of Health Services, Hauppauge, NY, Pub. No. 1084, p. 1.
5. "Guidelines for Prevention of Transmission of Human Immunodeficiency Virus and Hepatitis B Virus to Health-Care and Public Safety Workers," U.S. Department of Health and Human Services, Public Health Service, Centers for Disease Control, NIOSH, U.S. Government Printing Office (February 1989), pp. 2–4.
6. *Commercial Biotechnology: An International Analysis,* Washington, D.C.: U.S. Congress, Office of Technology Assessment, OTA-BA-218 (January 1984), p. 3.
7. Curtiss, R., III. "Genetic Manipulation of Microorganisms: Potential Benefits and Biohazards," *Ann. Rev. Microbiol.* 30:507–33 (1976).
8. Fuscaldo, A. A., B. J. Erlick, and B. Hindman. *Laboratory Safety Theory and Practice,* New York: Academic Press (1980), p. 162.
9. "Biosafety in Microbiology and Biomedical Laboratories," U.S. Department of Health and Human Services, Public Health Service, Centers for Disease Control, National Institutes of Health, U.S. Government Printing Office (1988), p. 3.
10. "Guidelines for Research Involving Recombinant DNA Molecules," FR 51:16958–16985 (1986).
11. Collins, C. H. *Laboratory-Acquired Infections,* London: Butterworths (1983), p. 3.
12. "Biosafety in Microbiology and Biomedical Laboratories," U.S. Department of Health and Human Services, Public Health Service, Centers for Disease Control, National Institutes of Health, U.S. Government Printing Office (1988), p. 1.
13. Pelczar, M. J., Jr., R. D. Reid, and E. C. S. Chan. *Microbiology,* New York: McGraw-Hill (1977), p. 500.
14. Smith, D. T., N. F. Conant, and J. R. Overman. *Microbiology, 13th ed.,* New York: Appleton-Century-Crofts (1964), p. 365.
15. Pelczar, M. J., Jr., R. D. Reid, and E. C. S. Chan. *Microbiology,* New York: McGraw-Hill (1977), p. 500.
16. *McGraw-Hill Concise Encyclopedia of Science and Technology,* New York: McGraw-Hill (1984), p. 1258.
17. "Guidelines for Research Involving Recombinant DNA Molecules," FR 51:16958–16985 (1986).

18. Pike, R. M. ''Laboratory-associated Infections: Summary and Analysis of 3,921 Cases,'' *Health Lab Sci.* 13:105–114 (1976).
19. ''Biosafety in Microbiology and Biomedical Laboratories,'' U.S. Department of Health and Human Services, Public Health Service, Centers for Disease Control, National Institutes of Health, U.S. Government Printing Office (1988), p. 2.
20. Darlow, H. M. ''Safety in the Microbiological Laboratory,'' in *Methods in Microbiology,* J. R. Norris and D. W. Robbins, Eds., New York: Academic Press (1969), pp. 169–203.
21. Chatigny, M. A. ''Protection Against Infection in the Microbiology Laboratory, Devices and Procedures,'' in *Advances in Applied Microbiology. No. 3.,* W. W. Umbreit, Ed., New York: Academic Press (1961), pp. 131–192.
22. Phillips, G. B. and S. P. Baily. ''Hazards of Mouth Pipetting,'' *Am. J. Med. Technol.* 32(2):127 (1966).
23. ''Laboratory Safety Monograph,'' U.S. Department of Health, Education, and Welfare, Public Health Service, National Institutes of Health (July 1978), pp. 32–37.
24. ''Biosafety in Microbiology and Biomedical Laboratories,'' U.S. Department of Health and Human Services, Public Health Service, Centers for Disease Control, National Institutes of Health, U.S. Government Printing Office (1988), p. 5.
25. ''Laboratories'' in *Handbook-HVAC Applications,* Atlanta, GA: ASHRAE (1991), p. 14.12.
26. ''Laboratories'' in *Handbook-HVAC Applications,* Atlanta, GA: ASHRAE (1991), p. 14.13.
27. ''Laboratories'' in *Handbook-HVAC Applications,* Atlanta, GA: ASHRAE (1991), p. 14.13.
28. ''Laboratories'' in *Handbook-HVAC Applications,* Atlanta, GA: ASHRAE (1991), p. 14.13.
29. ''Biosafety in Microbiology and Biomedical Laboratories,'' U.S. Department of Health and Human Services, Public Health Service, Centers for Disease Control, National Institutes of Health, U.S. Government Printing Office (1988), pp. 5–6.
30. ''Classification of Etiologic Agents on the Basis of Hazard,'' U.S. Department of Health, Education, and Welfare, Public Health Service, Centers for Disease Control (1969).
31. ''Biosafety in Microbiology and Biomedical Laboratories,'' U.S. Department of Health and Human Services, Public Health Service, Centers for Disease Control, National Institutes of Health, U.S. Government Printing Office (1988).
32. ''Guidelines for Research Involving Recombinant DNA Molecules,'' FR 51:16958–16985 (1986).
33. ''Biosafety in Microbiology and Biomedical Laboratories,'' U.S. Department of Health and Human Services, Public Health Service, Centers for Disease Control, National Institutes of Health, U.S. Government Printing Office (1988), pp. 11–28.

CHAPTER 3

Research Laboratory Ventilation Systems: A New Technology

Neil Hawkins and Stan Lengerich

INTRODUCTION

Ventilation systems in most laboratories consist of two components: (1) the laboratory fume hood and (2) the general supply and exhaust ventilation servicing the building. The primary purpose of the building supply and exhaust system is to keep occupants of the building ''comfortable'', i.e., within the ranges of acceptable temperatures, relative humidities, and other atmospheric requirements (adequate oxygen, removal of body odors, etc.).[1] In contrast to the general building ventilation system, the ''business end'' of a laboratory ventilation system is the laboratory fume hood. The primary function of the fume hood is to enclose and capture the vapors and aerosols which can be generated during laboratory use.

A fundamental notion in industrial hygiene is to control potential air contaminants *at their sources* before they can reach the breathing air of laboratory

0-87371-392-3/93/$0.00 + $.50

personnel. The laboratory fume hood, when functioning adequately, provides this protection.[2,3] Most organizations rely almost entirely on adequate hood performance, along with worker training, to ensure that laboratory workers are not being overexposed. Because most day-to-day work in research laboratories is non-routine, traditional industrial hygiene air monitoring programs usually are impractical — hence the reliance on control technologies to prevent overexposures to chemicals.[4,5]

While comfort ventilation and fume hoods have been separately introduced in Volume 1 of this series, it is the central theme of this chapter to highlight the interaction of these two ventilation system components, how this interaction has led to a new laboratory ventilation technology, and what issues need to be addressed when applying this new technology. Both the traditional systems and the new technology can be applied to protect laboratory workers from harmful chemicals — the key is to know the strengths and weaknesses of the systems.

TRADITIONAL APPROACHES

The simplest of the traditional approaches to laboratory ventilation is a laboratory fume hood with a dedicated blower (fan) on the roof of the building. This fume hood and associated blower may run continuously and may or may not have a bypass vent above the sash (door) of the hood (see Figure 1). On a hood without a bypass, the amount of air being pulled by the fan remains relatively constant; therefore, the higher the sash position, the lower the hood

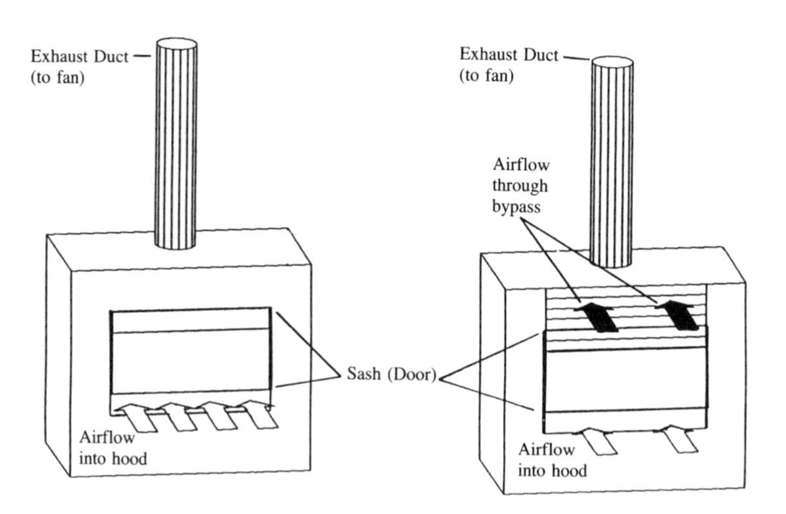

Figure 1. Two common laboratory fume hood ventilation systems. In the hood with no bypass (left), the face velocity varies with sash height. In the hood with the bypass (right), the velocity remains semi-constant regardless of sash height.

face velocity, and the lower the sash position, the higher the hood face velocity. The bypass feature is installed as an attempt to maintain a relatively constant face velocity at the fume hood opening. By allowing the bypass to open or close as the sash closes or opens, air velocity fluctuation through the hood face is somewhat controlled. A hood with a sash bypass is designed to keep a constant area opening, and therefore, a constant face velocity. However, there is not a one-to-one correlation between the face and bypass areas. The face velocity may become excessive since total air flow through the blower and the open hood area remains constant.[3]

Regardless of whether a hood is a bypass or non-bypass design, there is an optimal range of hood face velocities for controlling emissions from laboratory fume hoods often cited as ranging from 70 to 125 fpm.[6,7] If face velocities are too high, materials such as paper or foil inside the hood can be sucked up into the fan, flames can be extinguished, containers can be upended, and undesirable turbulent air currents can form within the fume hood leading to fume hood leakage. If face velocities are too low, emissions from experiments can escape the laboratory fume hood and pass into workers' breathing air.[6,7]

Pros and Cons of the Traditional Approaches

The single fan/single hood systems have the advantage of being relatively simple to maintain and operate. As long as users are properly trained and follow appropriate operating procedures, and as long as operating conditions remain relatively constant, the hoods should provide reliable protection for lab personnel. The traditional approaches can also have the advantage of a single blower servicing a single fume hood. Therefore, should a blower break down, only the individual fume hood connected to the malfunctioning fan would fail — not a group of laboratory hoods (as would be the case with manifolded or "ganged" systems, which are multiple hoods connected to single blowers). The single fan/single hood approach also has the advantage of preventing the mixing of exhaust fume streams from several laboratories, important when unpredictable reactive chemistry is potentially a problem. There is a long history of use of these types of systems, which demonstrates that when properly built, used, and serviced, laboratory workers can be adequately protected.

The primary disadvantage of single fan/single hood systems is one of economics because these systems are not cost-optimized relative to energy use. That is, the traditional systems have fans that pull a constant air flow regardless of the level of fume hood use. With the single fan/single hood systems, large volumes of conditioned air (i.e., heated or cooled and adjusted for comfortable relative humidity) may be unnecessarily exhausted out a roof stack (which is costly since all replacement air must be conditioned prior to distribution throughout the building). Ventilation system annual operating costs in large research laboratories can run into millions of dollars, making even modest percentage savings worthwhile.[4,10,11] Single fan/single hood systems also require a high initial capital expenditure, and that holes for ductwork be cut in the roof. These holes can often cause leaks and will raise the installation and maintenance costs for roofing.

Another drawback of traditional systems is the dependence on hood users to set sashes at or below the appropriate heights indicated for safe operation. With the traditional systems, hood operators can easily cause insufficient or excessive face velocities by improper setting of hood sash heights. Most of the single fan/single hood systems rely on duct static pressure monitoring (if any) for real-time indication of hood performance rather than face velocity measurements, the latter being more directly linked to hood performance.[6] Hoods without bypass systems may have wildly fluctuating face velocities which can produce swirling air currents — a source of fume hood leakage. Those hoods with a bypass also have face velocity fluctuation although usually to a lesser degree.[4] Finally, the traditional approaches to laboratory ventilation do not accommodate effective control of inter-room pressure differences, which is desirable for lab safety. It is good practice to maintain laboratories at slightly negative pressure relative to offices and hallways where chemicals are not in use, so that potential air contaminant releases are controlled within labs and do not spread into nonlab areas.[3,7,12]

Single fan/single hood systems are usually not directly connected with the general building exhaust/supply ventilation system, although building engineers do try to keep overall systems balanced. Conditioned supply air is directed to each section of the building in predefined amounts by general system air flows, but the presence of lab hoods can contribute to air flow imbalances. For example, if insufficient air flow is supplied to rooms with fume hoods, room doors may be difficult to open or close, and the blower for the fume hood in that room may be starved for air since an inadequate amount of air has been supplied to the room by the general building ventilation system. The single fan/single hood system is not able to compensate for inappropriate inter-room pressures (e.g., the pressure difference between a lab and a hallway).[8,9]

A number of different fume hood systems have been developed to address these problems. At a fundamental level, however, the problems are unsolvable unless a *comprehensive* ventilation system is designed which balances all the air flows into and out of a room within a predefined set of parameters (such as the range of acceptable hood face velocities or range of acceptable inter-room pressure differences).[4,10] The execution of this comprehensive system requires either a full-time building ventilation engineer or a microprocessor controlled system.

A NEW TECHNOLOGY: VARIABLE-AIR-VOLUME (VAV) SYSTEMS WITH MICROPROCESSOR CONTROL

A new technology which has emerged to address some of the problems of traditional approaches is the use of variable-air-volume (VAV) systems with microprocessor controls. Some of these VAV systems have been installed on individual hoods, but retrofitting of single fan/single hood systems can be prohibitively expensive. This chapter only considers the installation of a comprehensive VAV system throughout an entire laboratory.

Typically, a VAV system controls variables such as face velocities on all laboratory fume hoods, inter-room pressure differences, room air changes per hour (for comfort), and temperature and relative humidity of room air. These variables are monitored by automatic sensors which are built directly into laboratory fume hoods and mounted directly in walls between, for example, a hallway and a laboratory. These sensors operate on the basis of measuring either air velocity (e.g., a hot-wire anemometer) or pressure differential (e.g., a pressure transducer). Additionally, some systems operate using a sensor connected to the sash to electronically measure sash height. The general concept of these systems is shown in Figure 2.[4,9,10]

Often there are several large blowers drawing outdoor air into the building where it is first conditioned for temperature and relative humidity. This conditioned supply air (also known as makeup air) is then directed to all points in the building by ducts which have automatically controlled dampers to adjust the amount of air passing into different areas. The exhaust for these buildings will normally be controlled by several large blowers which primarily service laboratory fume hoods in the building. Typically, 20 to 30 hoods will be ganged to one blower; therefore, a laboratory with 100 hoods may have 4 or 5 primary exhaust blowers (see the next section for fan redundancy design).[9] The ganging of hoods in this way makes reactive chemical review necessary to ensure the safe placement of research activities into connected lab areas.

Each fume hood duct has an automatically controlled actuator which opens and shuts a damper to allow the proper amount of air flow for maintaining face velocity within a given range. For these ganged systems, the supply and exhaust blowers (both with variable motor speed control), duct damper actuators, and face velocity and room pressurization controllers are connected to a computer or microprocessor. The computer controls the dampers and the blower motor speeds to simultaneously maintain all predefined variables at their set points. For example, program criteria would frequently include the following: all hoods to operate within the range of 90 to 110 fpm face velocity, independent of their sash heights, whether hallway doors in the labs are open or shut, and the number of fume hoods in use at any given moment. This is a sizable programming and controller-technology challenge.

When a VAV system has been properly installed and adjusted, lab workers benefit by having hoods operating at face velocities within safe ranges (too low or too high is unacceptable). Workers in non-lab environments benefit from the pressure differential control which minimizes the release of lab contaminants into non-lab areas. Furthermore, building ventilation (HVAC) operating costs are minimized since conditioned air is exhausted only when hood sashes are open.[10]

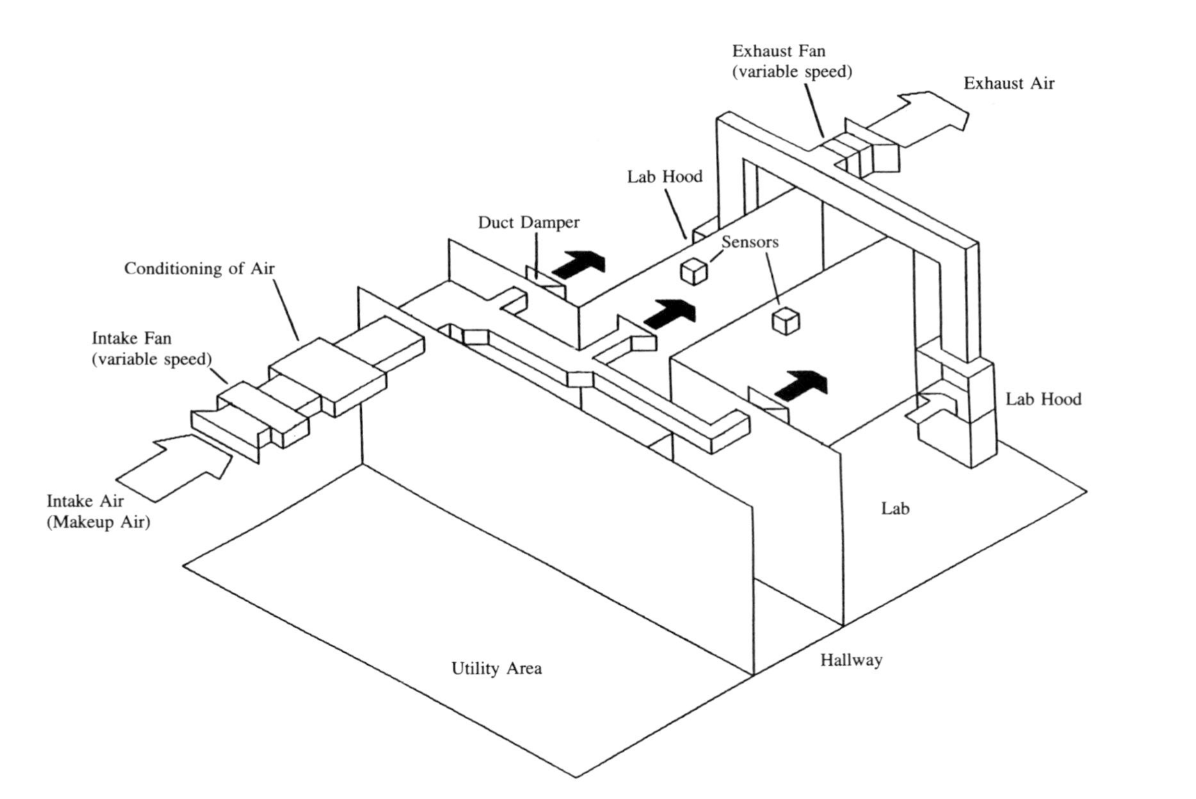

Figure 2. An example of a comprehensive variable-air-volume (VAV) system made up of a utility room, a hallway and two labs. Intake air from the outside is pulled in by a variable speed intake fan, then conditioned. Air is distributed to individual rooms (black arrows) and then exhausts, via the lab hoods and the variable speed exhaust fan. Sensors detect differences in pressure (between rooms, shown here mounted on the walls) or differences in velocity (at the hood face). A computer control balances overall airflow in response to the sensors by varying the fan speeds and using duct.

Using VAV Technology With Microprocessor Control

While the new VAV technology shows promise, there are a number of issues which must be addressed. Many of these issues are common to those encountered with traditional systems.

A VAV system is no better than each individual sensor installed for measuring hood face velocity or inter-room pressure difference. A great deal of maintenance and calibration is needed in operating these systems effectively, even after a sizable time investment is made in initially getting a system up and running as designed. Sensor/controller features of importance include the amount of maintenance and calibration required once installed, and from a functional point of view, the speed of sensor/controller action. When watching a duct damper operating in a system such as this, damper oscillation (also known as "hunting") while searching for the predefined set point is sometimes noticeable. Some sensors/controllers end the search much more quickly than others, and these are the most desirable for worker health protection. Sash height sensors react most quickly, but they have some disadvantages because they do not directly measure face velocity or pressure differential. One of the most effective sensors in terms of speed and accuracy has been the hot-wire type, mounted in the side wall of lab hoods. It is unclear what effect oscillating dampers have on hood performance during transient periods, but such oscillation is certainly not beneficial and should be minimized where possible by use of fast-acting sensors.[8,10,13]

Industrial hygiene personnel must have input early into any new lab project with respect to design and contract specifications (Table 1 provides a summary of important points to consider). For large R and D labs with this type of system, the ventilation system is an integral part of the overall building design. Engineers and architects may tend to be less responsive to industrial hygiene concerns as projects advance because changes in the ventilation system may cause radical changes in overall building plans (affecting look, layout, cost, utilities, etc.).

Exhaust fans on ganged fume hood systems (e.g., up to 30 hoods per fan) should be installed with a dedicated backup blower (or blowers). If an exhaust blower malfunctions with no dedicated backup, all (e.g., 30) hoods serviced by the blower will malfunction, resulting in a potentially negative impact on worker health and safety (and jeopardizing research, experiments, or production

Table 1. Considerations When Using VAV Technology

- Industrial hygiene input in early design stages
- Quick, accurate, and reliable control sensors are needed; hot-wire type appear to be the best
- Redundant fans for "ganged" hood systems
- Reactive chemical reviews for "ganged" hood systems
- Real-time indicators of hood performance on each hood
- Select well-designed lab hoods (do not rely solely on contractor selections)
- Ongoing, periodic hood evaluation program once the systems is installed (preferably by quantitative performance testing)

in progress). One solution is to install two or more smaller blowers instead of one large one. By wiring them so that they can operate individually or together, an extra boost can be provided when fume hood demands are high.[9] As an example, the fans may be sized so that each can pull at least 60 to 80% of the estimated maximum capacity for the system. This provides a number of advantages for installing, operating, and maintaining the system. First, if one fan malfunctions, the other fan(s) can pick up all or most of the load so that work in the hoods can continue safely while the malfunctioning fan is being repaired. Second, when less than normal capacity is required, fewer fans need to be operating. Third, smaller motors and blowers are easier to install and repair by virtue of their size. Fourth, two or more small fan assemblies can often be purchased for approximately the same cost as one large unit. Finally, the multi-fan design with redundant capacity allows for ventilation system expansion in the future without the need for upgrading to a larger fan.

Only some manufacturers of VAV systems provide a real-time indicator of air flow on the control unit at each hood. In these new VAV lab hood system applications there is a trend to not install manometers, face velocity meters, or other indicators of hood flow on the fume hoods when it is not part of the package. This frustrates and angers laboratory researchers, and it also removes the opportunity for proactive lab personnel to monitor the performance of their own hoods and notify building maintenance when problems are noticed.[6] Current hood guidelines already strongly recommend real-time hood monitoring devices, and pending government standards will likely require them in the near future.[3,5,7,12] When static pressure taps are installed for either a manometer or for an alarm system, they should not be installed downstream of the duct damper for the laboratory fume hood. Installation downstream of the damper will always indicate that static pressure is present on the hood, even when the damper of the hood is closed.

Industrial hygienists must be involved with the selection of laboratory fume hood models. Poorly designed hoods provide inadequate control of contaminants due to turbulent air current formation in the hood and vapor leakage from the hood. For example, hoods should have an air-foil design around the perimeter of the hood opening, including the bottom of the sash.[2,6] Often, hood selection becomes part of a large ventilation package offered by a contractor, who supplies the specifics on the hood types "at a later date". Hood types and models should be specified early in contract negotiations. Also, it would be a good idea to specify in the contract that all hoods meet predefined performance characteristics after the installation of the complete system. The requirements specified should include, as a minimum, meeting face velocity requirements on all hoods under various conditions of use. A more advanced contract requirement would require the application of hood performance testing (e.g., ASHRAE #110-1985[14] or modified ASHRAE performance testing procedures) when the job has been completed to evaluate vapor leakage from hoods under operating conditions.

Industrial hygienists must have an active hood testing program to identify those hoods not performing adequately. These surveys should at least consist of face velocity checks and smoke tests, but ideally should include quantitative performance tests for hood leakage.[15] The surveys should be performed at a specified frequency (e.g., annual) or at the request of laboratory personnel.[3,7] Lab personnel are in the best position to evaluate the day-to-day performance of their own hoods, and often are the first to realize that operating conditions have changed significantly.

SUMMARY

This chapter has presented a very brief introduction to the overall concepts of laboratory ventilation systems, some traditional approaches and their issues, and one new technology which has emerged to address these issues. The new technology, variable-air-volume (VAV) systems with microprocessor control, has the potential for excellent worker protection and optimized economy. Because of this, it is expected that more of these systems will be designed and installed. Caveats have been offered here to assist occupational health professionals (among others) in using these systems to provide adequate worker health protection. The key to making this happen is informed industrial hygiene input on the project from its inception and good communication with design engineers. These new, comprehensive VAV systems can only be fully optimized to protect worker health during initial design and contract specification stages.

REFERENCES

1. DiBerardinis, L., J. S. Baum, M. W. First, G. T. Gatwood, E. F. Groden, and A. K. Seth. *Guidelines for Laboratory Design: Health and Safety Considerations,* New York: John Wiley and Sons (1987).
2. Mikel, W. G. and L. R. Hobbs. "Laboratory Hood Studies," *J. Chem. Educ.* 58:A165–A168 (1981).
3. Chamberlin, R. I. and J. E. Leahy. "Laboratory Fume Hood Standards Recommended for the U.S. Environmental Protection Agency." Contract #68-01-4661 (January 15, 1978).
4. Monsen, R. R. "Practical Solutions to Retrofitting Existing Fume Hoods and Laboratories," *ASHRAE Transactions,* Vol. 95, Pt. 2 (1989).
5. OSHA. "Health and Safety Standards: Occupational Exposure to Toxic Substances in Laboratories, Proposed Rule," *Federal Register* (July 24, 1986).
6. Sanders, G. T. "Updating Older Fume Hoods," *J. Chem. Educ.* 62:A175–A181 (1985).

7. SAMA. ''Standard for Laboratory Fume Hoods,'' Scientific Apparatus Makers Association, LF-10-1980.
8. Lacey, D. R. ''Observed Performance of VAV Hood Controls,'' *ASHRAE Transactions,* Vol. 95, Pt. 2 (1989).
9. Shumaker, R. B. ''Users' Experiences with Variable Volume Laboratory Fume Hood Exhaust Systems,'' *ASHRAE Transactions,* Vol. 95, Pt. 2 (1989).
10. Wenz, R. G. ''A Practical Laboratory Ventilation Control System,'' *ASHRAE Transactions,* Vol. 95, Pt. 2 (1989).
11. Wiggin, M. E. and R. H. Morris. ''Electronic Control for Fume Hoods,'' *Heating, Piping and Air Conditioning* (February 1988).
12. Kempsell, H. F., Ed. ''Standard on Fire Protection for Laboratories Using Chemicals,'' *NFPA 45* (1986).
13. Garrison, R. P., Y. Dong, S. K. Lengerich, and T. M. Rabiah. ''Performance Characteristics of Control Devices to Maintain Constant Face Velocity for Laboratory Exhaust Hoods,'' *Am. Ind. Hyg. Assoc. J.* 50(9):501 (1989).
14. ASHRAE Standard 110-1985, ''Method of Testing Performance of Laboratory Fume Hoods,'' Atlanta, GA (1986).
15. Ivany, R. E., M. W. First, and L. J. DiBerardinis. ''A New Quantitative Method for In-Place Testing of Laboratory Hoods,'' *Am. Ind. Hyg. Assoc. J.* 50:275 (1989).

PART II

Healthcare Provider Exposures

Part II, **Healthcare Provider Exposures,** contains Chapters 4, 5, and 6:

Chapter 4, **Occupational Exposure to Bloodborne Pathogens, HBV and HIV,** deals with the most explosive of healthcare issues, hepatitis B (HBV) and human immunodeficiency virus (HIV, the precursor to AIDS). The chapter discusses Centers for Disease Control (CDC) recommendations regarding HBV and HIV (these documents are contained in Appendices B and C to this volume), and also discusses implementation of 29 CFR 1910.1030, ''Occupational Exposure to Bloodborne Pathogens; Final Rule'' (Appendix D). All facets in the prevention of exposure to bloodborne pathogens, including a comprehensive table on how to apply universal precautions, are included in this chapter.

Chapter 5, **Tuberculosis: Resurgence and Control,** discusses the recent increase in tuberculosis in the United States, including what populations are at risk (e.g., the homeless and HIV-positive individuals) and what occupational populations (e.g., healthcare providers such as social workers) are at risk. Included is a general explanation of what tuberculosis is, how it is transmitted, how to test for it, and how its transmission can be minimized (CDC guidance on tuberculosis is included as Appendix E).

Chapter 6, **Occupational Health Hazards in the Dental Office,** examines problems unique to dentists and dental hygienists. In addition to confronting the same potential exposures as faced by many in laboratories or healthcare (such as to bloodborne pathogens or to tuberculosis), there are additional, different hazards in dentistry. These include radiation from x-ray machines, ergonomic problems resulting from maintaining awkward positions for long periods, chemical exposures from dental amalgams, and others. This chapter gives clear examples for resolving problems specific to the dental profession.

CHAPTER 4

Occupational Exposure to Bloodborne Pathogens

Marc A. Gomez

INTRODUCTION

Throughout the 1980s, and now into the 1990s, occupational exposure to bloodborne pathogens has become an increasing concern. Bloodborne pathogens are defined by OSHA to mean "pathogenic microorganisms that are present in human blood and can cause disease in humans. These pathogens include, but are not limited to, hepatitis B virus (HBV) and human immunodeficiency virus (HIV)."[1] These bloodborne pathogens may be transmitted from an infected individual to other individuals when blood or certain other body fluids are exchanged. Biological hazards (biohazards) such as these bloodborne pathogens present new challenges to occupational health. It is important to realize that in contrast to hazardous chemical agents in the work place, hazardous biological agents such as bloodborne pathogens have the ability to replicate. Thus, "safe" doses of chemical and physical agents can be defined, however, there is no safe level of a bloodborne pathogen.

0-87371-392-3/93/$0.00 + $.50

UNDERSTANDING HEPATITIS B

HBV Etiology

Hepatitis means "inflammation of the liver" and is caused by various agents. The most common causative agent of hepatitis is viruses. A virus is a submicroscopic pathogen that consists essentially of a protein shell or envelope that protects a string of genetic material (DNA or RNA). Viruses live and reproduce within living cells.

Hepatitis B, formerly called "serum hepatitis", is caused by HBV. HBV has an inner core structure containing DNA, enzymes, and proteins. The most important protein is the hepatitis B core antigen. The outer shell is comprised of a lipoprotein called hepatitis B surface antigen.[2] In fact, the hepatitis B core antigen and the hepatitis B surface antigen are the important markers in the serodiagnosis of hepatitis B infection. Once into the body, this virus attacks and replicates itself in liver cells.

HBV Epidemiology

Hepatitis B infection does not occur uniformly throughout the U.S. population. The disease is encountered more often in certain ethnic and racial groups, and is especially prevalent in various groups related to occupation.

The percentage of the U.S. population who are hepatitis B carriers (that is, at risk of developing chronic liver disease and capable of transmitting the disease to others) is approximately 0.2% for whites, 0.7% for blacks, and as high as 13% for foreign born Asians.[3] In 1987, the Centers for Disease Control (CDC) estimated the total number of HBV infections in the general U.S. population to be 300,000 per year, with approximately 75,000 (25%) of infected persons developing acute hepatitis. Of the estimated 300,000 total infected individuals, 18,000 to 30,000 (6 to 10%) will become HBV carriers, at risk of developing chronic liver disease (chronic active hepatitis, cirrhosis, and primary liver cancer), and capable of transmitting the disease to others.[4]

In the 1970s the risk of HBV infection to healthcare workers became well defined by the CDC. In the 1980s, there were many studies completed which documented that HBV infection risk was directly related to various healthcare occupational groups. A subsequent section of this chapter ("Who is at Risk?") will discuss specific types of healthcare personnel that are at high risk.

HBV Pathology

When an individual becomes infected with the Hepatitis B virus, two responses are produced, self-limited *acute* hepatitis B and *chronic* hepatitis B infection.[5] The most frequent response seen in healthy adults is the development of self-limited acute hepatitis and the corresponding production of hepatitis antibody. The production of this hepatitis antibody coincides with the destruction of liver cells containing the virus, elimination of the virus from the body, and lifetime immunity against reinfection. Following this acute infection with HBV and corresponding production of an antibody, approximately one third of all infected individuals will experience no symptoms, one third will experience a mild flu-like illness (which is usually not diagnosed as hepatitis), and one third will experience more severe symptoms such as fatigue, anorexia, nausea, dark urine, abdominal pain, fever, and jaundice (a yellowing of the eyes and skin). These more severe symptoms occur because the destruction of liver cells, in the body's attempt to rid itself of the infection, often leads to this clinically apparent acute hepatitis B.

The second type of hepatitis disease outcome, which is even more severe, is chronic hepatitis B infection. Upon infection with HBV, approximately 6 to 10% of all cases cannot eliminate the virus from their liver cells. In these cases, hepatitis antibody is produced in the body for many years, usually for life. This 6 to 10% of all hepatitis cases become chronic HBV carriers and are at a high risk of developing chronic persistent hepatitis, chronic active hepatitis, cirrhosis of the liver, and primary liver cancer. Chronic persistent hepatitis is a relatively mild, non-progressive type of liver disease experienced by approximately 25% of these carriers (the 6 to 10% group). The chronic active hepatitis is a progressive, debilitating disease that can lead to cirrhosis of the liver after 5 to 10 years (this is experienced by another 25% of the carriers). This condition may lead to fluid accumulation in the abdomen, esophageal bleeding, coma, and death.

UNDERSTANDING AIDS

HIV Etiology

Acquired immunodeficiency syndrome (AIDS) was first recognized in the U.S. in the summer of 1981 and has now been reported in all 50 states, as well as in more than 80 countries worldwide.[6] This disease weakens the body's immune system leaving the infected individual susceptible to many life-threatening "opportunistic" diseases and cancers which are not ordinarily fatal. AIDS is caused by the appropriately named human immunodeficiency virus (HIV). HIV is a member of a group of viruses known as the human retroviruses. Structurally, this virus has an inner core containing its genetic material, called ribonucleic acid (RNA), which is surrounded by a shell consisting of lipids and proteins.[7] Once in the body, HIV gradually depletes the body's cells which are normally used for the immune response, thus rendering the infected individual increasingly susceptible to opportunistic infections.

Table 1. AIDS Cases by Transmission Group in the U.S., 1988

Transmission Group	Cumulative Total AIDS Cases (%)
Homosexual/bisexual individuals	62
Intravenous (i.v.) drug users	20
Homosexual/bisexual men who are also i.v. drug users	7
Heterosexual transmission	4
Transfusion recipients	3
Persons with hemophilia disorders	1
Undetermined source	3

HIV Epidemiology

Every year, the number of reported AIDS cases increases alarmingly. Over the first five and a half years of the disease's existence, nearly 30,000 AIDS cases were reported to the CDC.[8] By the beginning of 1989, the number jumped to nearly 89,000 cases.[9] It is interesting to note that the number of AIDS cases reported each year continues to increase, however, the rate of increase has steadily declined, except in 1987, when the case definition for AIDS was revised (resulting in an abrupt increase in reported cases). Most adults with AIDS in the U.S. can be placed into well-established transmission categories as shown in Table 1.[10]

The AIDS epidemic has expanded in scope and magnitude as HIV infection has affected different populations and geographic areas throughout the U.S. Although homosexual/bisexual individuals continue to account for most AIDS cases, cases associated with i.v. drug use are more common in several northeastern states. In 1990, the incidence of AIDS increased most rapidly among persons exposed to HIV through heterosexual contact. In addition, the rate of increase of AIDS cases was greatest in the southern states.[11]

Also in the U.S., over 1000 cases of AIDS have occurred in children. Most (78%) of these cases have been children born to mothers infected with HIV. These mothers transmit the virus to their children prenatally, during birth, or during infancy through breast feeding, although the latter is rare. Other transmission categories for children are hemophiliacs (6%) and transfusion recipients (13%).[12]

The significance of all these statistics, with respect to potential occupational exposure, resides in the fact that the CDC estimates between 1 and 1.5 million persons in the U.S. are now infected with HIV.[13] The risk to healthcare workers becomes apparent since these 1 to 1.5 million often unknowingly infected individuals eventually require medical treatment for related and unrelated conditions. On a global basis, the outlook is even worse. The World Health Organization (WHO) estimates that 8 to 10 million adults and 1 million children worldwide are infected with HIV. By the year 2000, 40 million persons may be infected with HIV.[14] More than 90% of these persons will reside in developing countries.

HIV Pathology

AIDS is primarily a disease of the body's immune system. The human immunodeficiency virus attacks the immune system leaving the infected individual vulnerable to a wide range of diseases that usually lead to death. The CDC has divided the progression of HIV infection into several stages, Groups I through IV, depending upon the type of signs or symptoms of infection:[15]

Group I — Upon infection with HIV, many individuals show no immediate symptoms. In some cases, however, within a month after exposure, the first evidence of HIV infection may appear as a mononucleosis type flu named "acute retroviral syndrome". The signs and symptoms of this illness include, fever, swollen lymph glands, diarrhea, fatigue, and rash.

Group II — Typically, it takes 6 to 12 weeks after HIV enters the body before antibodies to the virus are detectable in blood samples. However, individual reactions vary, and this seroconversion may take 8 months or even longer. As previously mentioned, the majority of cases show no symptoms for months to years after infection. It is apparent that one of the most dangerous considerations of the AIDS disease is that these individuals can transmit the virus to others throughout this time.

Group III — Some HIV-infected individuals will develop an unusual persistence of one or more of the clinical symptoms shown in Table 2. At this time, there may be no indication of abnormal function of the immune system. And in the absence of other explanations, unusual persistence of two or more of these symptoms for more than three months is cause for concern. These symptoms are typical of what is known as AIDS-related complex (ARC) or lymphadenopathy.

Group IV — The clinical manifestations of this "full-blown AIDS" group may vary extensively. Some of these patients may develop what is known as HIV "wasting syndrome" characterized by severe involuntary weight loss, chronic diarrhea, weakness, and a long term fever of a month or more. This syndrome itself may result in death. As mentioned previously, an individual with full-blown AIDS is susceptible to opportunistic infections that an individual with a normal immune system would only rarely experience. Therefore, there are specific diseases that are considered indicators of AIDS. Among these are parasitic diseases such as *pneumocystis carinii* pneumonia (the most common opportunistic infection and cause of death); fungal diseases such as candidiasis of the

Table 2. Clinical Symptoms of HIV-Infected Individuals

- Fatigue or listlessness
- Weight loss of 10–15 lbs or 10% of body weight
- Fever of 100°F
- Drenching night sweats
- Swollen lymph nodes in the neck and/or armpits in addition to the groin area
- Diarrhea

esophagus, trachea, bronci or lungs; viral diseases such as cytomegalovirus disease; cancer and neoplastic diseases such as Kaposi's sarcoma; and bacterial infections such as mycobacterium avium complex.

AIDS is a uniformly devastating disease. Since AIDS was first recognized and reported in 1981, more than 179,000 persons with AIDS have been reported to public health departments in the U.S. Of these, more than 113,000 (63%) are reported to have died. During this period, HIV infection has emerged as a leading cause of death among men and women less than 45 years of age and children 1 to 5 years of age in the U.S.[16]

ROUTES OF TRANSMISSION OF AIDS AND HEPATITIS B

In the discussion of how HBV and HIV are transmitted to humans, it is most important to remember that not everyone who comes into direct contact with these viruses becomes infected. The simple analogy is that not everyone exposed to a flu virus contracts the flu. The point here is that there is a difference between *exposure* (the opportunity for viral invasion) and *infection* (the virus actually enters a living cell). The question that now must be answered is what *type* of exposure leads to infection.

Modes of Transmission — General Public

The routes of transmission for HBV and HIV are known to be very similar for the general public. HBV and HIV are not transmitted by casual contact. The three generally accepted modes of transmission are as follows:[17]

1. Through parenteral (direct inoculation through the skin) injection of virus contaminated blood or blood products; this includes blood transfusions, needle-sharing among drug users, and contact with an open wound or non-intact skin.
2. Through sexual contact in which there is exchange of infected body fluids, i.e., semen, vaginal secretions.
3. From infected mothers to their babies, including in the uterus, during the birth process, and (less likely) through breast milk.

As public health officials sum it up, the major transmission routes for the general public are via blood, sex, and birth.

Modes of Transmission in the Workplace: Patient to Healthcare Worker

In the occupational setting, blood is the single most important source of HBV and HIV infection. Both of these viruses have been transmitted in the workplace by the following:[18]

1. Parenteral (direct inoculation through the skin) injection which includes contact with an open wound, contact with non-intact skin (chapped, abraded, weeping skin), and injections through the skin (needle sticks and cuts with sharp instruments).
2. Mucous membrane exposure which includes blood or blood containing body fluid contamination of the eye or mouth.

CDC has estimated that 12,000 healthcare workers whose job entail exposure to blood become infected with hepatitis B each year. Of this 12,000, 500 to 600 are hospitalized as a result of the infection, and 700 to 1200 of those infected become HBV carriers. Of the 12,000 hepatitis B infected workers, approximately 250 will die (12 to 15 from fulminant hepatitis, 170 to 200 from cirrhosis, and 40 to 50 from liver cancer).[19] These statistics support the fact that 10 to 30% of all healthcare workers show serologic evidence of past or present HBV infection.

On the other hand, occupational transmission of HIV has been documented in only a few healthcare workers. CDC has identified only 25 cases where HIV infection is *directly* related to occupational exposure.[20] These 25 cases represent a diverse group of healthcare personnel including nurses, laboratory personnel, and a dentist. In 16 of these cases, exposure to the blood of HIV-infected individuals occurred by needle stick. Two infections are believed to have resulted from cuts with sharp, HIV-contaminated objects. In the remaining seven cases, HIV exposure occurred via mucous membrane or non-intact skin.

It is important to realize that the potential for HBV transmission is greater than the potential for HIV transmission. The risk of hepatitis B infection (for an individual who has not had prior hepatitis B vaccination) following a parenteral exposure (such as a needle stick from a hepatitis B carrier) is approximately 6 to 30%.[21] The risk of infection with HIV, following a needle stick exposure to blood from a patient known to be infected with HIV, is approximately 0.35%.[22] This rate of transmission is considerably lower than that for HBV, probably as a result of the significantly lower concentrations of virus in the blood of HIV infected persons.

Besides blood, there are numerous other body fluids which are either known or suspected in the transmission of HBV and/or HIV. They include cerebrospinal fluid (CSF), synovial fluid, pleural fluid, peritoneal fluid, pericardial fluid, amniotic fluid, semen, and vaginal secretions. There are several other body substances to which the risk of HBV and/or HIV transmsision is extremely low or nonexistent including feces, nasal secretions, sputum, sweat, tears, urine, and vomitus *unless* they contain visible blood.[23]

Modes of Transmission in the Workplace: Healthcare Worker to Patient

It is also worth mentioning that not only is there documentation of HIV transmission from patient to healthcare worker, but in 1991 the first evidence to strongly suggest HIV transmission from healthcare worker to patient was discovered.[24] The CDC investigation of a Florida dentist with AIDS revealed it is likely that 3 to 5 patients became infected with HIV while receiving dental care. Neither the precise mode of HIV transmission to these patients nor the reasons for transmission to multiple patients in a single practice are known. However, HBV has also been transmitted to multiple patients in the practice of HBV-infected healthcare workers during invasive procedures.[25-27] Most reported transmissions of HBV to clusters of patients in the U.S. occurred before awareness increase of the risks of transmission of bloodborne pathogens and before emphasis was placed on the use of universal precautions and hepatitis B vaccine among healthcare workers. Factors that may be associated with this transmission of bloodborne pathogens from infected healthcare workers to patients include variations in procedures performed and techniques used by the healthcare worker, infection control precautions used, and the titer of the infecting agent.

WHO IS AT RISK?

From an occupational health standpoint, anyone who comes into contact with (or has the potential to come into contact with) another individual's blood or body fluids while working is at risk of occupational exposure to bloodborne pathogens. Healthcare workers are by far the largest group that fall into this ''at risk'' category. The CDC has defined healthcare workers as ''persons, including students and trainees, whose activities involve contact with patients or with blood or other body fluids from patients in a healthcare setting''.[28] There are many types of healthcare workers identified as being ''high risk'', these include (but are not limited to) those listed in Table 3.

Besides healthcare personnel, there are other non-healthcare operations that have a serious potential exposure to bloodborne pathogens for their workers, examples of these are listed in Table 4.

REGULATORY REQUIREMENTS

Hepatitis B virus has been recognized as a pathogen capable of occupational transmission causing serious illness and death for the last few decades. HIV, on the other hand, has only been recognized as a serious occupational health issue over the past few years. During the 1980s both CDC and OSHA became involved in making recommendations and setting regulatory requirements to control occupational exposure to bloodborne pathogens.

Centers for Disease Control

In 1983, the CDC published a report titled *Guidelines for Isolation Precautions in Hospitals*. One section of this document, *Blood and Body Fluid Precautions,*

Table 3. Examples of Healthcare Workers With High-Risk Exposure to Bloodborne Pathogens

- Physicians, i.e., surgeons and pathologists
- Dentists, i.e., periodontists, oral surgeons, endodontists
- Dental professionals, i.e., dental hygienists and assistants
- Nursing professionals, i.e., intravenous therapy nurses, critical care nurses
- Laboratory personnel, i.e., phlebotomists, blood bank technicians, medical technologists
- Operating room personnel
- Dialysis unit personnel
- Emergency room personnel
- Laundry and housekeeping personnel
- Emergency medical technicians

recommended that blood and body fluid precautions be taken when a patient was known or suspected to be infected with bloodborne pathogens. The CDC later published a report titled *Recommendations for Prevention of HIV Transmission in Health-Care Settings* (Appendix B to this volume). This 1987 report, in contrast to the 1983 document, recommended that blood and body fluid precautions be consistently taken for *all* patients regardless of their bloodborne infection status. This extension of blood and body fluid precautions to all patients is referred to as ''universal blood and body fluid precautions'' or ''universal precautions''. Under universal precautions, the blood and certain body fluids of all patients are considered to be potentially infectious for HIV, HBV, and other bloodborne pathogens.

In July 1991, the CDC released *Recommendations for Preventing Transmission of HIV and HBV to Patients During Exposure — Prone Invasive Procedures* (Appendix C to this volume).[29] These recommendations were developed to update their previous recommendations for prevention of HIV and HBV transmission in healthcare settings. The recommendations state that as long as the healthcare worker adheres to appropriate infection control procedures, the risk of transmitting HBV from an infected healthcare worker to a patient is small, and the risk of transmitting HIV is likely to be even smaller. However, the recommendations state the likelihood of exposure of the patient to the blood of a healthcare worker is greater for certain procedures designated as ''exposure-prone''. Thus, healthcare workers who perform exposure-prone procedures should know their HBV and HIV status. Healthcare workers who are infected with HIV or HBV should not perform exposure-prone procedures unless they notify prospective patients of their seropositivity and have sought counsel from an expert review

Table 4. Examples of Non-Healthcare Workers With "High Risk" Exposure to Bloodborne Pathogens

- Morticians' services personnel (post-mortem procedures)
- Firefighters
- Law enforcement personnel
- Correctional-facility personnel
- Personnel involved in infectious waste disposal
- Personnel involved in service and repair of medical equipment, i.e., biomedical technicians

panel and been advised under what circumstances, if any, they may continue to perform these procedures.[30] The CDC recommendations do not call for mandatory testing of healthcare workers for HIV or HBV. At this time, the CDC believes the current assessment of the risk that healthcare workers will transmit HIV or HBV to patients during exposure-prone procedures does not support the diversion of resources that would be required to implement mandatory testing programs.

Occupational Safety and Health Administration

In 1983, OSHA issued a set of voluntary guidelines designed to reduce the risk of occupational exposure to HBV. These guidelines were sent to healthcare employers throughout the U.S. Then in November of 1987, OSHA published an *Advanced Notice of Proposed Rulemaking* announcing the initiation of the rulemaking process to develop a standard for protecting employees from HBV and HIV. OSHA received a tremendous response from this request for information relevant to reducing occupational exposure to these bloodborne pathogens. OSHA then analyzed all the data received, and in May 1989 the agency published its proposed rule governing *Occupational Exposure to Bloodborne Pathogens* (Appendix D to this volume). In order to solicit feedback on the proposed rule, written comments and public hearings were accepted, and public hearings were held through October 1989. In December 1991, after more than two years of comments and controversy, OSHA issued its final standard to protect workers from bloodborne pathogens. This standard, which went into full effect in July 1992, represented the first regulation of occupational exposure to biological hazards from OSHA. The following is a summary of the key provisions of the final bloodborne pathogens standard as it applies to each employer having employees with occupational exposure to blood or other potentially infectious materials.[31]

Exposure control plan — Each employer must develop a written exposure control plan to eliminate or minimize employee exposure to bloodborne pathogens. The plan must include an exposure determination to identify potentially exposed employees. The plan should also include a schedule and method for implementing other provisions of the standard, such as hepatitis B vaccination, post-exposure evaluation and follow-up, communications of hazards to employees, and recordkeeping.

Methods of compliance — Engineering and work practice controls are designated as the primary means of eliminating or minimizing employee exposures. The standard defines engineering controls as ''controls that isolate or remove the bloodborne pathogens hazard from the workplace''. Also, universal precautions, which prevent contact with blood or other potentially infectious materials, are mandated by the standard.

Housekeeping — Employers must ensure that the worksite is clean and sanitary. All equipment and surfaces must be cleaned and decontaminated after contact with blood or other potentially infectious materials.

Hepatitis B vaccination — Employers must make the hepatitis B vaccination available at no cost to employees who have occupational exposure.

Communication of hazards to employees — Warning labels are required on all containers of regulated waste, refrigerators, and freezers containing blood or other infectious materials and other containers used to store, transport, or ship blood or other infectious materials.

Information and training — All employees with occupational exposure must participate in a training program provided at no charge. Training must be provided at the time of initial assignment and at least annually thereafter.

Recordkeeping — Each employer must establish and maintain employee medical records that include name, social security number, hepatitis B vaccination status, and results of examinations, tests, and follow-ups.

METHODS OF DISEASE CONTROL

The significant risk that bloodborne pathogens represent can be minimized (or eliminated) using a combination of vaccination (where applicable), engineering controls, work practice controls, personal protective equipment and clothing, training, and medical follow-up of exposure incidents.

Vaccination

The most effective way to control the risk of hepatitis B infection is through immunization of workers at risk with the hepatitis B vaccine. This vaccine induces protective antibody levels in healthy adults and, in most cases, provides the individual with immunity to hepatitis B. Unfortunately, no vaccine currently exists to protect workers from the risk of AIDS.

In 1982, the first hepatitis B vaccine was licensed in the U.S. The vaccine was recommended for healthcare employees with the risk of blood and body fluid exposures. Early efforts to immunize healthcare workers were slowed by fear that the blood plasma used in the vaccine might contain HIV causing AIDS. It was found that the procedure used to produce the vaccine inactivated the HIV virus.[32] Thus, it was shown that individuals receiving the vaccine did not develop AIDS.

In 1987, a second hepatitis vaccine, synthetically produced in yeast by recombinant technology, was licensed for use. This vaccine is not prepared from blood, therefore, it did not encounter the same resistance to use as its predecessor. This vaccine, called ''recombivax'', is given in three doses over a six month period. It induces protective antibody levels in 85 to 97% of healthy adults and is believed to provide protective immunity at least 7 years.[33]

In 1988, the CDC estimated that 1.4 million individuals in the U.S. had received the hepatitis B vaccine and that 85% of these people were healthcare workers. Hepatitis B vaccination is the single most important component of reducing occupational exposure to HBV. Unfortunately, it is estimated that only 30 to 40% of high-risk healthcare workers have been vaccinated.[34] It is widely agreed that *all* workers at high risk of exposure to HBV should receive the hepatitis B vaccine. As previously mentioned, in OSHA's final standard governing occupational exposure to bloodborne pathogens, the Agency requires that the vaccine shall be offered to all employees with occupational exposure to blood or other potentially infectious materials, unless it has been determined through antibody testing that an employee is immune or has previously received the vaccine.

Universal Precautions

It has now become widely accepted that because a medical history and examination cannot reliably identify all patients infected with HIV, HBV, or other bloodborne pathogens, blood and body fluid precautions should be consistently used for all patients. This approach, recommended by CDC in 1987 and then made law in 1991 by OSHA, is referred to as universal blood and body fluid precautions or universal precautions. These precautions should be followed in the care of *all* patients, especially in emergency care settings when the risk of blood exposure is increased and the infection status of the patient is usually unknown. This philosophy is advocated by the majority of healthcare associations including the American Hospital Association, the American Dental Association, and OSHA.

Personal Protective Equipment (PPE)

Inherent in the theory of universal precautions is the universal usage of PPE. Personal protective equipment refers to specialized clothing or equipment worn by an individual to protect them from a hazard. When dealing with bloodborne pathogens, PPE is used to prevent the entry of pathogens into the workers body. This includes entry via skin lesions, or entry through the membranes of the eye, nose, or mouth. Typical PPE for bloodborne pathogens includes gloves, eye and face protection (such as goggles, glasses, face shields, and masks), gowns, fluid-proof aprons, mouthpieces, resuscitation bags, or other ventilation devices. Table 5 includes an overview of the different types of personal protection equipment to be used during various activities and tasks.

Gloves

Examples of tasks which require the use of gloves include dentistry, surgery, phlebotomy, laboratory analysis of blood or body fluids, cleanup of blood or body fluid spills, and rendering emergency medical assistance to individuals with traumatic injury (see Table 5). The Center for Devices and Radiological

Health (within the Food and Drug Administration) is responsible for regulating the medical glove industry. Medical gloves include sterile surgical gloves and nonsterile examination gloves made of vinyl or latex. General purpose utility ''rubber'' gloves are also used in the healthcare setting, but they are not regulated by the FDA since they are not promoted for medical use. There are no reported differences in barrier effectiveness between intact latex and intact vinyl used to manufacture gloves.[35]

Careful consideration should be given to choosing the appropriate type of glove for the task being performed. The following general guidelines are recommended for glove selection:[36]

1. Use sterile gloves for procedures involving contact with normally sterile areas of the body.
2. Use examination gloves for procedures involving contact with mucous membranes, unless otherwise indicated, and for other patient care or diagnostic procedures that do not require the use of sterile gloves.
3. Change gloves between patient contacts.
4. Hands and other skin surfaces should be washed immediately and thoroughly if contaminated with blood or other body fluids; hands should be washed immediately after gloves are removed.
5. Do not wash or disinfect surgical or examination gloves for reuse. Washing with surfactants may cause ''wicking'', i.e., the enhanced penetration of liquids through undetected holes in the gloves; disinfecting agents may cause deterioration.
6. General purpose utility gloves (rubber household gloves) can be used for housekeeping applications involving potential blood contact and for instrument cleaning and decontamination procedures; utility gloves may be decontaminated and reused but they should be discarded if they are punctured, torn, cracked, discolored, peeling, or there is other evidence of deterioration.

Eye and Face Protection

Protective glasses, goggles, masks, and chin-length face shields should be worn whenever splashes, spray, spatter, or droplets of blood or other potentially infectious materials, i.e., saliva with blood and bone chips may be generated and there is potential for mucous membrane (eye, nose, mouth) contamination. If protective eye wear is chosen over the use of a face shield, then the eye wear must be worn in conjunction with a face mask since the aim of this requirement is to provide protection for the eyes, nose, and mouth. Some individuals have experienced problems when wearing protective eye wear such as goggles. These problems include discomfort, fogging, and blurred or limited range of vision. Examples of procedures where eye and face protection is particularly recommended include many types of surgery, medical procedures such as endotracheal intubation and bronchoscopy, and in most dental procedures. For further information on this subject, refer to Table 5.

Table 5. Occupational Bloodborne Pathogen Control

Task or Activity	Personal Protective Equipment				Work Place Control	Engineering Control
	Disposable Gloves	Protective Eye Wear	Mask	Gown		
Medical						
Measuring temperature	No	No	No	No	—	Disposable mouthpiece
Measuring blood pressure	No	No	No	No	—	—
Bathing patient	Yes	Yes[a]	No	No	—	—
Handling soiled instruments	Yes	No	No	No	SOP	Sharps disposal container
Giving an injection	Yes	No	No	No	SOP	Protective devices[b]
Starting an i.v.	Yes	No	No	No	SOP	Protective devices[b]
Blood drawing	Yes	No	No	No	SOP	Protective devices[b]
Oral/nasal suctioning	Yes	Yes	Yes	No	SOP	—
Manually cleaning airway	Yes	Yes	Yes	No	SOP	—
Endotracheal intubation	Yes	Yes	Yes	No	SOP	—
Bleeding control (minimal bleeding)	Yes	No	No	No	SOP	—
Bleeding control (spurting blood)	Yes	Yes	Yes	Yes	SOP	—
Surgery — invasive	Yes	Yes	Yes	Yes[a]	SOP	—
Childbirth	Yes	Yes	Yes	Yes	SOP	—
Needle/sharps handling and disposal	Generally yes	No	No	No	SOP	Sharps disposal container
CPR	Yes	No	No	No	SOP	Resuscitation device
Cardiac catheterization	Yes	Yes	Yes	Yes[a]	SOP	—
Dental						
Dental exam (detection of caries and pockets)	Yes	No	No	No	SOP	—
Oral injections	Yes	No	No	No	SOP	Protective devices
Taking X-rays	Yes	No	No	No	SOP	Protective devices
Dental cleaning	Yes	Yes	Yes	No	SOP	—
Instrument/equipment sterilization	Yes	Yes	Yes	No	SOP	—
Simple extraction	Yes	Yes	Yes	No	SOP	—
Major extraction	Yes	Yes	Yes	Yes	SOP	—
Root canal therapy	Yes	Yes	Yes	No	SOP	—
Biopsy	Yes	Yes	Yes	Yes	SOP	—
Oral surgery	Yes	Yes	Yes	Yes	SOP	—

Housekeeping						
Biohazardous trash handling	Yes	No	No	No	SOP	—
High-risk area cleaning	Yes	No	No	No	SOP	—
Blood spill clean-up	Yes	Yes	Yes	Yes	SOP	—
Soiled linen handling	Yes	Yes[a]	Yes[a]	Yes[a]	SOP	—
Laboratory						
Blood/body fluid specimen handling	Yes	Yes[a]	Yes[a]	Yes[a]	SOP	—
Pipetting	Yes	No	No	No	SOP	Mechanical device
Phlebotomy	Generally yes	No	No	No	SOP	Protective devices[b]
Mechanical specimen handling	No	No	No	No	—	Isolation by mechanical device
Morticians' services						
Autopsies	Yes	Yes	Yes	Yes	SOP	—
Handling deceased persons	Yes	No	No	No	Barrier wounds	—
Dialysis	Yes	No	No	No	SOP	—
Fire and emergency medical services	Yes	Yes[a]	Yes[a]	Yes[a]	SOP	—

[a] If blood/body fluid splashing or spattering is possible.
[b] Protective devices are available to prohibit needle sticks/sharps cuts. SOP requires a procedure to minimize exposure.

Gowns and Aprons

If there is a potential for soiling of clothes with blood or other potentially infectious materials then gowns, aprons, lab coats, or similar clothing should be worn. These items are commonly made of tightly woven or fused materials that will prevent the employees' underlying clothing from becoming contaminated. If splashing or spraying of blood (or other potentially infectious liquids) is probable, employees should wear more protective ''fluid-resistant'' or ''fluid-proof'' gowns, aprons, or coveralls. This is necessary because a larger volume of blood or other potentially infectious material creates a greater chance of soak-through and skin contact, thus, a more protective type of barrier clothing is required.[37] Due to the fact that these fluid-proof and fluid-resistant gowns provide more protection, some individuals complain of retention of body heat, accumulation of sweat, and skin rashes when wearing these gowns. If the overgarment does become contaminated, it should be removed when soiled or at the end of the workshift. The contaminated overgarment should remain within the work area for cleaning, laundering, or disposal.

Resuscitation Devices

Although currently there is no evidence of transmission of bloodborne pathogens from administering cardiopulmonary resuscitation (CPR), theoretically this is a real possibility. OSHA, CDC, and the American Federation of State, County, and Municipal Employees (AFSCME) all agree on the need to minimize exposure from mouth-to-mouth resuscitation.[38] These organizations believe the most effective way to do so is to require that ventilation devices be provided for resuscitation. These devices are very inexpensive with some models well under $10.00. Consistent with all PPE, potentially affected employees must be trained in the use of such equipment and the equipment should be strategically located in order to facilitate and maximize its use.

Engineering Controls

Engineering controls are designated to reduce workers' occupational exposure by either removing the hazard or by isolating the individual from the exposure. When dealing with bloodborne pathogens there are essentially three kinds of engineering controls which are illustrated in Figures 1, 2, 3, and 4. Whenever possible, engineering controls should be utilized as they are the most effective way (except for the vaccine for HBV) to prevent occupational exposure to bloodborne pathogens.

Work Practice Controls

Work practice controls reduce the potential for worker exposure to bloodborne pathogens by changing the manner in which a task is performed and by

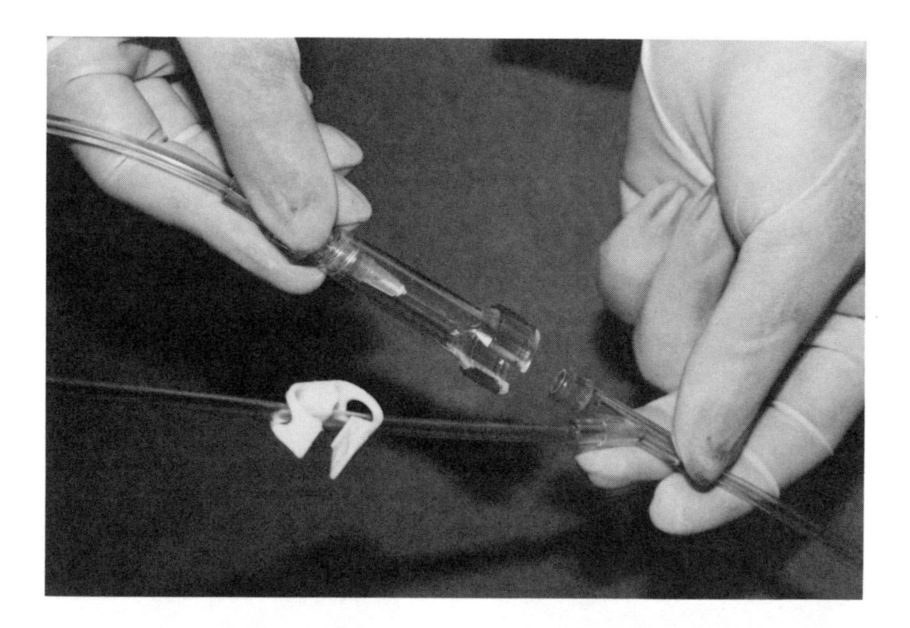

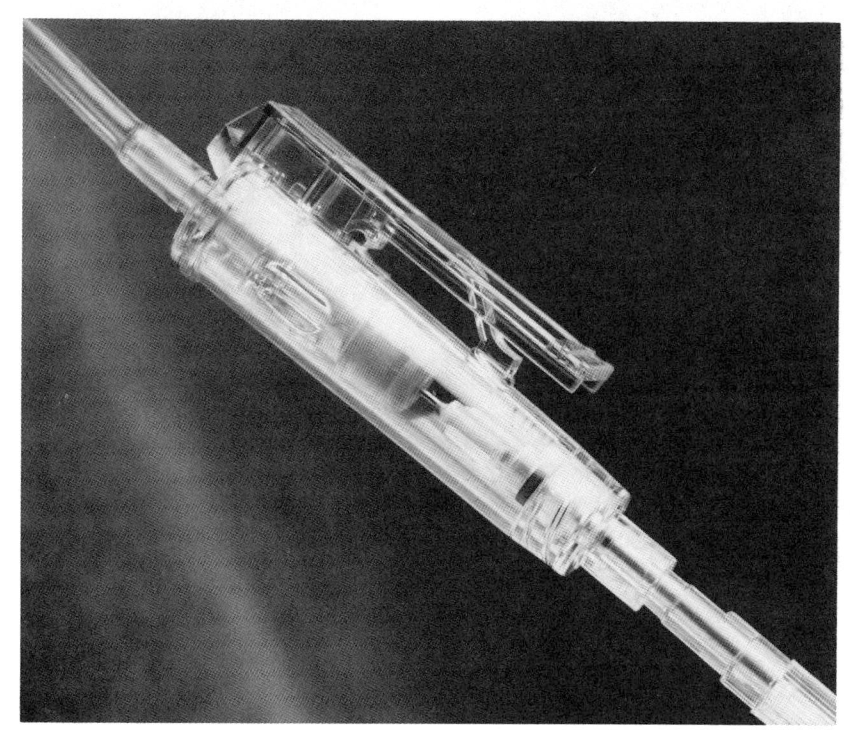

Figure 1. Process or equipment redesign, i.e., i.v./catheter and secondary i.v.s that connect with "protected" needles (needle recessed in plastic). Photos provided courtesy of ICU Medical, Inc., Irvine, CA.

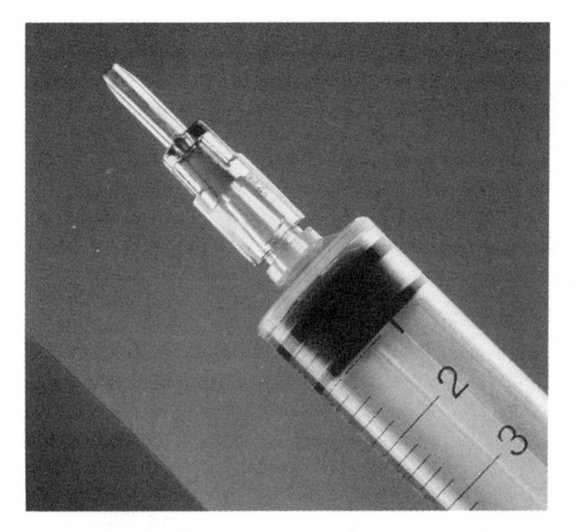

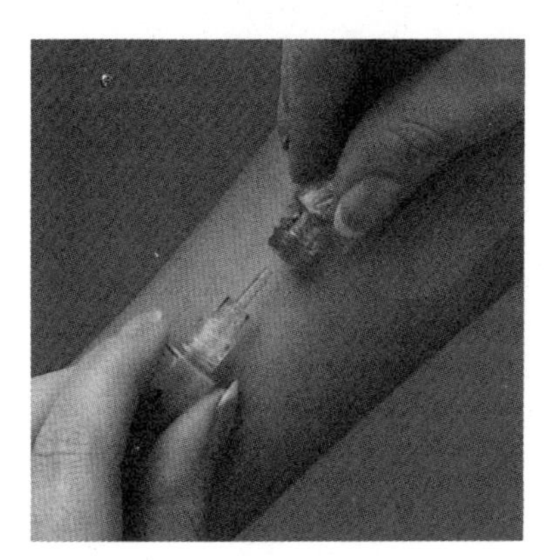

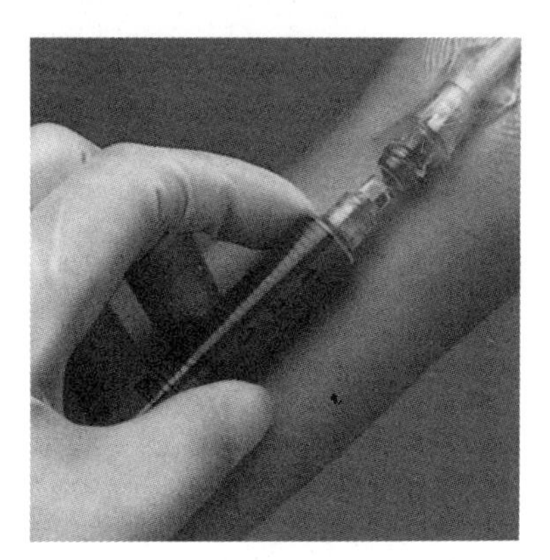

Figure 2. Process or equipment redesign, i.e., blood sampling without needles. Photos provided courtesy of Becton Dickinson.

designating a standard operating procedure for accomplishing the task. Work practice controls act on the source of the hazard but the protection they provide is based upon employer and employee behavior (rather than on the installation of an engineering control). Work practice controls for reducing exposure to bloodborne pathogens include:[39]

1. Adherence to the practice of universal precautions in situations of possible exposure.
2. Prohibiting the bending or breaking of needles and other sharps and not permitting recapping or manipulation of needles or sharps by hand.
3. Prohibiting pipetting or suctioning by mouth.

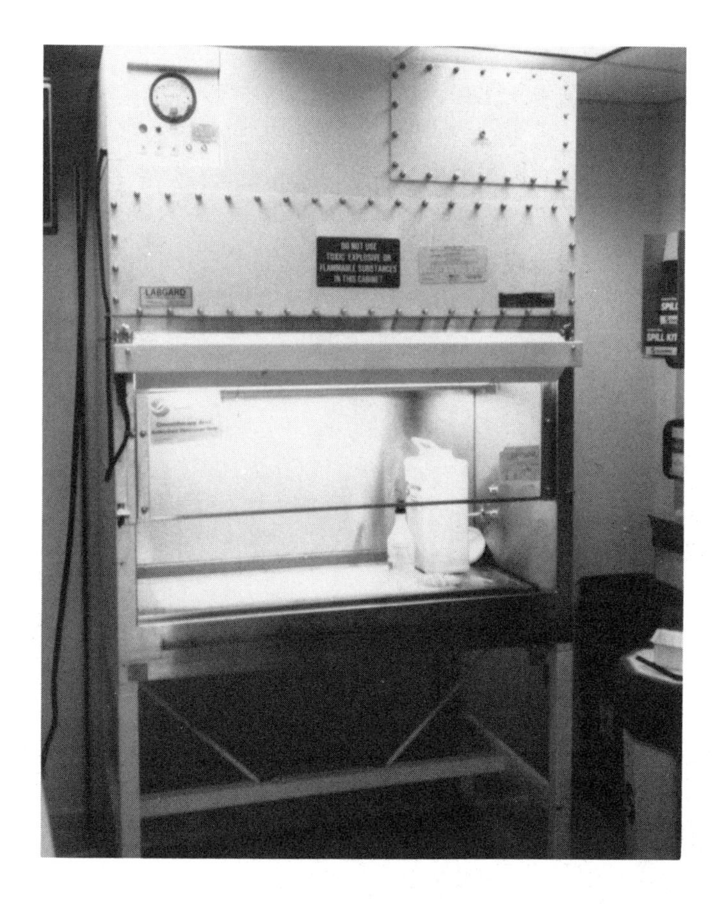

Figure 3. Employee isolation, i.e., biosafety cabinets used in laboratories designed to isolate the worker from the bloodborne pathogen.

In each of these examples, the possibility for exposure to blood or other potentially infectious materials has been eliminated or minimized simply by changing the way the employee performs the task.

Needles and Sharps

Needles and sharps are documented to be a primary mechanical means of employee exposure to both HIV and HBV. Of the 25 healthcare workers diagnosed as having HIV from occupational transmission, 18 of the exposures occurred by needle stick or cuts with sharp HIV-contaminated objects. In addition, as mentioned previously, the chance of becoming infected after a single needle stick from a hepatitis B patient ranges from 6 to 30%. In order to minimize the chance of needle stick or cut when dealing with needles and sharps, it is necessary to combine work practice controls with engineering controls. First and most

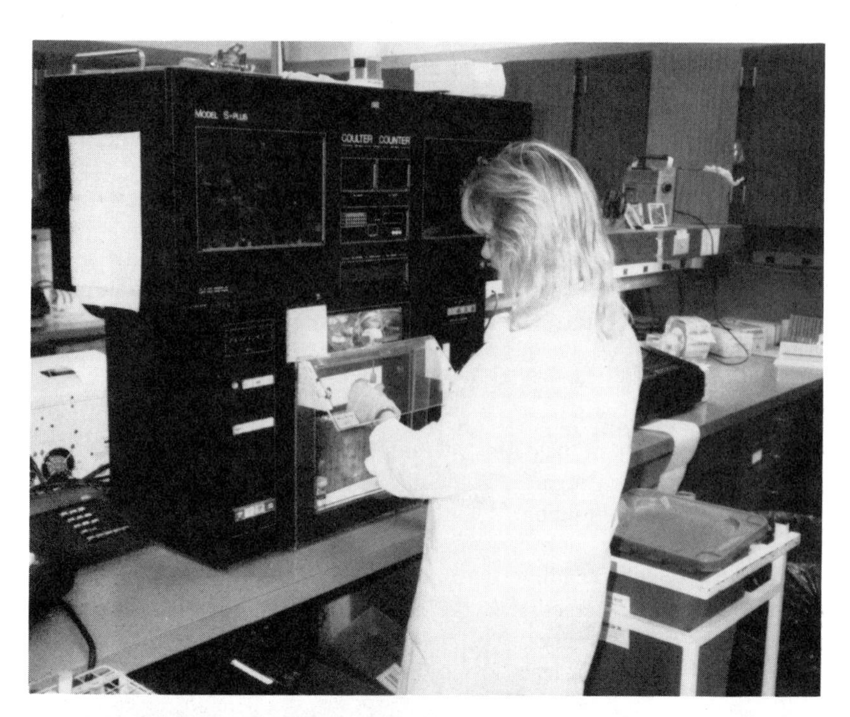

Figure 4. Process or equipment enclosure, i.e., installation of a plexiglass shield over laboratory blood testing equipment to prevent blood from spraying into the operator's face.

important is the fact that handling of needles/syringes and other sharps should be *minimized.* For example, disposable syringes with needles, scalpel blades, and other sharp items should be disposed of immediately in puncture-resistant containers located as close as is practical to the area where they are used. If reusable items are used, devices that minimize needle exposure/handling should be utilized, i.e., self-sheathing needles and secondary i.v.s that connect without needles. This combination of specific work practices and specially designed equipment help to reduce needle sticks and cuts that play a significant role in HIV or HBV transmission.[40]

Laundry

Although soiled linen has been identified as a source of large numbers of certain pathogenic microorganisms, the risk of actual transmission of bloodborne pathogen disease is negligible.[41] Soiled linen should be handled as little as possible, and with minimum agitation, to prevent microbial contamination of the air and of persons handling the linen. All linen soiled with blood or other body fluids should be immediately placed (and transported) in bags that prevent leakage. If hot water laundry cycles are used, the soiled linen should be washed with detergent in water at least 71°C (160°F) for 25 min. If lower temperature laundry cycles are used, chemicals suitable for low-temperature washing should be used at proper use concentrations.

STERILIZATION AND DISINFECTION TECHNIQUES

Viability of Hepatitis B Virus

It is generally agreed that the hepatitis B virus can survive in the open environment, at room temperature and in a dried state, for at least one week. Transmission of HBV to an individual from environmental surfaces has been documented to be a significant mode of virus spread. For example, it is believed that HBV-contaminated dried blood from the surface of dialysis machines can be carried to patients on the hands of medical personnel. Unsterilized or improperly sterilized needles used in acupuncture are also believed to be responsible for the transmission of hepatitis B infection. In the laboratory, HBV transmission can occur via common environmental surfaces such as test tubes, lab benches, lab accessories, and other surfaces contaminated with infected blood or other infectious body substances. The probability of disease transmission from a single exposure of this type may be remote, but the frequency of this type of exposure makes this mechanism of transmission a concern. Also, in the CDC's guidelines for dental procedures, potential problems of environmental HBV contamination have been addressed in regards to dental equipment and environmental surfaces in the dental operatory.[42] This subject is covered in more detail in Chapter 6.

Viability of Human Immunodeficiency Virus

To date, no environmentally mediated mode of HIV transmission has ever been documented. The most extensive study of HIV survival in the open environment after drying involved samples of HIV that were at least 100,000 times more concentrated than what is typically found in the blood of HIV-infected patients.[43] Even with this extreme concentration of HIV, the rate of virus inactivation after drying was rapid.

Sterilization and Disinfection

Both HBV and HIV are believed to be much less resistant to sterilization and disinfection procedures than are the microbial endospores and mycobacteria that are used as reference criteria. Thus, sterilization and disinfection products and procedures that are registered with the U.S. Environmental Protection Agency (EPA) as sterilizing agents or high-level disinfectants will kill HBV and HIV if used as directed. In addition to these commercial available chemical germicides, a solution of sodium hypochlorite (household bleach) prepared daily is an inexpensive and effective germicide. Concentrations ranging from approximately 500 ppm sodium hypochlorite (1:100 dilution of household bleach) to 5000 ppm sodium hypochlorite (1:10 dilution of household bleach) are effective depending upon the amount of organic material, i.e., blood, mucus, etc. present on the surface to be cleaned and disinfected.[44] Certain low-level germicides such as quarternary ammonium compounds are not considered to be effective against HBV. Soaking medical and dental instruments in low-level germicide solutions is a common and potentially dangerous procedure because healthcare workers may handle sharp instruments soaked in these solutions with a false sense of security.[45]

As a part of standard infection-control practice, instruments and other reusable equipment used in performing invasive procedures should be appropriately disinfected and sterilized as follows:[46]

1. Equipment and devices that enter the patient's vascular system or other normally sterile areas of the body should be sterilized before being used for each patient.
2. Equipment and devices that touch intact mucous membranes but do not penetrate the patient's body surfaces should be sterilized when possible or undergo high-level disinfection if they cannot be sterilized before being used for each patient.
3. Equipment and devices that do not touch the patient or that only touch intact skin of the patient need only be cleaned with a detergent or as indicated by the manufacturer.

Table 6. Training Outline For Occupational Exposure to Bloodborne Pathogens

1. Introduction

2. Understanding hepatitis B
   - Etiology
   - Epidemiology
   - Pathology

3. Understanding AIDS
   - Etiology
   - Epidemiology
   - Pathology

4. Routes of transmission of HIV and hepatitis B
   - Modes of transmission:
     - general public
     - workplace, patient to HCW
     - workplace, HCW to patient

5. Who is at risk?
   - HCW
   - Non-HCW

6. Regulatory requirements
   - Centers for Disease Control (CDC)
   - Occupational Safety and Health Administration (OSHA)

7. Methods of disease control
   - Vaccination
   - Universal precautions
   - Personal protective equipment
     - Gloves
     - Eye and face protection
     - Gowns and aprons
     - Resuscitation devices
   - Engineering controls
   - Work practice controls
   - Needles and sharps
   - Laundry

8. Sterilization and disinfection techniques
   - Viability of hepatitis B virus
   - Viability of human immunodeficiency virus
   - Sterilization and disinfection

9. The importance of employee training
   - Time of hire
   - Annual refresher

HCW = healthcare worker.

EMPLOYEE TRAINING

The training of workers is a fitting place to conclude our discussion of occupational exposure to bloodborne pathogens as it brings together what has been discussed. Effective training is a key element to employee protection from bloodborne pathogens. Training will ensure that workers understand the hazards these viruses present, their routes of transmission, and methods that can be used to minimize the risk of infection. The organization of this chapter is an excellent example of an outline for employee training and is presented in Table 6. To be most effective, employee training should be conducted at the time of hire and then annually thereafter. An example of the importance of training is apparent when one considers how critical it is that a new employee know what types of personal protective equipment are necessary for each particular task, where this equipment is located, and its proper use and disposal.

SUMMARY

The subject of occupational exposure to bloodborne pathogens is one that is evolving and changing almost every day. There still are many things we do not know about this subject that are crucial to our developing the appropriate measures of protection for both healthcare workers and the population in general. This is a very exciting time for the study and risk assessment of occupational exposure to bloodborne pathogens. It is certain that we will hear more from the CDC, OSHA, and other concerned organizations.

REFERENCES

1. "Proposed OSHA Rule Governing Occupational Exposure to Bloodborne Pathogens," OSHA proposed rule and notice of hearing, 54 FR 23042 (May 30, 1989), p. 2259.
2. Hoeprich, P. D., Ed. *Infectious Diseases,* 3rd ed., Philadelphia: Harper and Row (1983), p. 714.
3. "Proposed OSHA Rule Governing Occupational Exposure to Bloodborne Pathogens," OSHA proposed rule and notice of hearing, 54 FR 23042 (May 30, 1989), p. 2175.
4. "Guidelines For Prevention of Transmission of Human Immunodeficiency Virus and Hepatitis B Virus to Health-Care and Public Safety Workers," U.S. Department of Health and Human Services, Public Health Service, Centers for Disease Control, NIOSH, U.S. Government Printing Office (February 1989), p. 3.
5. "Proposed OSHA Rule Governing Occupational Exposure to Bloodborne Pathogens," OSHA proposed rule and notice of hearing, 54 FR 23042 (May 30, 1989), p. 2173.

6. "Managing AIDS Patients: The Healthcare Professionals' Survival Guide," Brown University STD Uptake, Manisses Communications Group, Providence, RI (1988), p. 1.
7. "Managing AIDS Patients: The Healthcare Professionals' Survival Guide," Brown University STD Update, Manisses Communications Group, Providence, RI (1988), p. 2.
8. "Managing AIDS Patients: The Healthcare Professionals' Survival Guide," Brown University STD Update, Manisses Communications Group, Providence, RI (1988), p. 1.
9. "Update: Acquired Immunodeficiency Syndrome — U.S. 1981–1988," *MMWR* 38:229–236 (1989).
10. "Proposed OSHA Rule Governing Occupational Exposure to Bloodborne Pathogens," OSHA proposed rule and notice of hearing, 54 FR 23042 (May 30, 1989), p. 2179.
11. "The HIV/AIDS Epidemic: The First 10 Years," *MMWR* 40:357–369 (1991).
12. "Proposed OSHA Rule Governing Occupational Exposure to Bloodborne Pathogens," OSHA proposed rule and notice of hearing, 54 FR 23042 (May 30, 1989), p. 2179.
13. "Proposed OSHA Rule Governing Occupational Exposure to Bloodborne Pathogens," OSHA proposed rule and notice of hearing, 54 FR 23042 (May 30, 1989), p. 2178.
14. World Health Organization. *In Point of Fact,* Geneva: World Health Organization (May 1991), No. 74.
15. "Proposed OSHA Rule Governing Occupational Exposure to Bloodborne Pathogens," OSHA proposed rule and notice of hearing, 54 FR 23042 (May 30, 1989), p. 2179.
16. "The HIV/AIDS Epidemic: The First 10 Years," *MMWR* 40:357–369 (1991).
17. "Managing AIDS Patients: The Healthcare Professionals' Survival Guide," Brown University STD Update, Manisses Communications Group (1988), p. 5.
18. "Guidelines For Prevention of Transmission of Human Immunodeficiency Virus and Hepatitis B Virus to Health-Care and Public Safety Workers," U.S. Department of Health and Human Services, Public Health Service, Centers for Disease Control, NIOSH, U.S. Government Printing Office (February 1989), p. 2–4.
19. "Guidelines For Prevention of Transmission of Human Immunodeficiency Virus and Hepatitis B Virus to Health-Care and Public Safety Workers," U.S. Department of Health and Human Services, Public Health Service, Centers for Disease Control, NIOSH, U.S. Government Printing Office (February 1989), p. 3.
20. "Proposed OSHA Rule Governing Occupational Exposure to Bloodborne Pathogens," OSHA proposed rule and notice of hearing, 54 FR 23042 (May 30, 1989), p. 2180.
21. Seeff, L. B., E. C. Wright, H. J. Zimmermann et al. "Type B Hepatitis After Needle Stick Exposure: Prevention With Hepatitis B Immune Globulin," *Ann. Intern. Med.*, 88:285–93 (1978).
22. Wormser, G. P., C. S. Rabkin, and C. Joline. "Frequency of Nosocomial Transmission Of HIV Infection Among Health Care Workers," *N. Engl. J. Med.*, 319:307–8 (1988).
23. "Update: Universal Precautions for Prevention of Transmission of Human Immunodeficiency Virus, Hepatitis B Virus, and Other Bloodborne Pathogens in Health-Care Settings," *MMWR* 37:378 (1988).

24. "Update: Transmission of HIV Infection During Invasive Dental Procedures — Florida," *MMWR* 40:377–381 (1991).
25. Grob, P., B. Bischoff, and R. Naeff. "Cluster of Hepatitis B Transmitted by a Physician," *Lancet* 2:1218–20 (1981).
26. Rimland, D., W. E. Parkin, G. B. Miller et al. "Hepatitis B Outbreak Traced to an Oral Surgeon," *N. Engl. J. Med.* 296:953–8 (1977).
27. Ahtone, J. and R. A. Goodman. "Hepatitis B and Dental Personnel: Transmission to Patients and Prevention Issues," *JAMA* 106:219–22 (1983).
28. "Recommendations For Prevention of HIV Transmission in Health-Care Settings," *MMWR* 36:3S (1987).
29. "Recommendations for Preventing Transmission of Human Immunodeficiency Virus and Hepatitis B Virus to Patients During Exposure-Prone Invasive Procedures," *MMWR* 40:RR-8:1–9 (1991).
30. "Recommendations for Preventing Transmission of Human Immunodeficiency Virus and Hepatitis B Virus to Patients During Exposure-Prone Invasive Procedures," *MMWR* 40:RR-8:1–9 (1991).
31. "Final OSHA Standard On Bloodborne Pathogens," 56 FR 235 (December 6, 1991), p. 64175–64182.
32. "Proposed OSHA Rule Governing Occupational Exposure to Bloodborne Pathogens," OSHA proposed rule and notice of hearing, 54 FR 23042 (May 30, 1989), p. 2176.
33. "Proposed OSHA Rule Governing Occupational Exposure to Bloodborne Pathogens," OSHA proposed rule and notice of hearing, 54 FR 23042 (May 30, 1989), p. 2176.
34. "Proposed OSHA Rule Governing Occupational Exposure to Bloodborne Pathogens," OSHA proposed rule and notice of hearing, 54 FR 23042 (May 30, 1989), p. 2176.
35. "Update: Universal Precautions for Prevention of Transmission of Human Immunodeficiency Virus, Hepatitis B Virus, and Other Bloodborne Pathogens in Health-Care Settings," *MMWR* 37:381 (1988).
36. "Update: Universal Precautions for Prevention of Transmission of Human Immunodeficiency Virus, Hepatitis B Virus, and Other Bloodborne Pathogens in Health-Care Settings," *MMWR* 37:381 (1988).
37. "Proposed OSHA Rule Governing Occupational Exposure to Bloodborne Pathogens," OSHA proposed rule and notice of hearing, 54 FR 23042 (May 30, 1989), p. 2244.
38. "Proposed OSHA Rule Governing Occupational Exposure to Bloodborne Pathogens," OSHA proposed rule and notice of hearing, 54 FR 23042 (May 30, 1989), p. 2243.
39. "Proposed OSHA Rule Governing Occupational Exposure to Bloodborne Pathogens," OSHA proposed rule and notice of hearing, 54 FR 23042 (May 30, 1989), p. 2241.
40. "Proposed OSHA Rule Governing Occupational Exposure to Bloodborne Pathogens," OSHA proposed rule and notice of hearing, 54 FR 23042 (May 30, 1989), p. 2241.
41. "Recommendations for Prevention of HIV Transmission in Health-Care Settings," *MMWR* 36:11S–12S (1987).

42. "Proposed OSHA Rule Governing Occupational Exposure to Bloodborne Pathogens," OSHA proposed rule and notice of hearing, 54 FR 23042 (May 30, 1989), p. 2176.
43. "Recommendations for Prevention of HIV Transmission in Health-Care Settings," *MMWR* 36:10S–11S (1987).
44. "Recommendations for Prevention of HIV Transmission in Health-Care Settings," *MMWR* 36:9S–10S (1987).
45. "Proposed OSHA Rule Governing Occupational Exposure to Bloodborne Pathogens," OSHA proposed rule and notice of hearing, 54 FR 23042 (May 30, 1989), p. 2176.
46. "Guidelines for the Prevention and Control of Nosocomial Infections: Guideline for Handwashing and Hospital Environmental Control." Centers for Disease Control, Atlanta, Georgia: Public Health Services (1985), pp. 1–20.

CHAPTER 5

Tuberculosis: Resurgence and Control

Fredrick T. Horn

INTRODUCTION

Tuberculosis is increasing, both in incidence and virulence. Once commonly assumed to be easily preventable and well under control, tuberculosis has recently begun to make a resurgence. Tuberculosis is a potential occupational health hazard for healthcare practitioners. This includes, for example, social workers, homeless shelter staff, counselors, nurses, physicians, long-term care providers, employees at drug treatment centers and correctional institutions, and others in contact with potentially infectious individuals or populations.

Tuberculosis is coming back for a variety of reasons. For example, social conditions are conducive to its transmission, such as homelessness or overcrowding in correctional facilities. Substance abusers with tuberculosis are frequently unable to maintain their treatment. Also, there has been the evolution of drug resistant strains of tuberculosis.

The focus of this chapter is the specific threat that tuberculosis presents to healthcare providers. This chapter will discuss tuberculosis as an occupational

0-87371-392-3/93/$0.00 + $.50

Table 1. Symptoms of Pulmonary Tuberculosis

Generalized symptoms	Pulmonary symptoms
· Fever	· Productive, prolonged cough (chronic)
· Night sweats	· Sputum production
· Malaise and easy fatiguability	· The spitting or coughing up of blood
· Loss of appetite	· Chest pain
· Weight loss without dieting	· Recurrent kidney or bladder infections
· Chills	

From National Tuberculosis Training Initiative "Core Curriculum on Tuberculosis," U.S. DHHS No. 00-5763 (1991), p. 21, CDC "Tuberculosis Medical Orientation Course," Instructor Guide, Department of Health and Human Services, Public Health Service, Centers for Disease Control, Center for Protection Services, Division of Tuberculosis Control, Atlanta, GA (September 1989), p. 24, and "Tuberculosis Control Program Manual," Fact Sheet #7, New Mexico Department of Health (September 1, 1987).

hazard, including how it is transmitted, what populations are particularly susceptible to it, and how workers in high risk occupations can protect themselves from contracting it.

TUBERCULOSIS: WHAT IS IT?

Tuberculosis is an infectious disease caused by the tubercle bacillus, *Mycobacterium tuberculosis* (*M. tuberculosis*). The name of the disease is derived from the word tubercle, meaning a small lump or nodule. Tuberculosis most commonly affects the respiratory system (but can infect other parts of the body such as the gastrointestinal tract, lymph nodes, nervous system, skin, etc.).

Tuberculosis is characterized by inflammatory infiltrations, formations of tubercles, caseation, necrosis, abscesses, fibrosis, and calcification.[1] The tubercle is a concentration of granulomatous inflammation consisting of lymphocytes, epithelioid cells, macrophages, and giant cells. The granulomas seen in tuberculosis are characterized by a form of tissue necrosis (tissue death). Prior to the time of necrosis the lesion may heal completely by resolution, however, once necrosis has occurred it heals only by fibrosis, encapsulation, calcification, and scar formation. Early in the primary infection, the organisms may be transported to the draining lymph nodes and may be widely spread throughout the body. Symptoms of pulmonary tuberculosis are summarized in Table 1.

Tuberculosis may occur in an acute, generalized form known as miliary tuberculosis or in a chronic, more localized form. Active cases should receive hospital care. In the late nineteenth century this consisted of placing the tuberculosis patient in a sanitorium, where treatment consisted of bed rest, good food, and fresh air; additionally, the patient was isolated from the rest of the community.[2] Currently, chemotherapeutic drugs are commonly used to combat against the tubercle bacillus, among them: streptomycin, para-amino-salicylic acid (PAS), rifampin (RMP), ethambutol (EMB), and isoniazid (INH). Treatment is long term and typically lasts for 9 to 24 months. Symptomatic medications may also be required for cough, hemoptysis, chest pain, and other symptoms. The course of chemotherapy in response to tuberculosis requires that the patient adhere to a self-administered treatment schedule, usually for a period of months. Where the patient is unable to maintain treatment, perhaps as in homeless or substance abusing populations, it is the intermittent administration of medication which can result in the development of drug resistant strains of tubercle bacillus (for example, in Chicago, 6 cases of multi-drug resistant cases in 1988 and 13 such cases in 1991).[3]

HISTORY

Tuberculosis has existed for centuries. Hippocrates described the disease circa 400 B.C., and until the nineteenth century a commonly used term for tuberculosis was ''consumption'' (from the Latin *consumare,* to take/waste).[2] During the industrial revolution of the 18th and 19th centuries the disease was also known as consumption and the white plague.[4] The disease has been observed in the U.S. from the mid-eighteenth century. In North America, the first mortality rates for tuberculosis were 300 deaths per 100,000 population per year (for whites in Salem, MA).[5] As more population growth took place in the U.S., increasing urbanization brought mortality rates of up to 1,000 per 100,000 population per year in New England in 1800. Because of westward settlement, urbanization of other parts of the U.S. came about later. Tuberculosis mortality peaked subsequently in New Orleans (approximately 1840) and on the West Coast of the U.S. (approximately 1880).[5] Tuberculosis was the leading cause of death in the U.S. in 1900, causing approximately 10% of all deaths.[2]

Tuberculosis in North America was originally not a significant problem for either blacks or Native Americans. Social changes affected the incidence of tuberculosis in these groups. For example, urbanization of blacks, resulting from emancipation during the Civil War, contributed to significant mortality increases which peaked at approximately 650 deaths per 100,000 population per year in 1890. The peak in tuberculosis mortality for Native Americans took place (approximately 1910) and occurred as a result of the crowded conditions on reservations.[5]

EPIDEMIOLOGY

Worldwide, despite the great success in the development and implementation of effective tuberculosis treatment, tuberculosis is still a serious medical problem in many developing countries today. The World Health Organization (WHO) estimates that 1.76 billion people worldwide are infected with the tubercle bacillus, that 8 to 10 million new cases occur each year, and that 2 to 3 million people die from tuberculosis each year.[6]

Table 2. Methods for Diagnosing Tuberculosis

- Patient history
- Tuberculin skin test
- Chest radiograph
- Bacteriologic examination

Adapted from the CDC "Tuberculosis Medical Orientation Course," Department of Health and Human Services, Public Health Service, Centers for Disease Control, Center for Protection Services, Division of Tuberculosis Control, Atlanta, GA (September 1989), p. 31.

When available, treatment for tuberculosis has been very effective, as have screening methods. Diagnosis and treatment are routine procedures (Table 2) and have been highly successful.[7] In the U.S., where the medical care available has been arguably better than in most other countries, the tuberculosis morbidity rate has generally decreased since the early 1900s. The U.S. Public Health Service Tuberculosis Program was first started in 1944 when 126,000 total cases were reported and the morbidity rate was 95 per 100,000.[7] Since that time the annual number of tuberculosis cases decreased steadily for several decades; in 1985 there were 22,201 cases and a morbidity rate of just under 10 per 100,000.

Unfortunately, this long-term decrease in the number of cases of tuberculosis flattened in the mid-1980s and has more recently begun to increase. For the years 1985 to 1990 in the U.S., Table 3 shows the number of tuberculosis cases reported, the number of cases expected, the excess tuberculosis cases, and the percent increase. Excess tuberculosis cases represent the additional cases observed when compared to the number of cases expected.

One study suggests that prolonged exposure to a homeless shelter environment leads to tuberculosis infection and active disease.[8] The elderly (those over 65) make up 12% of the population in the U.S. but account for 25% of the tuberculosis morbidity.[9] Other populations have greater risk; the South Dakota Pine Ridge Reservation, with 16,000 residents, has an incidence rate of 66.3 cases per 100,000 population (compared with 1.2 cases per 100,000 population for whites in South Dakota).[10]

Table 3. Tuberculosis Cases Associated With a Leveling of the National Trend in the 1980s

Year	Reported Cases	Expected Cases	Excess Cases	Percent Increase Associated With Change
1985	22,201	21,256	945	4.3
1986	22,768	20,070	2,698	11.8
1987	22,517	18,944	3,573	15.9
1988	22,436	16,885	4,554	20.3
1989	23,493	16,885	6,610	28.1
1990	25,701	15,818	9,883	38.5

Adapted from "Tuberculosis Elimination, U.S.A. (Facsimile of Slides)," Division of Tuberculosis Elimination, National Center for Prevention Services, Centers for Disease Control (August 1991).

DIAGNOSIS OF TUBERCULOSIS

Early identification of individuals having tuberculosis infection, and the application of preventive therapy, can be effective in preventing tuberculosis transmission (as well as in controlling it in the infected subject). This is true for transmission to others in the community as well as to healthcare providers who come in contact with infected individuals. The diagnosis of tuberculosis, outlined in Table 2, is further discussed below. Diagnosis of persons with tuberculosis is very significant in reducing occupational exposure to tuberculosis because of the preventative measures which can then be taken. Not all healthcare providers will have the luxury of screening or diagnosing tuberculosis in their patients, clients, or facility occupants.

Patient History

An extensive history of the patient is important. Key elements should include past or present potential exposures to tuberculosis, living arrangements (for example, long-term inpatient or homeless), and a good medical history; the historian must be accurate and the client must be cooperative. Anyone who has been exposed to an infectious case of tuberculosis within the last 24 months has a high risk of developing the disease and is referred to as a contact.[2]

Persons with tuberculosis may be symptomatic (i.e., have symptoms shown in Table 1) or asymptomatic (not have the symptoms). There may be localized symptoms, in addition to the generalized symptoms of pulmonary tuberculosis, if a particular organ is affected. Tuberculosis symptoms are seldom sudden in onset; most symptoms will be present for many weeks.

If the patient has a history of tuberculosis, an inadequate treatment regimen (or poor compliance by the patient) can make recurrence of the disease likely and could signal the possible presence of a drug resistant strain of tuberculosis.[11] Certain other medical conditions may predispose a person with tuberculosis infection to develop the disease. Certainly this includes persons with HIV or at risk for HIV. Table 4 describes some groups with risk factors for tuberculosis.

It is very important to recognize patients with active disease so that they can be placed under treatment to terminate infectivity and to protect the healthcare providers. An effective tuberculosis control program requires several key steps to be successful including early identification, isolation, and treatment of persons with active tuberculosis. Screening, investigation, and recognition of symptomatic cases at entry points to the medical care system, and surveillance of high risk groups are key in the prophylactic treatment of tuberculosis.

The Tuberculin Skin Test

The tuberculin skin test (Mantoux) is the only method currently available that demonstrates infection with *M. tuberculosis* in the absence of active tuberculosis symptoms.[2] The test uses tuberculin which is a purified protein extract of tubercle

Table 4. A Relative Ranking of Risk Factors For Tuberculosis

Group A
- Immunosuppressed individuals, such as persons having HIV
- Recent contacts with infectious tuberculosis cases
- Persons with abnormal chest radiographs (consistent with tuberculosis)

Group B
- Foreign born persons from countries with a high tuberculosis prevalence
- Homeless persons
- Persons from low-income populations
- Substance abusers, particularly intravenous drug users
- Residents of institutions (nursing homes, correctional facilities, etc.)
- Persons over the age of 70
- Hospital and mycobacteria laboratory personnel
- Healthcare workers (and others) who come in contact with high-risk persons
- Persons who administer aerosolized pentamidine
- Persons with medical risk factors including diabetes mellitus, prolonged corticosteroid therapy, hematologic and reticuloendothelial diseases, end stage renal disease, postgastrectomy, chronic malabsorption syndromes, silicosis, or being 10% or more below ideal body weight

Group C
- Individuals not listed above

Adapted from "Tuberculosis Medical Orientation Course," DHHS/CDC, Center for Protection Services, Division of Tuberculosis Control, Atlanta (September 1989).

bacilli which have been killed by heating. The test is performed by injecting purified protein derivative (PPD) between the layers of the skin, usually on the forearm, and the patient's arm is examined 48 to 72 hours later.

Analysis of skin test results require the interpretation of trained professionals and depends upon the size of the skin reaction that results. In general, the test will show that persons infected with tubercle bacilli will have an area of palpable swelling (an induration) around the site of the injection. Persons who have not been infected with tubercle bacilli will not have an induration or may have a small reaction (less than 5 mm). The test may miss some cases because it is affected by some viral illnesses such as common fevers. False positives may result from past exposure to other germs. For the different risk groups in Table 4, a positive skin test reaction is evidenced for Group A individuals with an induration of $\geq$5 mm, for Group B individuals with an induration $\geq$10 mm, and for Group C individuals with an induration $\geq$15 mm.[2] Again, this is a simplification of the application and interferences associated with the tuberculin skin test.

Chest Radiograph

The chest X-ray or radiograph is a useful diagnostic tool since about 85% of tuberculosis patients have pulmonary involvement. However other conditions (such as lung cancer, pneumonia, sarcoidosis, nontubercular mycobacterial disease, and a variety of other pulmonary disorders) may resemble tuberculosis on a chest radiogram. Prior to the development of chemotherapy, the prognosis of tuberculosis was based on the extent of the disease that could be seen on a chest radiograph; in fact, chest radiographs were commonly administered during the 1950s in a concerted effort to wipe out tuberculosis.

Now the forecast depends on the bacteriologic response to chemotherapy. Although an experienced radiologist can often discern that an abnormality seen on a film is due to tuberculosis, only a bacteriological culture which is positive for *M. tuberculosis* definitely proves that a patient has tuberculosis. At the present time the prognosis of an infected individual can be better followed through a series of sputum smears and cultures than through a series of chest radiographs. Most individuals infected with tubercle bacilli (but without clinical disease) have normal chest radiographs, and so as a result the chest radiograph is an inadequate tool for detecting infection before it has progressed to disease.

Bacteriologic Examination

The tuberculosis control program depends heavily on information provided by the microbiology laboratory. Growth of tubercle bacilli in the laboratory, from a specimen obtained from an infected individual or one suspected to be infected, constitutes the only definitive proof that the individual does have tuberculosis. However, this does not mean that an individual whose cultures are negative does not have tuberculosis.

Most commonly the individual simply coughs up sputum which is collected in a sterile container for processing and examination. This type of sample has the most public health significance, since the individual who can cough up tubercle bacilli is the most likely to be infectious. This procedure must be done under strict control (particularly with regard to isolation of other susceptible individuals).

If an individual cannot cough up an adequate sputum specimen, other techniques are occasionally used to obtain a specimen in the form of induced sputum production from inhalation of a saline mist, pulmonary secretions via bronchoscopy, and gastric washing. About 15% of tuberculosis patients have extrapulmonary tuberculosis (that is, active tuberculosis outside the lungs).[2] Rates of extrapulmonary tuberculosis are higher in HIV-infected individuals. Specimens other than sputum are obtained for extrapulmonary tuberculosis. For example, urine would be examined if tuberculosis of the kidney were suspected, and cerebrospinal fluid would be obtained to diagnose tuberculosis meningitis.

TUBERCULOSIS TRANSMISSION

It is unfortunate, to say the least, that the tuberculosis rate in the U.S. is increasing. But it is *why* the increase is occurring, and in what parts of the population, that is most important for healthcare providers to understand since the additional tuberculosis cases often are in groups which are in regular contact with healthcare providers. This includes an understanding of how tuberculosis is transmitted.

When an uninfected individual is exposed to someone who has the disease, then transmission and the pathogenic process potentially can begin. If the infected individual is dispersing large numbers of tubercle bacilli in the same environment (perhaps by coughing), the possibility for transmission increases. This is because the tubercle bacilli are approximately 1 to 5 μ in size and so are easily inhaled. If present in the air, the infectious droplet nucleus can be inhaled through the nose or mouth, or both, and carried through the airways to the lung and the alveoli.

If the tubercle bacilli are inhaled and reach the alveoli, they are initially capable of multiplying freely because it usually takes the body's defenses 2 to 10 weeks to become effective in stopping bacilli multiplication. Particularly during this time the tubercle bacilli can spread unopposed (from their initial location in the lung tissue) to the lymph nodes (in the center of the chest) then to other parts of the body by way of hematogenous (blood) pathways. In this manner it is possible for tubercle bacilli to develop in places such as the kidneys, brain, and bone. In healthy individuals, the body's immunologic defenses can be activated in time to prevent this serious spread.

It is important to note that even in healthy individuals who become infected with the tubercle bacillus, the immune system is mostly successful but is not able to kill all the tubercle bacilli in the body. This residual bacteria in an otherwise healthy host is called a latent or inactive infection. The tuberculosis bacilli may be able to overcome the defenses of the body at any time in the future, producing active tuberculosis disease even after up to 50 years of dormancy.

For individuals who are already not well when exposed to the bacteria, there can be a great difference in the body's ability to defend against tuberculosis (depending upon the presence and type of other medical afflictions). For example, in individuals infected with the HIV virus the immune system is suppressed. Because the immune system is suppressed, the body is unable to control the multiplication of tubercle bacilli, therefore the HIV-positive individual is at much greater risk of developing and transmitting tuberculosis (active disease). Because of this inability to fight the tubercle bacilli, persons with both HIV and tuberculosis can present a serious tuberculosis threat to healthcare providers. In fact, tuberculosis has greatly increased in geographic areas and demographic groups with large numbers of cases of acquired immune deficiency syndrome (AIDS), which suggests that the spread of HIV has begun to influence tuberculosis morbidity.[12] Put simply, persons working in proximity to HIV-positive individuals may be at much greater risk of contracting tuberculosis since the HIV individual's body is unable to defend against the tuberculosis infection. It is possible to get HIV from HIV-positive individuals when there is transmission of body fluids such as blood; however, it may be easier to get tuberculosis from an HIV-positive individual who also has tuberculosis since transmission can take place through the air. While it may be emotionally more supportive of the HIV-positive individual for healthcare workers (such as social workers) to stress physical proximity with the client, this proximity may enhance the transmission of tuberculosis.

INFECTED vs. INFECTIOUS

Approximately 5% of newly infected individuals in the U.S. will develop the disease in the first year or two following infection. The remaining 95% of newly infected individuals are usually not even aware that they have been infected.[2]

Persons who have been infected are not necessarily infectious (again, this is referred to as latent disease). If a person's immune system becomes weakened and unable to control the bacteria, their infection can progress and develop into tuberculosis. Table 5 defines terminology used to describe different stages in the progression of the disease.

Table 5. The Stages of Exposure and Infection to Tuberculosis

- **Exposed.** A person is said to be exposed if there has been contact with anyone having active tuberculosis in a transmissible form. The degree of exposure can be household, close, or casual.
- **Primary Infection.** The initial infection within the lung caused by exposure to an infectious individual; its symptoms are flu-like. Most recover from the primary infection without treatment. Preventative treatment given at this time will stay future development of active disease, and those with primary infection are usually not infectious.
- **Recently Infected.** Anyone in the first three years following their primary infection. Also known as a **Recent Converter.**
- **Latent Infection.** Also known as **Inactive Infection,** this condition exists post primary infection when the body has built up antibodies to keep the tuberculosis bacteria inactive. These individuals will have a positive tuberculin skin test, a normal X-ray, and will not be ill.
- **Active Disease.** Active disease will develop if for any reason the immune system does not keep the latent tubercle bacteria in check. Persons with active disease are infectious to others. If an individual with latent tuberculosis remains healthy, active disease will not develop.
- **Chronic Active Disease.** This refers to individuals who cannot or will not take their medication and who will symptomatically appear to get better without treatment but in whom the disease is still active. Persons with chronic active disease, a condition which may last for years, are infectious to others.
- **Inactive Disease.** Anyone formerly having an active disease but who no longer has an active infection.
- **Reactivation.** Reactivation occurs when the individual does not take all of the medication prescribed for the complete duration of treatment. A reactivation case of tuberculosis is often more difficult to treat because by taking only a part of the medication, the patient's germs may have mutated and so will be resistant to that medication in the future.

Adapted from the "Tuberculosis Control Program Manual," Fact Sheet #8, New Mexico Department of Health (September 1, 1987).

Tuberculosis transmission is a recognized risk in traditional healthcare settings; for example, airborne nosocomial infections, including tuberculosis, have been observed in hospitals.[13,14] Resurgent tuberculosis, including drug resistant strains, is surfacing in other high risk populations. Tuberculosis is a major problem in correctional facilities, where tuberculosis cases occur at least three times more often than in the general adult population.[15] As a result of close quarters and recirculated air, almost 200 inmates became infected in one prison after the first case had been discovered, including eight active cases.[16] Tuberculosis is also prevalent in homeless shelters. Seventeen cases of active pulmonary tuberculosis occurred in residents of homeless shelters in three Ohio cities.[17] Tuberculosis morbidity among the homeless is increasing elsewhere; Tennessee has established separate space within existing homeless shelters to accommodate homeless persons being treated for tuberculosis.[18] HIV infection and social factors such as homelessness have contributed to increases in tuberculosis cases reported in New York City since 1979.[19] The 1990 tuberculosis case rate in New York City of 49.8/100,000 is more than five times the national case rate.[20] High incidence groups for tuberculosis include people born in high-prevalence countries, medically underserved low income populations, especially blacks, Hispanics and Native Americans, and residents of long-term care facilities.[21]

Transmission of tuberculosis in chronic care institutions such as nursing homes may be facilitated because diagnosis is delayed and because of close contact between patients and staff. In addition to healthcare providers, other personnel such as housekeeping, maintenance, clerical, janitorial staff, and volunteers are also high risk work populations and should be included in a preventative program. The magnitude of the risk varies considerably by the type of work setting and population in which a person works. The risks may be higher in clinical settings likely to have individuals with active tuberculosis (but in whom a diagnosis has not been made). Risks may also be higher where procedures such as bronchoscopy, endotracheal intubation and suctioning with mechanical ventilation, open abscess irrigation, and autopsy are performed.

Tuberculosis is primarily delivered by the airborne transmission of droplet nuclei. The transmission of tuberculosis from an infectious individual to an uninfected individual is perpetuated by the exhalation and inhalation of infected droplet nuclei. In an uncontrolled environment the droplet nuclei can be released by infected individuals (that is, individuals with active disease) through coughing, sneezing, singing, spitting, etc. A "typical" sneeze can potentially release millions of tuberculosis carrying microscopic droplets.[22] Uninfected individuals can inhale those nuclei, become infected, and ultimately release infectious droplet nuclei. Exactly what degree of contact is required to contract tuberculosis is debatable, with consensus being that tuberculosis usually requires prolonged, intense exposure for transmission to occur.[2] Obviously, healthcare workers are encouraged to take preventative steps when possible.

PREVENTING OCCUPATIONAL EXPOSURE TO TUBERCULOSIS

The potential for contracting tuberculosis in the healthcare setting can be reduced if proper diagnostic and screening methods are followed (previously discussed). Figure 1 presents screening guidelines developed for correctional facilities but which are largely applicable to other healthcare settings. As shown in Figure 1, it is important to routinely screen both employees (such as healthcare providers or other staff) as well as clients (patients, inmates, etc.) in order to reduce the likelihood of tuberculosis transmission, or, to treat it if present.

A vaccine for tuberculosis was developed in 1922 called Bacillus of Calmette and Guerin (BCG). BCG contains live vaccine (which itself can cause tuberculosis-like illnesses) and is not reliable.[7] As a result, BCG vaccine is only used to prevent tuberculosis in populations with high risk. BCG vaccine is recommended by WHO, however, it is not used in the U.S. because it is not accepted as a reliable preventative against tuberculosis.[7] Persons who have had the BCG "vaccination" may show positive reactions in the tuberculin skin test.[2]

In high-risk healthcare settings, certain techniques can be applied to prevent or reduce the spread of infectious droplet nuclei into the general air circulation. There are four types of control methods for preventing or reducing exposures to tuberculosis (all discussed below): engineering controls, personal protective equipment (PPE), work practices, and source control methods. None of the methods alone or in combination can completely eliminate the risk of tuberculosis transmission, however, they can substantially reduce the exposure risk (see Chapter 4 and Appendix D).

Engineering Controls

Once infectious droplet nuclei have been released via a cough or sneeze, etc., they can be eliminated or reduced by ventilation. Part of the reason that some environments are tuberculosis prone (such as prisons) is that the air routinely gets recirculated. Remember that the original method of treatment for tuberculosis, the sanitorium, involved plenty of fresh air; increasing fresh air throughout the work environment may help reduce occupational exposure to infectious droplet nuclei. Ventilation may be augmented by trapping organisms using high efficiency filtration or by killing the organisms with germicidal UV irradiation (100 to 290 nm). When possible, maintain an air pressure gradient between the worker and the patient. That is, air should flow from the uninfected individual (the healthcare worker) toward the potentially infectious individual. This can sometimes be accomplished simply by rearranging the furniture and helps to keep infectious droplet nuclei away from the healthcare worker. Always make certain that recirculated air is sufficiently diluted with "fresh" outside air; otherwise, the recirculated ventilation can accelerate the spread of the tuberculosis bacteria.

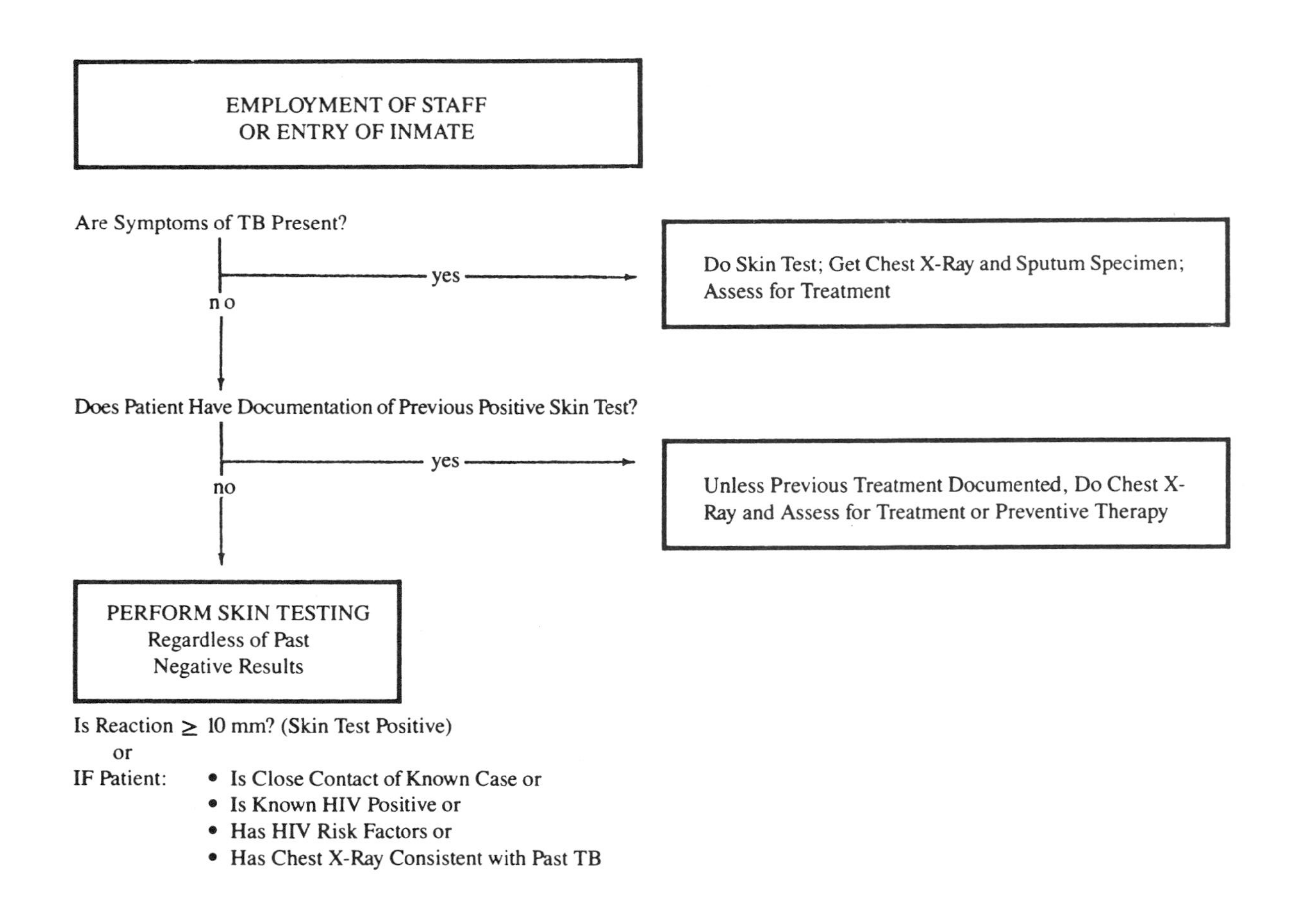
EMPLOYMENT OF STAFF
OR ENTRY OF INMATE
Are Symptoms of TB Present?
yes
Do Skin Test; Get Chest X-Ray and Sputum Specimen;
Assess for Treatment
no
Does Patient Have Documentation of Previous Positive Skin Test?
yes
Unless Previous Treatment Documented, Do Chest X-
Ray and Assess for Treatment or Preventive Therapy
no
PERFORM SKIN TESTING
Regardless of Past
Negative Results
Is Reaction ≥ 10 mm? (Skin Test Positive)
or
IF Patient:
• Is Close Contact of Known Case or
• Is Known HIV Positive or
• Has HIV Risk Factors or
• Has Chest X-Ray Consistent with Past TB

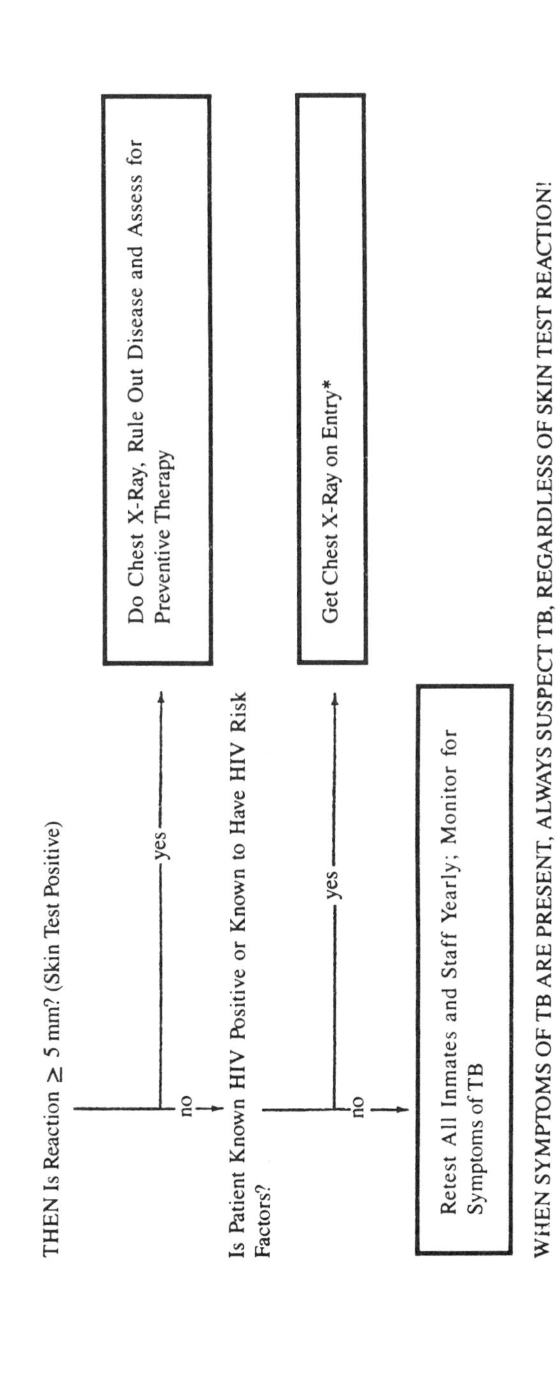

Figure 1. Guidelines for screening staff and inmates. Developed to apply to correctional facility staff and inmates, this guideline is applicable to other long-term care facilities. (Adapted from "Control of Tuberculosis in Correctional Facilities, A Guide for Health Care Workers," U.S. DHHS, CDC National Center for Prevention Services, Division of Tuberculosis Elimination, Atlanta, 1992.)

General Ventilation

General ventilation in healthcare facilities serves two functions: to provide dilution ventilation (the introduction of presumably fresh outside air) into the work environment and to remove airborne contaminants from the work environment. Ventilation standards for indoor air quality have been published by the American Society of Heating, Refrigeration and Air Conditioning Engineers (ASHRAE).[23] This includes 15 to 20 cfm/person (cubic feet of air per minute per person).[23] Other recommendations, specific for healthcare facilities as well as for other institutions where high risk populations exist, have been published by ASHRAE and by the Federal Health Resources and Services Administration.[24,25] Meeting these standards should reduce the probability of tuberculosis transmission in most areas, however, highly infectious individuals can still transmit infection even when these ventilation standards are met. Ventilation effectiveness can be enhanced by the addition of adequate makeup air to the system, by proper direction of the air flow through the room (usually with air supply at the ceiling and exhaust inlets near the floor). In an area occupied by a patient with infectious tuberculosis, air should flow from ''clean'' areas (such as a nurses' station) into the patient's room, then be exhausted.

For general use areas (emergency rooms, waiting rooms, examination rooms, and other general areas where there is an at-risk population), recirculated air is often used for general ventilation for reasons of economy (instead of constantly heating, cooling, humidifying or dehumidifying outside air). If air is to be recirculated, care must be taken to ensure that infectious agents are not also recirculated and transmitted in the process. Tuberculosis has been transmitted in long-term healthcare facilities[9] and prisons[26] where air was recirculated without disinfection or filtering.

In addition to the dilution ventilation and air flow control methods discussed above, several options exist for handling air which is potentially tuberculosis laden (and for the removal of infectious airborne particles). These include the use of High Efficiency Particulate Air (HEPA) filters and the use of germicidal UV light.

HEPA Filters

A HEPA filter is a high efficiency particulate filter which, by definition, will remove at least 99.97% of airborne particles >0.3 microns (μ) in diameter. HEPA filters have been shown to be effective in cleaning the air of Aspergillus spores which are on the order of 1.5 to 6 μ; tuberculosis bacillus droplet nuclei are approximately the same size and therefore HEPA filtration theoretically should remove infectious droplet nuclei. If properly designed and installed, systems using HEPA filters can be central units, filtering all of the air circulating through a duct. Portable HEPA units can be used to filter and circulate the air within a room.

Of course, as they continuously filter and get full of whatever particulates are in the air, presumably including tubercle bacilli, as a part of their maintenance the HEPA filters must be changed periodically. If HEPA filters are used as a control measure for tuberculosis, changing the filters could be potentially risky for maintenance personnel. Proper precautions should be taken (including training in the use of HEPA filtered vacuums and the use of PPE) to reduce the employee's exposure as well as the inadvertent release of contaminated material into the ventilation system or elsewhere.

Ultraviolet (UV) Light

Germicidal UV lamps have been used to prevent tuberculosis transmission, and can be considered as an adjunct to other control methods. Germicidal UV light can be contained within ductwork, as an overhead air disinfection system, or more rarely in unshielded environments where the employees wear the proper PPE to protect them from the adverse effects of UV radiation. There are advantages and limitations to the use of UV light for disinfection of airborne tuberculosis bacilli. UV light at a wavelength of 254 nm is fatal to most infectious droplet nuclei.[27] In one controlled study, UV lamps installed in the exhaust air ducts from rooms of individuals with infectious tuberculosis were shown to prevent infection of guinea pigs (which are highly susceptible to tuberculosis).[28] On this basis, and from the experience of tuberculosis clinicians and mycobacteriologists during the last three decades, the CDC has continued to recommend UV lamps (with appropriate safeguards to prevent short term overexposure) as a supplement to ventilation in settings where the risk of tuberculosis transmission is high.[29]

Adverse health effects can result from UV exposure, and particularly from germicidal UV exposure.[30] Short-term overexposure to UV irradiation can cause photokeratitis (inflammation of the cornea) in the eye and erythema (or sunburn). Long-term exposure to UV irradiation is associated with increased risk of basal cell carcinoma of the skin and with cataracts. It is therefore not possible to simply place UV light sources in the workplace to control bacteria; for all practical purposes patients and healthcare workers cannot be directly exposed to UV light.

UV lighting should be placed inside ventilation ducting or air plenums such that it can disinfect the air but not cause a health hazard in and of itself. When used in ducts or plenums for disinfection purposes, UV sources must operate continuously. Like any light bulb, UV bulbs have a recommended bulb life. Particularly when used for germicidal disinfection, UV bulbs should be changed using the schedule directed by the manufacturer, otherwise, there is a risk of the UV bulb(s) burning out (and not providing the needed disinfection during that time). This is particularly true since the bulbs inside ducting, etc. are out of sight. Maintenance workers responsible for changing the UV bulbs, or for performing maintenance on ductwork or on equipment adjacent to the UV bulbs, should be made aware of the potential hazards.

Table 6. Features to Consider Before Recommending Overhead UV Air Disinfection

If the answer to any statement is "NO," then overhead UV disinfection might not be suitable for the proposed situation.

YES	NO	
☐	☐	A potentially high risk of tuberculosis transmission in the population that cannot be controlled adequately by conventional interventions.
☐	☐	Sufficiently high ceilings (≥9 ft) that are relatively free of obstructions.
☐	☐	Relative humidity ≤70%.
☐	☐	The capability of maintaining and operating the UV lamps effectively and safely.
☐	☐	A means of measuring the reduction in airborne infections among the occupants.

From Kizer, K. W., "Using Ultraviolet Radiation and Ventilation to Control Tuberculosis," California Indoor Air Quality Program, Air and Industrial Hygiene Laboratory, and The Tuberculosis Control and Refugee Health Programs Unit, Infectious Disease Branch, Department of Health Services (1990).

In some cases it may be possible to install shielded UV lamps directly in the workplace. If this is desired, baffles must be installed around the UV lamps to prevent anyone from looking directly at the UV light.[31] Table 6 presents a checklist of items to consider if contemplating the installation of overhead UV air disinfection.

Whenever they are utilized, UV disinfection lamps should be used according to manufacturer's specifications and should meet applicable safety guidelines to prevent overexposure to UV light. NIOSH advises that warning labels be installed on the lamp and housing, but not in the work area.[32] Monitoring for exposure to UV radiation is advisable if UV lamps are installed. The American Conference of Governmental Industrial Hygienists (ACGIH) recommends that exposure to radiation of this wavelength not exceed 0.01 J/cm^2 or 0.1 uW/cm^2 for any 8 hour period.[33]

It may be desirable to use exposed UV lights in a facility (such as a laboratory working with dispersible infectious materials) to enhance the disinfection qualities. If this is to be done, consult a Certified Industrial Hygienist (CIH) or Certified Safety Professional (CSP) for advice on proper protective eye wear and skin wear for individuals who will be exposed to the UV light.

Personal Protective Equipment (PPE)

Different types of PPE can be used with respect to preventing the spread of tuberculosis. PPE control measures discussed here include surgical masks and respirators, skin protection, and work practices.

Surgical Masks and Respirators

At-risk work populations can reduce the risk of inhaling contaminated air by utilizing personal respirators capable of filtering >1 μ organisms. The CDC recommends that disposable particulate respirators be used by healthcare providers to prevent the inhalation of infectious respirable particles (Appendix E).[29] In general, a particulate respirator is a device worn to prevent the inhalation of any respirable particles which may be hazardous. Disposable particulate respirators look like the cup-shaped surgical masks commonly used in hospitals.[34]

Surgical masks (not the same as particulate respirators) are often worn by healthcare workers to provide protection against airborne tuberculosis droplet nuclei. Because surgical masks do not fit as tightly around the face as a respirator, air (and droplets) can penetrate into the lungs by leaking through the gaps between the surgical mask and the face.[35,36] Standard surgical masks provide protection as a shield against sprayed droplets (from a sneeze for example) but in general they are not as effective in preventing inhalation of droplet nuclei because they are not designed to provide a tight seal with the face. From the standpoint of healthcare worker protection, a better alternative is the personal disposable respirator. Respirators and fit testing are discussed in detail in the first volume of this series; consult a CIH for assistance in the selection of an appropriate respirator and/or respirator cartridge, respirator program, and fit testing.

Skin Protection

Universal precautions are intended to protect both patient and provider from HBV and HIV, both of which are bloodborne (see Appendix D and Chapter 4 for more on universal precautions). Universal precautions do not provide much protection against tuberculosis, which is primarily transmitted by air.

While tuberculosis is usually transmitted by the inhalation of infectious airborne particles, infection can occur from needle sticks. In at least one instance, an operating room technician (who had recently tested negative for tuberculosis) was stuck by a needle used in a biopsy of a patient in whom tuberculosis was subsequently diagnosed. The technician developed a subcutaneous abscess which tested positive for *M. tuberculosis* and was placed on medications (INH and RMP).[37] Nonpermeable gloves should be worn when a high-risk work population is in an environment where an infectious tuberculosis individual may be contacted. Glove changing should occur in between individuals.[38]

WORK PRACTICES

Personal hygiene should be practiced in the workplace to avoid the inhalation, ingestion or dermal contact with hazardous materials. Good hand washing must occur at all times. As always, eating, drinking, chewing of gum or tobacco, application of makeup, etc., should be avoided in the workplace in order to reduce the transmission of harmful agents. In addition to individual work practices such as good personal hygiene, Table 7 summarizes the components of a tuberculosis control program for reducing occupational exposures to healthcare workers and others who are occupationally at high risk.

SOURCE CONTROL METHODS

Source control methods (so-called here because they entrap infectious droplet nuclei as they are emitted by the individual or "source") are especially important during any aerosol generation. To prevent the introduction of tuberculosis bacilli into an environment, infectious patients may be induced to wear surgical masks or personal respirators in certain circumstances. In this type of situation a surgical mask or respirator acts as a barrier to prevent the generation of airborne nuclei and their transmission into the environment. The personal respirator used by these individuals must be valveless so that infectious particles do not escape through the exhalation valve. At the very least, individuals should be encouraged to cover their mouths when coughing or sneezing.

DECONTAMINATION

Airborne transmission is the most important route of exposure for tuberculosis, but infection can occur from other paths. Guidelines for cleaning, disinfecting, and sterilizing equipment have been published.[29,39,40] Procedures for each item depend upon its intended use. The rationale for cleaning, disinfecting, or sterilizing can be more readily understood if equipment were divided into three general categories: critical items, semicritical items, and noncritical items.[29,39]

Table 7. General Actions to Reduce the Risk of Tuberculosis Transmission in a Healthcare or Other Public Facility

- Screening suspect individuals for active tuberculosis and tuberculosis infection through a combination of patient history, physical examination, skin test* and chest radiograph.
- Regular screening of healthcare workers, and others at-risk at occupational populations, for tuberculosis infection and tuberculosis.
- Providing timely diagnostic services.
- Prescribing appropriate curative and preventative therapy.
- Maintaining physical measures to reduce the microbial contamination of the air, including engineering controls, PPE, and proper decontamination procedures.
- Providing isolation rooms (isolated with respect to ventilation and airborne transmission) for persons having (or suspected of having) infectious tuberculosis.
- Promptly investigating and controlling tuberculosis outbreaks.

* If the skin test is <5 mm and the HIV test is positive, consider evaluating for anergy.

Critical items — Instruments such as needles, surgical instruments, and cardiac catheters (anything introduced directly into the bloodstream or any appliance that would invade normally sterile areas of the body). Critical items should be sterilized.

Semicritical items — Come in contact with mucus membranes of the body. Examples are endoscopes, bronchoscopes, endotracheal tubes (or any appliance which comes in contact with the mucous membranes) should be disinfected (sterilization is preferred).[29]

Noncritical items — Any item which is not invasive and which will touch only the individual's unbroken dermis. Noncritical items need only to be cleaned with an approved detergent.

Routine cleaning of walls and other surfaces is recommended.[40] Although microorganisms are found on floors and walls, as well as on other surface areas, they are not normally considered to be a significant factor in the airborne transmission of tuberculosis (since inhalation is required for tuberculosis transmission). Walls and other surfaces should undergo routine cleaning with an approved germicide/disinfectant.[29]

SUMMARY

Tuberculosis is on the increase, and this increase has several causes. In part, social conditions which enhance tuberculosis transmission are more prevalent. Tuberculosis can spread as a result of overcrowding and poor ventilation in prisons, homeless shelters, drug treatment clinics, and long-term care environments such as geriatric care facilities. Certain populations are at high risk; it can be contracted and readily transmitted by HIV infected individuals.

In conjunction with these environmental factors has been the evolution of drug resistant strains of tuberculosis. Drug resistant strains of tuberculosis have developed, particularly among substance abusers, because patients cannot or will not follow the long-term medical therapy required to cure active tuberculosis.

These reasons, and others, have put certain professions at a higher risk of contracting tuberculosis. The work environments of concern for occupational exposure to tuberculosis are healthcare settings, correctional institutions, homeless shelters, long-term care facilities, and drug treatment centers. Particularly at risk are professions working in close contact with high-risk populations (such as nurses, physicians, social workers, counselors, long-term care providers, guards, and any others similarly engaged).

Proper training can mitigate and control high-risk occupational situations.[41] Engineering controls, including ventilation gradients, the addition of sufficient fresh outside air to recirculated air, HEPA filtration, and germicidal UV light, and personal protective equipment, such as respirators, can reduce the potential for occupational exposure to infectious droplet nuclei. Proper screening for symptoms of active tuberculosis (for example, among high-risk groups such as the homeless) can help to reduce transmission of tuberculosis to the healthcare professional. Routine screening of patients and employees is encouraged, and the use of controls in the healthcare facility can make the work environment less hospitable to the tubercle bacteria, and therefore, safer.

REFERENCES

1. Thomas, C. L., Ed. *Taber's Cyclopedic Medical Dictionary, Edition 16 Illustrated,* Philadelphia: F. A. Davis (1989).
2. CDC. "Tuberculosis Medical Orientation Course," Department of Health and Human Services, Public Health Service, Centers for Disease Control, Center for Protection Services, Division of Tuberculosis Control, Atlanta, GA (September 1989).
3. Kuharik, J. "Multidrug-Resistant TB Cases in Chicago Up 117 Percent," Department of Health and Human Services, Center for Prevention Services, Centers for Disease Control, Atlanta, GA (Summer 1991), pp. 10–11.
4. Wolinsky, E. "Tuberculosis," in *Textbook of Medicine,* J. B. Wyngarden and L. H. Smith, Eds., Bethesda, MD: Cecil (1985), pp. 1620–1630.
5. Stead, W. W. and J. H. Bates. "Epidemiology and Prevention of Tuberculosis," in *Pulmonary Diseases and Disorders,* A. P. Fishman, Ed., McGraw-Hill: New York (1980) pp. 1234–1254.
6. Snider, D. E. *TB Notes, Winter 1990,* Department of Health and Humans Services, Center for Prevention Services, Centers for Disease Control, Atlanta, GA (Summer 1991), pp. 1–2.
7. CDC. National Tuberculosis Training Initiative "Core Curriculum on Tuberculosis," U.S. DHHS No. 00-5763 (1991).
8. McAdam, J. M., P. W. Brickner, L. L. Scharer, J. A. Crocco, and A. E. Duff "The Spectrum of Tuberculosis in a New York City Men's Shelter Clinic (1982–1988)," *Chest* 97:798–805 (1990).
9. Lefrak, S. "Tuberculosis Screening in a Nursing Home: Indications for Preventative Therapy," *JAMA* 266(14):2000–2003 (1991).
10. Breault, J. "TB Control at Pine Ridge Reservation," in *TB Notes, Summer 1990,* Department of Health and Human Services, Center for Prevention Services, Centers for Disease Control, Atlanta, GA (Summer 1990), pp. 7–8.
11. Altman, L. K. "Public Health Service Moves to Curb Spread of Drug-Resistant Type of TB," *The New York Times* (May 1, 1992).
12. Snider, D. E. *TB Notes, Summer 1991,* Department of Health and Human Services, Center for Prevention Services, Centers for Disease Control, Atlanta, GA (Summer 1991), p. 1.
13. Schaal, K. P. "Medical and Microbiological Problems Arising from Airborne Infection in Hospitals," *J. Hosp. Infect.* 18(Suppl. A):451–459 (1991).
14. Danis, D. M., Ed. "The MMWR File," *J. Emer. Nurs.* 17(4):244–245 (1991).
15. CDC. "Control of Tuberculosis in Correctional Facilities, A Guide for Health Care Workers," U.S. Department of Health and Human Services, Public Health Service, Centers for Disease Control, National Center for Prevention Services, Division of Tuberculosis Elimination, Atlanta, GA (1992).

16. "Court Rules Against Minnesota Prisons in Tuberculosis Case," Associated Press Report of 5 December 1990, cited in Department of Health and Human Services, Center for Prevention Services, Centers for Disease Control, Atlanta, GA (Spring 1991), p. 9.
17. "Tuberculosis Among Residents of Shelters for the Homeless — Ohio, 1990," *MMWR,* 40(50):869–872 (December 20, 1991).
18. Gately, K. W. and G. Tamames. "Tennessee Establishes a Shelter for Homeless TB Patients as an Alternative to Hospital Inpatient/Nursing Home Care," Department of Health and Human Services, Center for Prevention Services, Centers for Disease Control, Atlanta, GA (Spring 1991), pp. 4–5.
19. Brudney, K. and J. Dobkin. "Resurgent Tuberculosis in New York City," *Am. Rev. Resp. Dis.* 144(4):745–749 (1991).
20. "Tuberculosis in New York City 1990, Information Summary," Bureau of Tuberculosis Control, New York City Department of Health (1991).
21. "Tuberculosis Takes a Turn for the Worse," *Infec. Con. Hosp. Epidemiol.* 12(8):506–507 (1991).
22. Woodard, C. "The Return of TB," *Newsday* (March 8, 1992), p. 5.
23. ASHRAE Standard 62-1989. *Ventilation for Acceptable Indoor Air Quality,* Atlanta, GA (1989).
24. ASHRAE. *1987 ASHRAE Handbook: Heating, Ventilating, and Air-conditioning and Applications,* Atlanta, GA (1987).
25. Health Resources and Services Administration. *Guidelines for Construction and Equipment of Hospital and Medical Facilities.* Rockville, MD: U.S. DHHS, PHS publication no. (HRSA) 84-14500, 1984.
26. Cruise, P. "TB in a Wisconsin Maximum Security Correctional Facility," Department of Health and Human Services, Center for Prevention Services, Centers for Disease Control, Atlanta, GA (Winter 1990), pp. 7–8.
27. Riley, R. L. and E. A. Nardell. "Clearing the Air: The Theory and Application of Ultraviolet Air Disinfection," *Am. Rev. Respir. Dis.* 139:1286–1294 (1989).
28. Riley, R. L. "The Changing Scene in Tuberculosis," in *Pulmonary Diseases and Disorders,* A. P. Fishman, Ed., New York: McGraw-Hill (1980), pp. 1229–1254.
29. CDC. *Guidelines for Preventing the Transmission of Tuberculosis in Health-Care Settings, with Special Focus on HIV-related Issues, MMWR* 1990:39 (No. RR-17). (Appendix E to this volume).
30. Murray, W. E. "Ultraviolet Radiation Exposures in a Mycobacteriology Laboratory," *Health Phys.* 58(4):507–510 (1990).
31. Kizer, K. W. "Using Ultraviolet Radiation and Ventilation to Control Tuberculosis," California Indoor Air Quality Program, Air and Industrial Hygiene Laboratory, and the Tuberculosis Control and Refugee Health Programs Unit, Infectious Disease Branch, Department of Health Services (1990).
32. National Institute for Occupational Safety and Health. *Occupational Exposure to Ultraviolet Radiation,* Washington, DC: NIOSH (1972).
33. Threshold Limit Values for Chemical Substances and Physical Agents and Biological Exposure Indices, 1991–1992, Cincinnati, OH: ACGIH (1991), pp. 123–126.
34. Hutton, M. "What is a Disposable Particulate Respirator?" in *TB Notes, Summer 1991,* Department of Health and Human Services, Center for Prevention Services, Centers for Disease Control, Atlanta, GA (Summer 1991), pp. 12–13.

35. Charney, W. "The Inefficiency of Surgical Masks for Protection Against Droplet Nuclei Tuberculosis," *J. Occup. Med.* 33(9):943–944 (1991).
36. Pippin, D. J., R. A. Verderame, and K. K. Weber. "Efficacy of Face Masks in Preventing Inhalation of Airborne Contaminants," *J. Oral Maxillofac. Surg.* 45:319–323 (1987).
37. Fleenor, M. E. and J. W. Harden. "Case Report: Tuberculosis Lymphadenitis from Needle Stick," in *TB Notes, Spring 1991,* Department of Health and Human Services, Center for Prevention Services, Centers for Disease Control, Atlanta, GA (Spring 1991), pp. 11–12.
38. 29 CFR 1910.1030, "Occupational Exposure to Bloodborne Pathogens; Final Rule." (Appendix D to this volume.)
39. Rutala, W. A., E. P. Clontz, D. J. Weber, and K. K. Hoffman. "Disinfection Practices for Endoscopes and Other Semicritical Items," *Infec. Con. Hosp. Epidemiol.* 12(5):282–288 (1991).
40. Garner, J. S. and M. S. Favero. "Guidelines for Handwashing and Hospital Environmental Control," Atlanta, GA: US DHHS, Public Health Service, CDC (1985).
41. CDC. "Tuberculosis Elimination, U.S.A. (Facsimile of Slides)," Division of Tuberculosis Elimination, National Center for Prevention Services, Centers for Disease Control (August 1991).

CHAPTER 6

Occupational Health Hazards in the Dental Office*

Audra S. Gomez and Marc A. Gomez

INTRODUCTION

There have always been occupational health hazards present in the dental office, but it is only recently that these hazards are receiving the concern and attention they deserve. Occupational hazards take many forms in the dental office. They range from exposure to the various chemicals utilized, to the biohazards associated with blood and body fluids, to exposure from radiation of dental X-ray machines, to the musculoskeletal disorders that develop as a result of the repetition and body positioning required during dental procedures. This chapter will address each of these types of hazards and will recommend ways to eliminate or control each.

* In order to more accurately demonstrate the posture of the operator, we have deliberately eliminated use of a lab coat/apron in the photographs in this chapter. The use of this type of protective device is now mandated by the OSHA Bloodborne Pathogens Standard (see Chapter 4).

0-87371-392-3/93/$0.00 + $.50

CHEMICAL HAZARDS IN THE DENTAL OFFICE

Within the typical dental office, one will usually find 30 to 50 different chemicals utilized for a variety of purposes, from filling teeth to sterilizing instruments. These chemicals range from extremely toxic chemicals like mercury to relatively nontoxic chemicals such as alcohol. This section will classify these chemicals by the function they serve in the dental office and then discuss the hazards and control measures for each category.

Dental Materials — Impression Materials

In a dental office, frequently it is necessary to make a model of a patient's teeth so that dental appliances such as crowns, fillings, and bite guards can be prepared. Impressions are a negative likeness of the teeth (or other structures of the oral cavity), made of a setting plastic material which is later filled with plaster of Paris, thus obtaining an exact copy of the oral structures.[1] These impressions are most commonly made using hydrocolloids which are a powder or gel (suspension of particles) that when mixed with water create a jelly-like substance. The health hazards associated with impression materials are to the respiratory system due to the dust generated when working with the hydrocolloids and the plaster of Paris. Potassium alginate particles are used in making the hydrocolloids. Some of this potassium alginate dust is small enough that it can be inhaled into the lungs. This respirable dust will typically contain up to 40% free crystalline silica (cristobalite). Cristobalite is one of three major crystalline forms of silica dust. Following inhalation of silica dust, some of the particles are trapped in the lung. This can begin the long process of development of a silicotic nodule, or growth, in the lung. Thus, long-term, unprotected exposure to impression material dust may cause lung disease (silicosis).[2] Another dust generated in the dental office, as previously mentioned, is plaster of Paris. Plaster of Paris is a white gypsum powder that when mixed with water sets to a smooth solid. Dust is generated while mixing the plaster of Paris (in preparation to pour into the mold/impression) and while smoothing, grinding and finishing the resulting model of the teeth. When working with potassium alginate and plaster of Paris, the dust generated may be controlled by several methods. First, and most important, is careful technique by the dental staff. For example, careful mixing and simple housekeeping can minimize excess dust. In offices where models/impressions are frequently made, it may also be necessary to install a local exhaust ventilation system to capture and control the dust as it is generated. If this is not possible or not effective in controlling the dust, then it may be necessary for the dental staff to wear personal protective equipment (PPE) such as a dust mask approved by the National Institute of Occupational Safety and Health (NIOSH). Both local exhaust systems and PPE are discussed in Volume 1.

Dental Materials — Filling Materials

Most of us have had a cavity sometime in our life. These cavities need to be removed and the teeth restored with a filling material. The most commonly used restoration material for this purpose is amalgam. In dental restorations, an alloy of silver, tin, copper, and zinc is mixed with pure mercury to form the desired amalgam.[3] Freshly mixed amalgam has a plasticity that permits it to be inserted into a prepared hole in the tooth and then carved to restore the normal contours of the tooth. Of all of the metals contained in amalgam, mercury is the most toxic since it is a general protoplasmic poison. The principal effect of mercury is upon the central nervous system, the mouth, and gums. Short-term exposure to mercury may cause headaches, chest pains, and difficulty in breathing. It may also cause chemical pneumonia, soreness of the mouth, loss of teeth, nausea, and diarrhea. Liquid mercury may irritate the skin. Mercury is also a poison through inhalation. Long-term exposure to mercury causes allergic skin rash, sores in the mouth, sore and swollen gums, insomnia, excess salivation, loss of memory, and intellectual deterioration.[4]

When mixing amalgam, it is common for dental personnel to handle mercury. For example, if the proper proportion of alloy and mercury (a 1:1 ratio) is not achieved, the excess mercury must be squeezed out through a "cheese cloth" like material. This procedure results in the direct absorption of mercury into the dental personnel's tissues. In response to this problem, premeasured disposal capsules of mercury and alloy were developed. Even with premeasured capsules (that reduce the need to touch excess mercury) dental personnel may still be exposed through the inhalation of vapors that are emitted by the volatilization of mercury.[5]

The current OSHA limit for occupational exposure to mercury vapor is 50 $\mu g/m^3$ (micrograms per cubic meter of air) time weighted average in any 8 hour workshift over a 40 hour work week.[6] Any dental personnel whose measured exposure exceeds this limit is considered to be overexposed and so the dental staff must undertake various measures to control mercury vapor exposure. Table 1 is a review of basic guidelines to be followed when working with mercury.[7] All dental offices should monitor their personnel for potential exposure to mercury vapor. Monitoring should be performed by an industrial hygienist at least annually and whenever a change is made in the way mercury is handled. This may be accomplished in several ways including using dosimeter badges or real time monitoring devices. Urinalysis may also be performed to measure dental office personnel's total body burden of mercury (resulting from skin absorption and inhalation).

Dental Materials — Acrylics

Due to advances in preventative dentistry, fewer of us lose our teeth to cavities. However, periodontal (gum) disease still prevails and is the most common reason

Table 1. Controlling Mercury Exposure

1. Avoid direct skin contact with mercury or freshly mixed amalgam (wear gloves).
2. Provide proper ventilation in the workplace by having appropriate fresh air exchangers and periodic replacement of air conditioning filters which may act as traps for mercury.
3. Use precapsulated alloys to eliminate the possibility of a bulk mercury spill; otherwise store bulk mercury in unbreakable containers with air tight covers on stable surfaces.
4. Use high volume suction (a form of exhaust ventilation) when finishing or removing amalgam.
5. Clean up visible beads or pools of spilled mercury properly by using trap bottles, or use commercial clean up kits that contain absorbents to minimize vaporization. Do not use household vacuum cleaners.
6. Dispose of mercury contaminated items in sealed bags according to applicable regulations.

for tooth loss. Some of the materials used to help replace these missing teeth can present a chemical hazard to dental personnel. Acrylics are used to prepare partial or complete dentures, temporary crowns, and bridges. Acrylic is a thermoplastic resinous material of acrylic acid and is the principal ingredient of many plastics used in dentistry. The most hazardous component of acrylics is methyl methacrylate which is a transparent resin, liquid at room temperature, and which is polymerized by the use of a chemical initiator.[8] The current OSHA limit for occupational exposure to methyl methacrylate is 100 ppm (parts per million). Methyl methacrylate is moderately toxic by inhalation, mildly toxic by ingestion, and is also a skin and eye irritant.[9] Table 2 lists the precautions to be followed when working with these substances.

Chemical Hazards of Disinfection and Sterilization

Sterilization is the process by which all forms of living organisms are completely destroyed. In the dental office, it is critical that all bacteria, fungi, viruses, and bacterial spores be destroyed. All instruments that are used in the mouth must be sterilized to protect the patient and the practitioner. With the recent focus upon bloodborne pathogens such as hepatitis B virus (HBV) and human immunodeficiency virus (HIV), sterilization has never been more critical.

The three forms of sterilization most commonly used in the dental office are autoclaving (pressurized steam), dry heat, and chemical vapor (pressurized chemical steam). Of these three, the only one that is potentially toxic is chemical vapor sterilization because it uses chemicals such as formaldehyde or ethylene oxide.

Table 2. Controlling Exposure to Acrylics

1. Avoid skin contact.
2. Avoid excessive inhalation of vapors.
3. Work in well-ventilated areas.
4. Keep containers tightly closed when not in use.

Both formaldehyde and ethylene oxide are suspected human carcinogens. Formaldehyde is a very toxic substance and is a poison by inhalation, skin contact, and ingestion. Short-term exposure to formaldehyde causes irritation to the eyes, skin, and respiratory tract while long-term exposure can cause hypersensitivity leading to contact dermatitis (severe skin irritation upon contact).[10] The current OSHA limit for occupational exposure to formaldehyde is 1 ppm. Ethylene oxide (ETO) is also a very toxic substance as it is a poison by inhalation and ingestion. Human systemic effects resulting from inhalation of ethylene oxide are convulsions, nausea, and vomiting. It has also been found to have effects on the reproductive system. Ethylene oxide is an irritant to mucous membranes of the respiratory tract and in high concentrations can cause pulmonary edema (fluid in the lung).[11] The OSHA occupational exposure limit for ethylene oxide is 1 ppm.

The major advantage to chemical vapor sterilization is the fact that this type of sterilization does not corrode carbon steel instruments or melt plastic instruments. For this reason, chemical vapor sterilization is still commonly used. Because of the toxicity associated with the chemical vapor sterilization process it is necessary to exhaust the chemical vapors out of the work area using a ventilation exhaust system and to locate the sterilizer in a well-ventilated area. Another work practice required when using ETO is the aeration of ETO-sterilized instruments. This is necessary to reduce the likelihood of mucous membrane irritation and the possible build up of off gassing in enclosed areas. It is also necessary to wear gloves when handling the chemicals and the freshly sterilized instruments.

Disinfection is the destruction of pathogenic microorganisms by chemical or physical methods. The purpose of disinfection is to reduce the microbial population when sterilization is not possible. For a disinfecting solution to be acceptable for use in dentistry it must be labeled tuberculocidal, bactericidal, viricidal, and fungicidal and carry the acceptance label of the Environmental Protection Agency (EPA) and the American Dental Association (ADA). Current methods of dental care cause widespread contamination of the treatment area. For example, droplet splatter of saliva and blood resulting from the operation of automated dental instruments such as handpieces, ultrasonic scalers, and air-slurry polishers, create the need for treatment room disinfection. Areas such as the dental chair, light handles, counter tops and cabinet handles should be disinfected before and after each patient's treatment. Also, anything that goes into the mouth that cannot be sterilized must be disinfected or discarded if disposable.

Six chemicals frequently used for disinfection in the dental office are glutaraldehyde, chlorine dioxide, iodophors (iodine), phenol compounds, sodium hypochlorite, and ethanol. When used as disinfecting agents, these chemicals are diluted with water or can be purchased already diluted into a disinfectant solution.

Glutaraldehyde is a high-level disinfectant used for instruments that cannot be sterilized with heat. The chemical is used in a disinfecting solution (typically

Table 3. Controlling Exposure to Glutaraldehyde

1. Gloves should always be worn when working with the solution and when removing instruments from the solution.
2. Glutaraldehyde vapor should be controlled by strict adherence to standard operating procedures including keeping the solution covered at all times or by installing a local exhaust ventilation hood to pull the vapors out of the work area.
3. Do not use glutaraldehyde as a surface disinfectant.
4. Be sure to rinse instruments thoroughly before use in the patient's mouth.

in a 2 to 3.2% concentration) in which instruments are soaked. Glutaraldehyde is capable of sterilization if there is long enough exposure to the instruments (from 6 to 10 hours, depending on the strength of the product).[12] Due to this time factor, glutaraldehyde is not a recommended form of sterilization for any instruments that can withstand heat. Glutaraldehyde, in solution, produces vapors that are very toxic to tissues. The current OSHA occupational exposure ceiling limit for glutaraldehyde is 0.2 ppm. The chemical is a suspected carcinogen, is moderately toxic by inhalation and skin contact, and is a severe human skin and eye irritant. In order to minimize dental office personnel exposure to glutaraldehyde, several control measures are recommended in Table 3.

Chlorine dioxide is a surface disinfectant that is effective on clinical surfaces when used for 1 to 3 min in conjunction with a thorough cleaning procedure. Chlorine is moderately toxic by inhalation and is a strong eye irritant. Typically, chlorine dioxide is used in a low concentration solution (approximately 0.15%). Due to the chlorine vapor that can be generated, it is recommended that the chemical be used in a well-ventilated area with safety goggles and gloves.[13]

Iodophors are solutions containing iodine. They are available as surgical scrubs (antiseptics) and hard surface disinfectants. When used properly, they are effective for surface disinfection in 3 to 30 min.[14] As long as iodophors are used in a well-ventilated area, generally, no precautions are necessary since they are minimally irritating to tissue.

Phenol is usually mixed together with other disinfectants such as glutaraldehyde or ethanol to form a synthetic phenol compound. The concentration of phenol in these disinfectants range from approximately 0.1 to 1.76%.[15] Phenol is a severe eye and skin irritant and is moderately toxic by skin contact. Dermatitis (skin irritation) resulting from contact with phenol or phenol-containing products is fairly common. The OSHA occupational exposure limit for phenol is 5 ppm. When working with disinfectants containing phenol, it is recommended that gloves and goggles be worn and that the chemical be used only in a well-ventilated area.

Sodium hypochlorite (household bleach) if prepared daily is an inexpensive and effective germicide and disinfectant. Concentrations ranging from approximately 500 ppm sodium hypochlorite (1:100 dilution of household bleach) to 5000 ppm sodium hypochlorite (1:10 dilution of household bleach) are effective depending upon the amount of organic material, i.e., blood, mucous, etc. present on the surface to be cleaned. These dilutions of bleach can be irritating to the eyes, skin, and nose, therefore safety glasses and gloves used in a well-ventilated area are recommended.

Ethanol (ethyl alcohol) is commonly combined with other disinfectants such as phenolics and is an effective chemical disinfectant. Typically it is found in high concentrations such as 50 to 80% and is used as an aerosol propellant. Ethanol is mildly toxic by inhalation and skin contact. Ethanol is an eye and skin irritant.[16] It is recommended that ethanol-containing disinfectants be used only in a well-ventilated area and that gloves and safety goggles be worn. When using these disinfectants, it is very important to spray the chemical close to the surface or object being disinfected. This will minimize any excess aerosolization of vapors that could be inhaled. This is true for all disinfectants previously mentioned.

Chemical Hazards of Anesthetics/Sedation

It has been estimated that somewhere between 2 and 20% of the population in the U.S. have such a high level of anxiety over dental visits that they are unable to seek necessary care.[17] The fear of pain or the expectation of pain causes most of the distress. Many dentists offer anti-anxiety medications and sedatives to help these patients. The most common of these is nitrous oxide, also known as laughing gas. Nitrous oxide is a general anesthetic which causes a decreased pulse rate without a drop in blood pressure or body temperature.[18] The combination of nitrous oxide and oxygen given to a patient induces an altered psychological state that helps eliminate the fear and pain of the dental experience. Nitrous oxide is administered to the patient through the use of a mask placed over the nose. The gases flow to the mask through attached tubes and are inhaled. Upon exhalation, the gases are captured by a gas scavenging system (vacuum suction) which prevents these gases from escaping into the work environment. Studies have shown that repeated exposure by dental personnel to nitrous oxide (from waste anesthetic gases and vapors) is moderately toxic by inhalation, and experimentally has produced reproductive effects. The recommended control measures for nitrous oxide (waste anesthetic gases and vapors) are to utilize a properly fitting mask to minimize leaking, maintain adequate fresh air ventilation in the dental office, and to install a gas scavenging system.

Chemical Hazards of X-Ray Solutions

Radiographs, commonly known as X-rays, are a vital diagnostic tool in dentistry. They help dentists evaluate structures that cannot be seen by clinical observation. Once an X-ray is taken, the exposed radiograph must be processed using either an automatic film processor or a manual system. Both systems require the use of developer and fixer. The developer solution reacts with the silver bromide salts of the film which have been partially activated by the

X-ray radiation to produce the image. The fixer solution dissolves the light sensitive silver bromide crystals that are not a part of the image and were not preserved by the developer solution.[19] The exposed X-ray film is first placed into the developer solution and then rinsed and put into the fixer solution. Essentially, the developer is an alkaline solution and the fixer is an acidic solution. Inhalation of vapors from these solutions can cause irritation to the nasal mucous membranes and the respiratory tract. The solutions may also cause irritation and damage to the eyes and skin. When working with developer and fixer, dental personnel should work in a well-ventilated area and should wear a fluid-resistant apron, safety goggles, and gloves.

In order to summarize the recommended control measures for the chemical hazards reviewed in this chapter, please refer to Table 4.

RADIATION HAZARDS

As mentioned previously, radiographs are an important diagnostic tool in dentistry. However, working with X-ray radiation is not without danger. X-ray radiation is a type of ionizing radiation (capable of ionizing water and other substances in the body). The human body is made up of approximately 80% water. Thus, when X-rays are absorbed by the body, they can disrupt cellular metabolism and cause the death of a cell. The absorption of X-rays by the human body may cause permanent damage to living cells and tissues. The extent of the effect upon the body is measured by:

1. The quantity (amount) of radiation exposure.
2. The quality (intensity) of radiation exposure.
3. The length (time) of exposure.
4. The type of tissue irradiated.

In the dental office there are two specific radiation hazards, primary and scatter radiation. Primary radiation is the central beam emitted from the X-ray tube. This central beam travels in a straight line and exposes the film to produce the diagnostically useful radiograph. The operator should not stand in line of this primary beam. Scatter radiation is produced when the central beam is deflected from its path during the impact with tissue. This scatter radiation is deflected in all directions and impossible to confine. The resultant scatter radiation has less energy, however, it is still harmful to those exposed to it. Without adequate protective barriers, dental personnel and persons standing or sitting nearby may be exposed to the scatter radiation.[20]

The ALARA concept (as low as reasonably achievable) describes the recommended method of radiation control. Adherence to the ALARA principle means that every available method of reducing exposure to X-ray radiation is used to minimize potential risks for dental personnel and patients. In order to achieve ALARA, certain control measures are recommended. In any dental suite

Table 4. Control Measures for Chemical Hazards

Hazard	Gloves	Goggles	Mask	Well-Ventilated Area	Specific Controls
Mercury	Y	Y		Y	Use precapsulated alloys and exhaust ventilation system
Potassium alginate/crystalline silica		Y	Y	Y	Exhaust ventilation and dust mask
Plaster of Paris		Y	Y	Y	Exhaust ventilation and dust mask
Methyl methacrylate	Y	Y		Y	Exhaust ventilation system
Chemical vapor sterilizer					
Formaldehyde	Y	Y		Y	Exhaust ventilation system
Ethylene oxide	Y	Y		Y	Exhaust ventilation system
Glutaraldehyde	Y	Y		Y	Exhaust ventilation system
Chlorine dioxide	Y	Y		Y	
Iodophors	Y	Y		Y	
Phenol compounds	Y	Y		Y	
Sodium hypochlorite	Y	Y		Y	
Ethanol (aerosol)	Y	Y	Y	Y	
Nitrous oxide				Y	Gas scavenging system
Developer solution	Y	Y		Y	
Fixer solution	Y	Y		Y	

where X-rays are taken, the scatter radiation is controlled by building the walls of the room with the ability to absorb the radiation produced. The walls should be constructed of at least two 5/8-inch thicknesses of gypsum sheetrock (or 1/16-inch thick sheet of lead embedded in the wall) in order to control scatter radiation. This will protect all individuals outside of the room from the radiation produced in the room. The patient (in the room) should be made to don a lead apron with a cervical collar (to protect reproductive organs and thyroid), and the operator should stand outside the suite in a safe area.

ERGONOMIC PROBLEMS

As anyone who has ever visited a dental office knows, the dental chair that the patient sits in has many controls that make it possible to place the patient in a range of positions from sitting upright to reclining. The purpose of this special chair is to position the patient so that dental personnel can easily access the mouth. Although the patient may not realize it, the dental chair is only relatively successful at accomplishing its purpose. This leads us to the hazard that produces the most frequent problems for dental personnel, i.e., musculoskeletal disorders. Day after day, dental personnel attempt to access their patient's mouths. In order to accomplish this, they must position and angle their bodies, specifically their backs, necks, elbows, and wrists into somewhat awkward positions.

Musculoskeletal Disorders — Back and Neck

Figure 1 is an example of how the back and neck must often be positioned in order to provide dental care. As can be seen in the picture the individual's back is somewhat bent over and twisted. Positioning the back in this manner is not uncommon. Frequently dental personnel will need to remain in this position for long periods of time during their day. Long-term positioning of the back like this puts a static load on the muscles of the back. Eventually, the muscles of the back will fatigue, leaving the ligaments in the back left to support the spine. This can cause significant pain and discomfort to dental personnel. In fact, this can lead to the inability to work. This usually can be corrected through rest, exercise to strengthen muscles and proper posture. If proper care is not taken, it is possible that the vertebrae of the lumbar and/or the thoracic spine may be allowed to be pushed out of line, causing a bulging of the disk as seen in Figure 2. When the bulging disk pushes far enough out it will touch the peripheral nerves and cause pain in the back. Medical treatment for this condition is usually required.

It is also not uncommon for the neck to be put into an awkward position. As can be seen in Figure 3, the individual's neck is twisted so that the dental staff

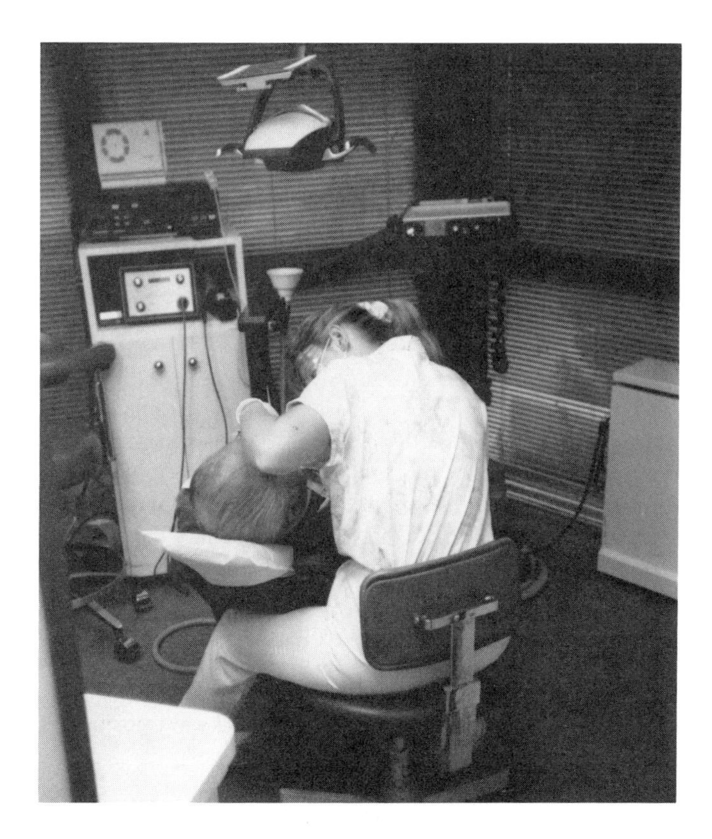

Figure 1. Improper working posture in the dental office. Positioning the back in this manner puts a static load on the muscles of the back.

person can see into the mouth. This twisting of the neck, similar to the back, can also put a static load on muscles. If allowed to continue, the muscles will fatigue and pain will occur. As previously mentioned, if not treated appropriately, the eventual result will be stretched ligaments that will allow the vertebrae in the cervical spine to move out of line, causing a bulging disk. When the bulging cervical disk pushes far enough out it will touch the peripheral nerves and cause pain in the neck and require medical treatment.

Although these problems are not easy to control, there are several possible actions that can be taken to minimize the pain. One solution to this problem is to position the patient in the subsupine position as seen in Figure 4. Note that when the head of the patient is positioned lower than the legs, the neck and back of the dental staff person need not be twisted as much to access the mouth. In addition to utilizing subsupine positioning of the patient, it is recommended that dental personnel attempt to position themselves as suggested in Table 5.

A

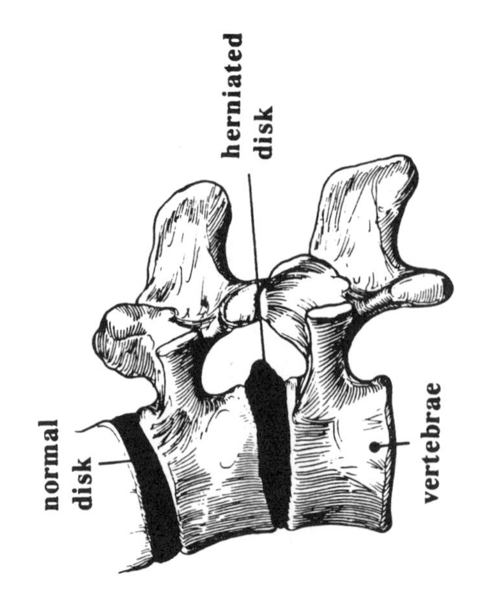

B

Figure 2. Improper working posture in the dental office, side view. Note the bending posture of the spine (A), leading to lumbar and/or thoracic disc compression, and perhaps to other adverse effects (B).

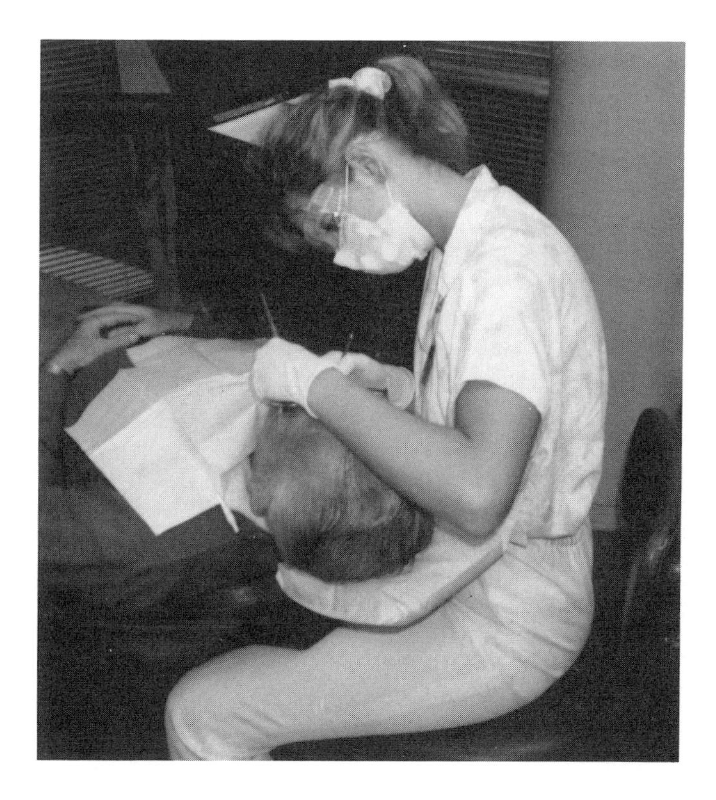

Figure 3. This photo shows proper positioning of the lumbar and thoracic spine, however, the neck is bent. This type of posture can ultimately lead to pain in the neck as the result of ligament stretching and compression of the cervical discs.

The chair/stool that the dentist and dental hygienist sit in to provide care is extremely important as it must be designed to provide support, stability, mobility, and comfort. As seen in Figure 5, the stool should have the following features:

1. An adjustable seat so that the knees remain at approximately a 90 degree angle.
2. An adjustable chair back so that the back cushion supports the lumbar area of the lower back.
3. The legs of the stool must have a broad base of support and casters to allow the stool to move.

Although not readily available, it is logical that a dental stool should be developed that also provides support when leaning forward or to the side. This could be accomplished using a tension bar that would allow dental personnel additional comfort and support while leaning forward or sideways while providing care to the patient as is found with a dental assistants' stool. Dental assistant must sit

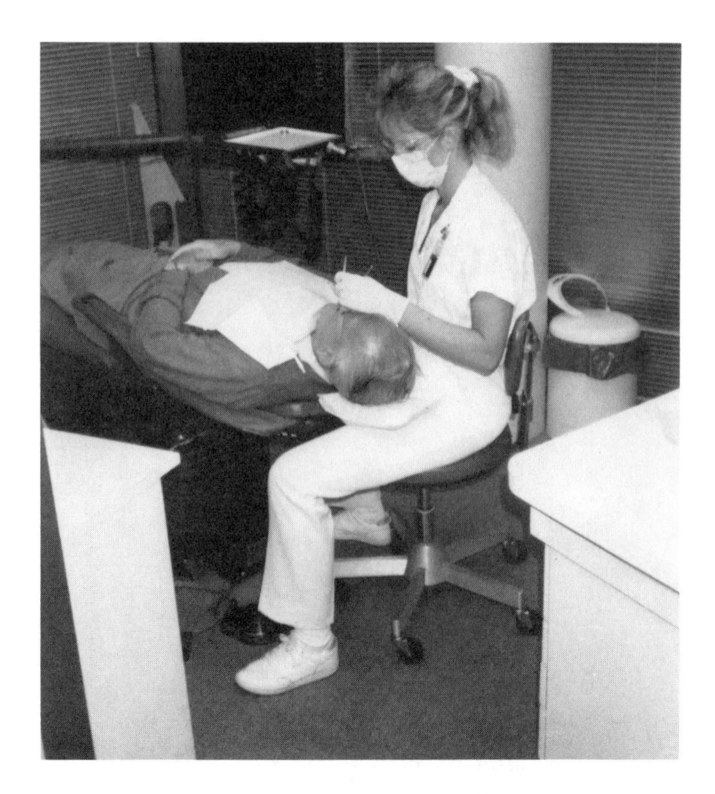

Figure 4. A subsupine positioning of the patient allows for dental personnel to take a more comfortable posture. Note the ability to see and move freely without bending the neck and back.

up higher than the dentist or dental hygienist as they are required to access the patient's mouth while the dentist is working, thus, the dental assistant's stool needs to be modified so that the seat may be raised much higher. Also, the assistant's stool should have a platform on which to rest the feet, enabling them to maintain a 90 degree angle at the knee. As seen in Figure 6, an arched extension (which is optional) from the back of the stool to the front provides support at the level of the assistant's abdomen when leaning forward toward the patient.

Table 5. Ergonomic Considerations For Dental Personnel

1. The operator should be seated in a relaxed position, elbows close to their sides.
2. The knees should be slightly above hip level with feet flat on the floor.
3. The back and shoulders should be kept as straight as possible.
4. The neck should be kept straight so that the head is held relatively erect. In some situations it is necessary to bend the neck to be able to see into the patient's mouth. When this is done, the neck should not be bent more than approximately 15° in any direction (forward or to the side).

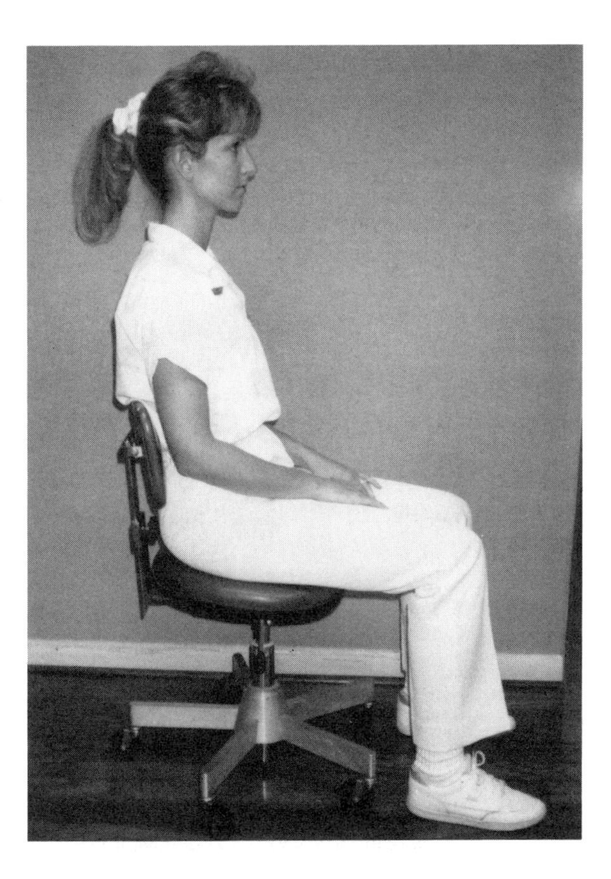

Figure 5. Proper seating for dental personnel.

Musculoskeletal Disorders — Hand and Wrist

It is important to note that much of the work performed in dentistry is accomplished manually. In fact, the hand and wrist can be considered an extension of the tools used by dental personnel. It is not uncommon for the wrist to be put into awkward positions while providing dental care in the mouth. Although most dental personnel learn proper hand positioning and movement while in school, in reality it is difficult to keep from bending and tensing the hand and wrist. Primarily, there are two wrist/hand positions that cause problems. The first, as shown in Figure 7, is flexion or bending of the wrist. This occurs to dental hygienists while performing dental prophylaxis (teeth cleaning) and to dentists while holding the various tools they work with. The second problem position, as seen in Figure 8, is the tensing of fingers around a dental instrument.

Although dental personnel are taught to use a light grip on their instruments, it is often necessary to grip the instruments tightly in order to adequately perform

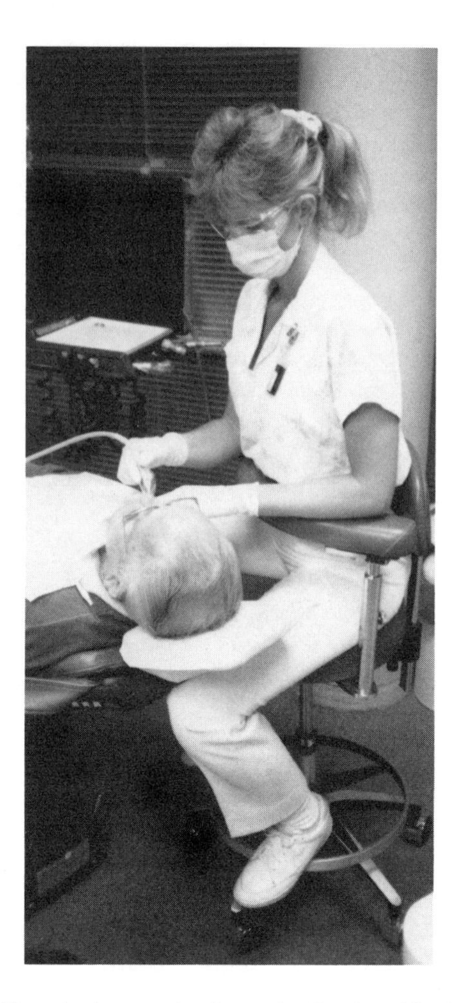

Figure 6. Seating with arched arm extension under the dental hygienist's left arm. This provides support to the upper body during leaning as well as support to the arm itself. Also note the footrest which allows legs to be supported at a comfortable angle.

dental care. If the wrist is kept in the flexed or bent position and/or the fingers are tensed around an instrument repeatedly for long periods of time, several conditions can occur. Dental personnel may first experience muscle pain and fatigue in the hand. A loss of strength and possibly tingling and/or numbness in the hand may follow. Simple treatment such as rest, strengthening exercises, and proper hand positioning can prevent a condition known as carpal tunnel syndrome (CTS) from occurring. As can be seen in Figure 9, inside the wrist there is an opening (or tunnel) through which the median nerve of the hand passes.

When the wrist is in the flexed or bent position, this tunnel is reduced in size and the nerve passing through the carpal tunnel is compressed. This compression

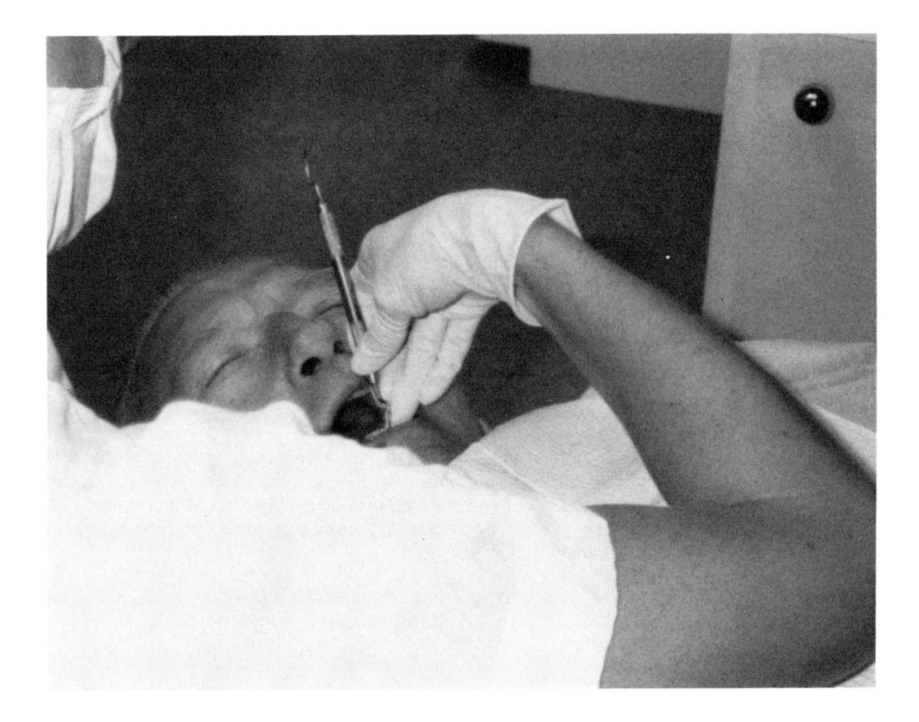

Figure 7. Wrist flexion or bending is common in the practice of dentistry.

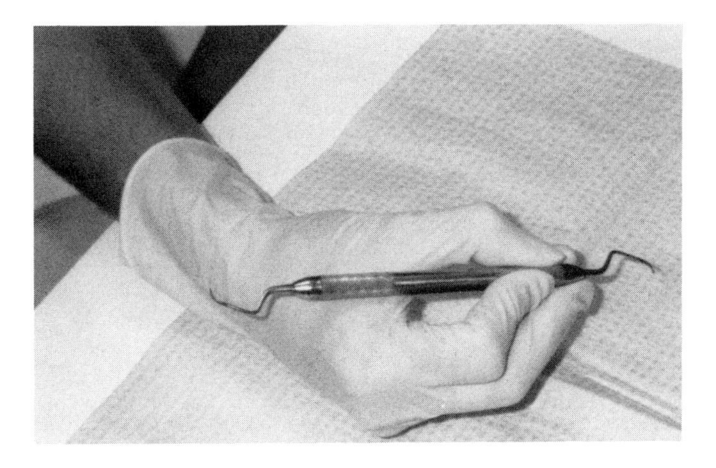

Figure 8. The precision grip required to maneuver dental instruments, if maintained for long periods, can lead to tensing, muscle fatigue, and stiffness. Unfortunately, this can also lead to cumulative trauma disorders such as carpal tunnel syndrome (CTS).

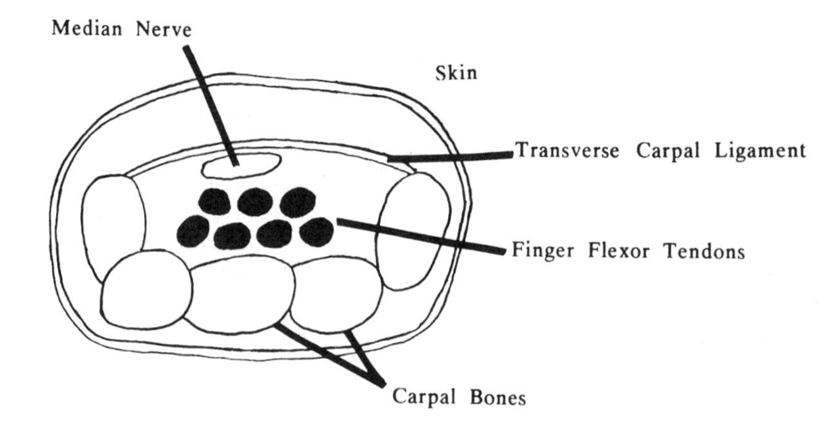

Figure 9. The carpal tunnel. This shows a representation of a wrist cross section. The carpal tunnel is constituted by the carpal bones and transverse carpal ligament. If the wrist is bent and/or if the fingers are tensed around an instrument, particularly for long periods, the carpal tunnel becomes smaller and thus compresses the median nerve, potentially causing pain, etc.

or pinching of the nerve causes pain, numbness and/or tingling of the fingers. The duration and degree of nerve compression determines the severity of CTS. As the problem progresses, there may be atrophy of muscles in the hand resulting in difficulty when gripping objects. Unfortunately, the repetitive motion and the delicate/fine manual work performed in the dental office put dental personnel at risk of developing CTS. Preventive measures to minimize the risk of developing CTS are listed in Table 6. If CTS is diagnosed early, it can be treated and controlled. If CTS is allowed to progress to an advanced stage, surgery may be required.

BIOHAZARDS — BLOODBORNE PATHOGENS

Chapter 4, Occupational Exposure to Bloodborne Pathogens, provides a comprehensive review of workplace exposure to pathogenic microorganisms such as

Table 6. Work Practices to Prevent Carpal Tunnel Syndrome (CTS)

1. Avoid the use of heavy instruments/handpieces.
2. Be sure that all air-driven or electrical dental instruments have sufficient cord length and range of motion.
3. Use dental instruments with a thick handle or shaft as this helps to eliminate finger/hand tension.
4. Dental hand cutting instruments should be kept sharp and well maintained as the sharpness of the blade determines the pressure/force required.
5. Whenever possible, the wrist should be kept straight and not bent or flexed.
6. If some of the symptoms of CTS develop, a wrist brace may be used to help keep the wrist straight.

hepatitis B virus (HBV) and human immunodeficiency virus (HIV). HBV is the causative agent of hepatitis (inflammation of the liver) and HIV is the causative agent of acquired immunodeficiency syndrome (AIDS). In this chapter, we will focus on exposure to these bloodborne pathogens in the dental office. The reason there is concern over these biohazards is that dental personnel frequently come into contact with the blood and saliva of their patients. Due to this, dental offices are included under the OSHA Bloodborne Pathogens Standard, fully effective in July 1992. Blood is the single most important source of HBV and HIV infection in the workplace. Both of these viruses have been transmitted from patient to dental staff by:

1. Parenteral injection (direct inoculation through the skin) which includes contact with an open wound, contact with non-intact skin (chapped, abraded, weeping skin), and injections through the skin (needle sticks and cuts with sharp instruments).
2. Mucous membrane exposure which includes blood or blood-containing body fluid contamination of the eyes or mouth.

It is important to realize that the potential for HBV transmission is greater than the potential for HIV transmission. The risk of hepatitis B infection (for an individual who has not had prior hepatitis B vaccination) following a parenteral exposure (such as a needle stick from a patient with hepatitis B) is approximately 6 to 30%.[21] By comparison, the risk of infection with HIV is approximately 0.35%.[22] This rate of transmission is considerably lower than that for HBV, probably as a result of the significantly lower concentrations of virus in the blood of HIV infected persons. The risk that bloodborne pathogens represent to dental personnel can be minimized (or eliminated) using a combination of vaccination (where applicable), engineering controls, and work practice controls such as PPE and clothing.

Vaccination

The most effective way to control the risk of hepatitis B infection is through immunization of dental personnel with the hepatitis B vaccine. This vaccine induces protective antibody levels in healthy adults and, in most cases, provides the individual with immunity to hepatitis B. Unfortunately, no vaccine currently exists to protect dental personnel from the risk of AIDS.

Universal Precautions and Personal Protective Equipment (PPE)

It has now become widely accepted that because a medical history and examination cannot reliably identify all patients infected with HIV, HBV, or other bloodborne pathogens, blood and body fluid precautions should be consistently used by dental personnel for all dental patients. This approach, referred to as universal blood and body fluid precautions or universal precautions, should be followed in the care of all patients. Inherent in the theory of universal precautions

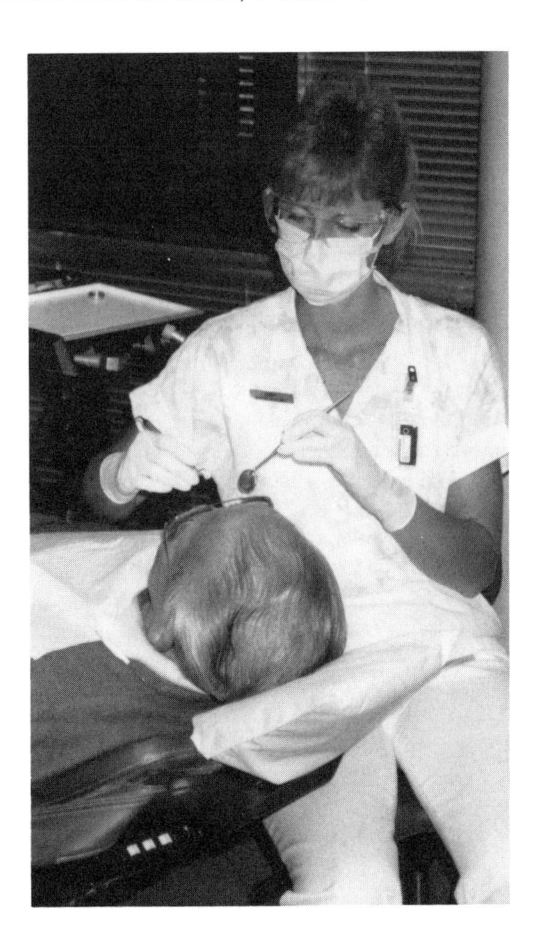

Figure 10. Personal protective equipment (PPE) used as a part of the universal precautions which should be followed in the care of all patients. Also see Table 5, Chapter 4 for further information on the universal precautions.

is the universal usage of PPE (see Figure 10). Personal protective equipment refers to specialized clothing or equipment worn by the dental staff to protect them from a hazard. When dealing with bloodborne pathogens, PPE is used to prevent entry of pathogens into the dental worker's body. This includes either entry via skin lesions or entry through the membranes of the eye, nose, or mouth. During many dental procedures there is a chance of saliva and/or blood spray. Typical PPE that must be worn includes gloves, lab coats/aprons, and eye and face protection such as goggles and masks.

Engineering Controls

Engineering controls are designed to reduce a worker's occupational exposure by either removing the hazard or by isolating the individual from the exposure.

There are several effective engineering type controls that have been developed to reduce worker exposure to bloodborne pathogens and these are reviewed in Chapter 4 of this volume. Unfortunately, in the dental office the available engineering controls are limited; more research and development needs to be completed to further protect dental personnel.

Needles and Sharps

Needles and sharps are documented to be a primary means of worker exposure to both HIV and HBV. In order to minimize the chance of a needle stick or cut when dealing with needles and sharps, careful and specific work practices should be combined with eingineering controls. First and most important is the fact that handling of needles/syringes and other sharps should be minimized. For example, when handling disposable syringes with needles or other sharps, they should be disposed of immediately in puncture-resistant containers located as close as is practical to the area where they are used. Also, the practice of recapping needles using the two handed technique must be discontinued. If reusable items are used, devices that minimize needle exposure/handling should be utilized, i.e., self-sheathing needles or devices that recap needles. This combination of specific work practices and specially designed equipment help reduce needle sticks and cuts that play a significant role in HIV and HBV transmission in the dental office.

SUMMARY

The subject of occupational health hazards in the dental office is an area that is finally receiving the concern and attention that it deserves. Although not usually thought of as a hazardous environment, the dental office presents a wide variety of potentially very serious exposures to dental personnel. The most evident hazards include radiation exposure and the chemicals used for disinfection. Other "hidden" hazards vary from dental material chemicals, to blood and body fluid biohazards to musculoskeletal disorders. It is important to note that all of the health hazards outlined in this chapter can and should be controlled.

REFERENCES

1. Dox, I., B. Melloni, and G. Eisner. *Melloni's Illustrated Medical Dictionary,* Baltimore: Williams & Wilkins (1979), p. 217.
2. Casarett, L. and J. Doull. *Toxicology, The Basic Science of Poisons,* New York: Macmillian (1975), p. 215.
3. Torres, H. and A. Ehrlich. *Modern Dental Assisting,* 4th ed., Philadelphia: W. B. Saunders (1990), p. 215.

4. Sax, N. and R. Lewis. *Dangerous Properties of Industrial Materials,* 7th ed., New York: Van Nostrand Reinhold (1984), p. 2197.
5. Dox, I., B. Melloni, and G. Eisner. *Melloni's Illustrated Medical Dictionary,* Baltimore: Williams & Wilkins (1979), pp. 215–217.
6. "Air Contaminants Final Rule," Occupational Safety and Health Administration, FR 54:2332 (1989).
7. "Mercury, Summary of Recommendations in 1990," *JADA,* 122:112 (1991).
8. Dox, I., B. Melloni, and G. Eisner. *Melloni's Illustrated Medical Dictionary,* Baltimore: Williams & Wilkins (1979), p. 411.
9. Sax, N. and R. Lewis. *Dangerous Properties of Industrial Materials,* 7th ed., New York: Van Nostrand Reinhold (1984). p. 2334.
10. Sax, N. and R. Lewis. *Dangerous Properties of Industrial Materials,* 7th ed., New York: Van Nostrand Reinhold (1984), p. 1764.
11. Sax, N. and R. Lewis. *Dangerous Properties of Industrial Materials,* 7th ed., New York: Van Nostrand Reinhold (1984), p. 1630.
12. Dox, I., B. Melloni, and G. Eisner. *Melloni's Illustrated Medical Dictionary,* Baltimore: Williams & Wilkins (1979), p. 155.
13. Sax, N. and R. Lewis. *Dangerous Properties of Industrial Materials,* 7th ed., New York: Van Nostrand Reinhold (1984), p. 768.
14. Torres, H. and A. Ehrlich. *Modern Dental Assisting,* 4th ed., Philadelphia: W. B. Saunders (1990), p. 155.
15. "Disinfectants, Instrument Immersion," *Clinical Research Associates Newsletter,* 15:3 (May 1991).
16. Torres, H. and A. Ehrlich. *Modern Dental Assisting,* 4th ed., Philadelphia: W. B. Saunders (1990), p. 155.
17. Torres, H. and A. Ehrlich. *Modern Dental Assisting,* 4th ed., Philadelphia: W. B. Saunders (1990), p. 256.
18. Sax, N. and R. Lewis. *Dangerous Properties of Industrial Materials,* 7th ed., New York: Van Nostrand Reinhold (1984), p. 2525.
19. Torres, H. and A. Ehrlich. *Modern Dental Assisting,* 4th ed., Philadelphia: W. B. Saunders (1990), p. 438.
20. Torres, H. and A. Ehrlich. *Modern Dental Assisting,* 4th ed., Philadelphia: W. B. Saunders (1990), p. 373.
21. Seeff, L. B., E. C. Wright, H. J. Zimmermann et al. "Type B Hepatitis After Needle Stick Exposure: Prevention with Hepatitis B Immune Globulin." *Ann. Intern. Med.* 88:285–93 (1978).
22. Wormer, G. P., C. S. Rabkin, and C. Joline. "Frequency of Nosocomial Transmission of HIV Infection among Health-Care Workers." *N. Engl. J. Med.* 319:307–8 (1988).

PART III

Laboratory Spills and Medical Waste Disposal

Part III, **Laboratory Spills and Medical Waste Disposal,** contains Chapters 7 and 8:

Chapter 7 addresses **Spills in Laboratories** (which can happen in any type of laboratory or clinic where biological, chemical or radiological agents are used). Depending upon the type of the spill or release, certain employees may have to be specifically trained to comply with 29 CFR 1910.120, "Hazardous Waste Operations and Emergency Response" (the so-called "Hazwoper Standard"). This chapter helps to identify when and if lab employees must be so trained. In addition, how to do a hazard evaluation and consequence assessment are explained, as well as the importance of a written emergency response plan. The chapter also provides assistance in determining the routine types of emergency and spill response training that everyone should have in order to allow them to safely deal with incidental releases. The importance of accident anticipation and emergency planning are stressed.

Chapter 8, **Medical Waste Disposal,** begins with a discussion of what infectious medical waste is, including how it is defined and regulated. Unfortunately, waste is something that all healthcare facilities and laboratories generate and this chapter helps to identify the scope of the medical waste problem. The chapter provides an excellent overview of medical waste disposal treatment technologies. It discusses the advantages and drawbacks of each method of medical waste disposal, including (but not limited to) methods as diverse as volume reduction, sterilization, disinfection, shredding, recycling, etc. The development of a comprehensive medical waste program is also discussed, including examples. Illustrations of medical waste disposal economics, site specific, off-site and regional approaches, are also presented.

CHAPTER 7

Spills in Laboratories

Thomas J. Cuthel

INTRODUCTION

The purpose of this chapter is to assist laboratory personnel who are employed in healthcare facilities or in clinical, academic, or research laboratories, and who work with relatively small amounts of hazardous materials, to safely deal with spills of chemicals that they routinely use. This chapter contains suggestions on how to plan for incidental releases of hazardous materials, how to train personnel, when to attempt an in-house clean up, and, when it might be prudent to contact HAZWOPER trained responders.

REGULATIONS DEALING WITH LABORATORY SPILLS

Federal regulatory requirements for dealing with hazardous materials have become increasingly stringent over the last few decades. Table 1 contains a partial listing of such legislation, particularly federal regulations dealing with hazardous materials and emergency response.

0-87371-392-3/93/$0.00 + $.50

Table 1. Overview of Federal Regulations Dealing With Emergency Response

Subject	Section	General Description
U.S. Nuclear Regulatory Commission, Instructions to Workers	10 CFR 19.12	Requires written SOPs for use of radioactive materials, which include spill response
Standards for Protection Against Radiation	10 CFR 20	Sets amounts for spills reportable to Nuclear Regulatory Commission (or to state)
Employee Emergency Plans and Fire Prevention Plans	29 CFR 1910.38	Applies to all emergency action plans required by a particular OSHA standard
Process Safety Management of Highly Hazardous Chemicals	29 CFR 1910.119	Requirements for substances with catastrophic release potential
Hazardous Waste Operations and Emergency Response ("HAZWOPER")	29 CFR 1910.120	Sets levels of response and training requirements for Emergency Responders; also a written Emergency Response Plan, PPE Program, and medical surveillance
Bloodborne Pathogens	29 CFR 1910.1030	Requires written program and training for the handling of body fluids, including specific training for the clean up of spills
Hazard Communication ("HAZCOM")	29 CFR 1910.1200	Requires written HAZCOM Plan and training for the use of all hazardous materials; minimum training requirement for anyone cleaning up a chemical spill
Occupational Exposure to Toxic Substances in Laboratories ("Laboratory Standard")	29 CFR 1910.1450	Requires written Chemical Hygiene Plan and training for the use of toxic materials in laboratories, which includes emergency response requirements for hazardous spills

One piece of legislation is of particular interest for persons who are responsible for dealing with chemical spills or releases in laboratories. Specifically, accidental releases of chemicals is covered by 29 CFR 1910.120, *Hazardous Waste Operations and Emergency Response*. With the introduction of this rule (also commonly known as "HAZWOPER") by the Occupational Safety and Health Administration (OSHA), specific requirements for hazardous materials spill responses have been defined. In addition to emergency response, HAZWOPER also covers workers routinely employed at hazardous waste sites, treatment storage and disposal facilities, and others, however, those situations are beyond the scope of this chapter. Table 2 outlines HAZWOPER (29 CFR 1910.120) particularly in terms of how it refers to any accidental releases.

It is important to know when, if, and to whom OSHA 1910.120 applies. HAZWOPER (i.e., 29 CFR 1910.120) itself addresses this issue.[1]

> Responses to incidental releases of hazardous substances where the substance can be absorbed, neutralized, or otherwise controlled at the time of release by employees in the immediate release area, or by maintenance personnel are not considered to be emergency responses within the scope of this standard.

The phrase ''incidental releases'' is very important in defining when HAZWOPER applies, particularly for healthcare and laboratory employees. Certain requirements of HAZWOPER (advanced training, etc.) are meant to apply to workers who respond to hazardous materials incidents as a part of their job function, such as fire and rescue personnel. As shown in Table 2, this does not usually include incidental releases and clean ups. In-house spill clean up, as referred to in this chapter, means the clean up of an incidental spill or chemical release by laboratory personnel who work with the material, on a normal basis,

Table 2. OSHA's HAZWOPER (29 CFR 1910.120) and Its Applicability to Laboratories

Classifying Accidental Chemical Releases or Spills	Who May Respond and Appropriate Training
Large Releases Emergency response operations for releases of, or substantial threats of releases of, hazardous substances without regard to the location of the hazard . . . a response effort by employees from outside the immediate release area or by other designated responders . . . to an occurrence which results, or is likely to result, in an uncontrolled release of a hazardous substance[a]	**Emergency Responders** (sometimes referred to as HazMat Teams) · First Responder Awareness Level[b] · First Responder Operations Level[b] · Hazardous Materials Technician[b] · Hazardous Materials Specialist[b] · Incident Commander[b]
Incidental Releases Responses to incidental releases of hazardous substances where the substance can be absorbed neutralized, or otherwise controlled at the time of the release *by employees in the immediate release area, or by maintenance personnel* are not considered to be emergency responses within the scope of the standard[c]	**Generic Employees** or **Maintenance Personnel** No training is specified for responding to incidental releases by this standard, however, other standards and guidelines do require laboratory worker training which is applicable to incidental spill responses (Table 1)

[a] From 29 CFR 1910.120 [a](1) Scope (v).
[b] As defined in 29 CFR 1910.120 [q](5) *specialist employees*.
[c] From 29 CFR 1910.120 [a](3) Definitions, emphasis added.

in that particular area of the workplace. A laboratory or healthcare worker who has received other training (such as Hazard Communication, Laboratory Standard, Bloodborne Pathogen or Radiation Safety Training, etc.) in the use of a hazardous material, and who normally works with a hazardous material, can clean up an incidental release or spill of the material within their normal work area without HAZWOPER training. However, if the release is such that the worker requests assistance from an outside source to clean up a spill (even an emergency response team that works in the same building), the specific emergency response procedures described by HAZWOPER take effect, and anyone who participates in the clean up from that point on must have the training described in 29 CFR 1910.120 for the tasks they are requested to do, or work under the control of the emergency response team as a ''specialist employee''. For example, if a railroad tank car overturns and ruptures, the persons responding must be trained at the advanced emergency response levels specified by OSHA in 1910.120.

This chapter does not train workers to be emergency responders at the OSHA 29 CFR 1910.120 levels. Instead, it specifically addresses the resolution of incidental in-house spills and clean ups.

PREPLANNING FOR CHEMICAL SPILLS OR RELEASES

The key to limiting employee exposure and property damage during a hazardous materials spill is prompt, safe, and efficient response in controlling and cleaning up a spill. The best way to actually do this is by first preplanning spill response so that laboratory personnel know what do do, where to go to find the materials that will be needed, etc.

Preplanning for spill response should include an Emergency Plan as a written document; training of personnel with respect to the Emergency Plan and in the proper emergency response procedures, and the assembling of the materials that will be needed into formal spill response kits (or at least centralizing the location of the materials in the area where they are most likely to be needed). The extent to which these things need to be done relies in large part on an evaluation of the potential hazards which exist in the workplace. Depending upon the complexity of the workplace and on the chemicals involved, it may become necessary to consult with a Certified Industrial Hygienist (CIH) or a Certified Safety Professional (CSP) to do effective emergency response preplanning.

Hazard Evaluation and Consequence Assessment

Emergency planning for a hazardous materials spill should start with a complete evaluation of the potential hazards in the laboratory. This can begin with an inventory of hazardous materials in the laboratory, followed by a hazard evaluation or consequence assessment of the potential outcomes of any possible chemical spill or release. This can also be referred to as Job Safety Analysis.[2]

Table 3. The Hazard Evaluation Process[a]

1. **Identify the activities and operations.** Identify and list all activities which can take place in the workplace.
2. **Identify the steps of each activity.** For each activity, identify the basic steps which comprise that activity.
3. **Identify the hazards.** Identify the hazards associated with each basic job step.
4. **Establish procedures.** For each hazard, establish the procedures to eliminate or minimize the hazard.

[a] Hazard evaluation is not a trivial matter and so professional assistance, such as from a CIH or CSP, is frequently required.

Adapted in part from National Safety Council, "Job Safety Analysis in an Industrial Laboratory Environment," Data Sheet I-706 Rev. 89 (1989).

The systematic performance of a hazard evaluation should be conducted for each activity or process in the workplace. Table 3 outlines the general procedure for conducting a hazard analysis.

Be careful to consider nonchemical potential hazards as well, such as electrical utilities, for any possible impact during a release. The identity of hazardous materials, the quantities present, the conditions of use, methods of storage, etc., must all be considered in this evaluation. Hazard evaluation is the estimation of the damage which could result from a leak, spill, or other release into the workplace or the environment. It is different for every workplace.

The inventory of chemicals in the laboratory or healthcare facility should be the first part of the hazard assessment. Material Safety Data Sheets (MSDS) will be helpful in the evaluating of potential chemical hazards. MSDSs are available from the manufacturers of chemicals or chemical products and must be accessible to all workers in the workplace who are actually or potentially exposed.[3] Also, containers must be clearly labeled.

The amounts and locations where the hazardous materials are used and stored should be noted so that the areas where spills are most likely to occur can be identified. This is useful in identifying high risk areas, for deploying spill clean up materials, and for training. Also, the possible worst case magnitude of any release can be determined based on quantities present. This information will allow the Emergency Plan to be site specific since it will have taken into account the chemicals (and quantities) present, the types of releases, possible locations where chemical releases are likely to occur, and the possible impact of those releases.

Written Emergency Response Plan

OSHA's General Industry Emergency Response and Preparedness Requirement requires specific elements as outlined in 29 CFR 1910.38(a) as a minimum be included in emergency plans if employees assist in the response or handling

Table 4. Emergency Plan Requirements

- Pre-emergency planning and coordination with outside parties.
- Personnel roles, lines of authority, training and communication.
- Emergency recognition and prevention.
- Safe distances and places of refuge.
- Site security and control.
- Evacuation routes and procedures.
- Decontamination.
- Emergency medical treatment and first aid.
- Emergency alerting and response procedures.
- Critique of response and follow-up.
- Personal protective equipment (PPE) and emergency equipment.

Note: If employees assist in response to a hazardous materials incident, the Emergency Plan must address these issues.

of a hazardous materials emergency. This includes planning for any large scale spills, which would require the assistance of trained emergency responders (as discussed above), as well as for incidental releases. These requirements are contained in Table 4. The level of detail required for the Emergency Plan will also vary from workplace to workplace. The Emergency Plan's technicalities and specifics will depend upon the complexity of the situation as revealed by the inventory and hazard analysis. Many workplaces already may have *de facto* emergency response procedures. Such procedures can serve as a starting point for the written plan.

As an aside to the Emergency Plan requirements listed in Table 4, it is important to note that any employer can elect, as a part of the written Emergency Plan, to simply call a local emergency response team (such as a local fire department) and evacuate all employees whenever there is an accidental chemical release. This option greatly reduces the obligation on the employer with respect to training and maintaining a spill, however, it is unconditional (the premises must be evacuated and the call made for every release). If an employer selects this option and plans on evacuating all employees from the workplace when an emergency occurs (and does not permit any employee to assist in handling of a hazardous materials incident), a written Emergency Plan is still needed, but it only must contain the elements shown in Table 5.

Table 5. Emergency Plan Requirements (If Employees Evacuate and Do Not Assist in the Response to a Hazardous Material Incident)

- Alarm/notification system to alert employees of an emergency.
- Emergency escape procedures and emergency escape route assignments.
- Procedures to be followed by employees who remain to operate critical plant operations before they evacuate.
- Procedures to account for all employees after emergency evacuation has been complete.
- Rescue and medical duties for those employees who are to perform them.
- The preferred means of reporting fires and other emergencies.
- Names or regular job titles of persons or departments who can be contacted for further information of duties under the plan.

For emergency response professionals (i.e., not for the typical healthcare or laboratory worker who is responding to their own incidental release), special training is required for all employees who respond to hazardous materials incidents. The requirements for emergency responder training are very stringent. For example, an employee would require an initial 8 hours of training just to initially respond and to take protective action from a distance (First Responder Operations Level). A total of 24 hours of training would be required to actually clean up a hazardous materials spill (Hazardous Materials Technician) and annual refresher training is required for all levels of response. Emergency responder training under HAZWOPER can be costly and takes time.

In either case, the Emergency Plan should also include information on when an in-house clean up is allowed, when emergency responders should be contacted and what to do if an in-house clean up is started but then found to be beyond the capacity of the staff. Generally, if an in-house clean up cannot be completed, the hazardous material should be contained as best as possible, the area evacuated and emergency responders contacted. The written Emergency Plan should address this type of contingency.

The distinction between an incidental release and large spill or release is subjective. It is based on the relative amounts of a material that a worker would normally use at one time and on the relative hazard to human health and life that the material poses. As a general rule, a laboratory worker who has received training in the handling and safe use of a material should be able to deal with a spill in amounts approximating that which is normally used. If a much larger amount of chemical is spilled, additional precautions may be necessary that the worker might not be familiar with. The difference between a small and large spill is based in part on the familiarity of the worker handling that relative amount. However, if a small amount of material presents a high hazard to human health and life, even though a laboratory worker can safely work with the material under controlled conditions, the worker may not be able to handle the routine quantity safely once the material is no longer under controlled conditions (as occurs during a spill). Spills of highly toxic materials may have to be considered an emergency (i.e., more than an incidental release and therefore requiring the assistance of trained emergency responders) when the amount that has been or is about to be released will cause harm to health or life and cannot be safely controlled and cleaned up by the laboratory worker. Calculations can be done to estimate potential airborne concentration resulting from releases of various amounts of volatile materials.[4,5] The calculated potential airborne concentrations should be compared to OSHA's Permissible Exposure Limits (PEL),[6] the American Conference of Governmental Industrial Hygienists' (ACGIH) Threshold Limit Values (TLV),[7] American Industrial Hygiene Association's (AIHA) Emergency Response Planning Guidelines (ERPG),[8] and other toxicological data to determine what size spills can cause harm to health or life. See Table 6 for an example of how to estimate potential airborne concentration for volatile chemical spills. The relative amounts of what constitutes an emergency response for

Table 6. The "Broken Bottle" Calculation

$$C = \frac{VS \times d \times 22.4 \times T \times 760 \times 10^6}{VR \times MW \times 273 \times P}$$

where C is the airborne concentration potentially resulting from the spill of VS (C has units of ppm); VS is the volume of spilled material (ml); d is the density (grams/ml); T is the absolute temperature in Kelvins (K), since standard room temperature is usually assumed to be 25°C, standard room temperature is about 298 K (0°C = 273 K); P is the atmospheric pressure (mmHg), assume 760 mmHg for standard atmospheric pressure; VR is the volume of the room or enclosure, etc. (liters, l). MW is the molecular weight of the volatile chemical spilled (grams per mole or g/mole); 22.4 is a constant representing the volume (in liters) occupied by 1 mole of any gas at standard temperature and pressure (25°C, 760 mmHg).

Example: If 500 mls of xylene (a solvent; d = 0.86 g/ml and M.W. = 106.2 g/mole) were spilled in a 20 ft × 10 ft × 8 ft room at standard temperature and pressure, find C (ppm).

Answer: The volume of the room must first be calculated; 1 ft^3 = 28.3 l. VR = (20 × 10 × 8 ft^3) × 28.3 l/ft^3 = 45,280 l

$$C = \frac{500 \text{ ml} \times 0.86 \text{ g/ml} \times 22.4 \text{ l} \times 298 \text{ K} \times 760 \text{ mmHg} \times 10^6}{45{,}280 \text{ l} \times 106.2 \text{ g/mole} \times 273 \text{ K} \times 760 \text{ mmHg}}$$

$$C = \frac{2.1815 \times 10^{15}}{9.9772 \times 10^{11}}$$

$$C = 2{,}186 \text{ ppm of xylene}$$

Note: The *"Broken Bottle"* calculation yields the potential airborne concentration resulting from the release of volatile chemicals. This assumes that 100% of the material spilled will vaporize instantaneously and that there is no dilution ventilation or local ventilation.

Adapted from NIOSH, *The Industrial Environment — Its Evaluation and Control,* Washington, D.C.: U.S. Government Printing Office (1973), p. 18.

specific materials or groups of hazardous materials should be outlined in the Emergency Response Plan.

Training

The staff should be trained in and be made familiar with all provisions of the Emergency Plan. Lack of training, or improper training, can turn routine, incidental releases into headaches. As an example, a medical center cleaning woman used a mop to clean up a small tritium spill. The mop (i.e., the same mop head) was then used to clean other floors of the same building. Because this spread the tritium throughout the building it was necessary to test 70 employees for contamination; of those, one employee's shoes and the skin of two employees had been contaminated.[9]

Once the Emergency Plan is complete, a copy should be made available at all times while the area is occupied. Training should be provided to the laboratory staff to ensure they are aware of the existence, location, and availability of the Emergency Plan, along with its specific emergency procedures and provisions. This training can be incorporated into existing training programs that are required by various federal standards dealing with hazardous materials (see Table 2). OSHA's Occupational Exposure to Toxic Substances in Laboratories (the Laboratory Standard) has similar requirements for training in emergency procedures.

Spill Response Kits

The exact types and amounts of supplies that should be included in a spill response kit will depend on the results of the inventory and on the hazard evaluation of the materials in the laboratory. The MSDSs for the hazardous chemicals present should be consulted to determine the types of Personal Protective Equipment (PPE) and absorbents and neutralizing agents that should be present in the kit; look for Spill and Leak Procedures and Special Protective Information sections of the MSDS. If the information in the MSDS is not complete or if the information or instructions are not clear, contact the manufacturer directly for additional information so that the spill response kit will be adequately stocked (OSHA requires that a telephone number be listed on the MSDS).

Personal Protective Equipment (PPE) should be present in all spill response kits. Since there is almost always the chance of material splashing during a clean up procedure, impermeable clothing and eye/face protection should be included. The impermeable clothing should include as a minimum a pair of appropriate work gloves but may also include lab coats, aprons, shoe covers, head coverings, sleeve covers, scrub suit, and coveralls depending on the material spilled. MSDSs and other reference materials should be consulted to determine the types of materials that will be impermeable to the hazardous materials specific to each area of the workplace. Multiple PPE, especially gloves and aprons of various materials, may have to be available for spill response in a single laboratory depending on the types of hazardous materials present. Eye/face protection should include chemical splash goggles but may also require a chemical face shield. One very important consideration concerns whether the PPE is to be disposable or reusable. If disposable, provisions for its safe disposal must be made. If reusable PPE is selected, provisions for proper PPE decontamination must be made.

The selection of PPE must be made by balancing considerations such as breakthrough time against the worker's convenience and ability to do the job while wearing it. Gloves (and similar PPE) protect people by acting as a barrier between person and chemical. Breakthrough time is the length of time it takes for a measurable amount of the chemical to permeate the protective material under test conditions. Permeation (or permeation rate) represents the rate of diffusion of a chemical through PPE materials. Degradation of PPE materials can sometimes result from chemical exposures; different materials offer different barrier effectiveness depending on the PPE material and the chemical agent.[10] Penetration, such as would result from a puncture, defeats any PPE barrier.

Aside from the measured effectiveness of any PPE material against any target chemicals, other factors must be considered. For example, for any given material, thicker gloves will have a longer breakthrough time than will thinner but the added thickness may decrease the wearer's dexterity and sense of touch. Selection of gloves, aprons and other personal protective materials can be done using published reference sources[10] or even using expert systems on the microcomputer.[11] Again, if in doubt, contact the manufacturer or a health professional such as a CIH or CSP. Whether for responding to incidental releases or for routine wear, always make sure that PPE is being maintained or replaced on an adequate schedule.

Respiratory protection will be required in many laboratory or healthcare spill situations since hazardous dust, vapors, gases, mists, and droplets can be present during many types of spills or chemical releases. Respiratory protection can include surgical masks, negative-pressure filtered respirators, powered air purifying respirators and supplied air units including Self-Contained Breathing Apparatus (SCBA), and airline respirators. Some devices, such as surgical masks, cannot be used for toxic materials but sometimes are helpful as a barrier when dealing with certain biological materials. Surgical masks should be available in spill response kits that are used for cleaning up blood spills. Use of true respirators, including supplied air units, are regulated by OSHA and are only briefly discussed here (see Volume 1 of this series for a complete discussion of respirators and respirator programs). In general, regulations for respirator use requires a written respirator program, a maintenance program, medical evaluation of all employees who will wear respirators, and training and fit testing for all individuals issued a respirator. Unless there is an individual or a department that is responsible for seeing that the requirements for a respirator program are met, the improper use of respirators could result in injury to personnel attempting a spill clean up and/or regulatory problems. Unless a complete respirator program is in effect for in-house personnel, the requirement for the use of a respirator during a spill clean up will exclude an in-house clean up (and so will necessitate the assistance of emergency responders).

Spill response kits should contain absorbent and neutralizing materials in sufficient quantities and types to deal with the chemicals in the laboratory as well as disinfectants for biological materials and decontamination solution for radioactive materials.[12] The MSDS and the manufacturer can be consulted for guidance in this area. See Table 7 for an overview of absorbents and neutralizers. See Table 8 for a synopsis of disinfectants. Other proprietary products are available for spill mitigation and cleaning; some will solidify or gel certain types of spilled liquids, thus limiting the scope of the liquid spill.

Disposable hand brushes, dust pans, and tongs should be included in the kits along with appropriate waste containers. Polyethylene bags, 0.006 inches or thicker, can be used for waste disposal. Hazard Identification Labels that can be attached to the waste containers should be available so that the waste can be easily identified for disposal. Barrier tape with "CAUTION" or other appropriate warnings written on it, self standing warning signs and/or orange cones should be available to block off the area of a spill and to warn that there is a hazardous spill in the area.

Table 7. Suggested Sorbents and Neutralizing Agents For Chemical Spills

Material	General Comments
Sand	Poor absorbent, expensive to dispose of and respiratory protection recommended; often used for diking large spills.
Vermiculite	Absorbs better than sand, especially for solvent and oil spills; respiratory protection recommended.
Activated carbon	Absorbs flammable solvent spills well and helps reduce vapors. Do not use with oxidizers; activated carbon itself can be hard to clean up.
Diatomaceous earth	Absorbs solvents, oils and water-based substances well.
Any weak base, e.g., sodium bicarbonate, sodium carbonate, sodium sesquicarbonate	Used to neutralize acid spills (excluding strong oxidizing and highly toxic acids such as perchloric and hydrofluoric acid).
Any weak acid, e.g., citric acid	Used to neutralize inorganic bases, alkali, and caustic spills.
Sodium bisulfite (1% solution)	Used to clean up small formaldehyde spills or to help reduce odor problems after a spill.
Sulfur powder	Used to clean up mercury spills; reacts slowly.
Zinc powder	Used to clean up mercury spills.
Gelling agents	Used to cause gelling or coagulation of spilled liquids.

Table 8. Common Disinfectants For Wiping and Cleaning Up Spills

Category of Disinfectant	QAC[a]	PC[b]	CC[c]	IC[d]	ALC[e]
Concentration of active ingredient (%)	0.1–2	0.2–3	0.01–5	0.47	70–85
Contact time (min)	10–30	10–30	10–30	10–30	10–30
Effective against[f]					
Tubercle bacilli	NR	P	P	P	LP
HIV	P	P	P	P	P
HBV	NR	LP	P	LP	LP

[a] QAC — Quaternary ammonium compounds
[b] PC — Phenolic compounds
[c] CC — Chlorine compounds
[d] IC — Iodophor compounds
[e] ALC — Alcohol (ethyl or isopropyl)
[f] P — Indicates a very positive response
LP — Indicates a less positive response
NR — Indicates a negative response (use not recommended)

Adapted from National Research Council, *Biosafety in the Laboratory: Prudent Practices for the Handling and Disposal of Infectious Materials,* Washington, D.C.: National Academy Press (1989), p. 40.

Proprietary spill response kits are available in small DOT approved drums or pails containing tools and absorbents; the spilled chemical (and clean up materials) can be disposed of in the drum or pail. Plastic dust pans, labels, indelible markers, razor blades, and tape (duct and/or filament) are also useful to have in kits. Commercially available spill response kits can be convenient and useful, however, make sure that the contents are adequate to meet the needs of the laboratory.

The supplies for the spill response kit can be placed into containers such as a box or small open top drum or stored in a drawer, on a shelf, or in a cabinet. The kit can be stored in a central location, however, the optimum locations are those identified during the inventory as places where spills are likely to occur (just be sure that the kit will be accessible when a spill does occur). Wherever the materials are stored, the location should be clearly indicated by a sign with "SPILL RESPONSE KIT" or other similar phrasing that is visible throughout the laboratory. All laboratory personnel should be aware where all spill response kits and other emergency equipment (e.g., fire extinguishers), and supplies are located.

GENERAL RESPONSE TO SMALL HAZARDOUS MATERIAL SPILLS

If a spill involves the contamination of an individual or individuals, the primary consideration should be the treatment of the individual(s). If assisting in the decontamination of an individual, take the proper steps to protect yourself from also being contaminated. The location and use of emergency showers is important. For emergency decontamination of personnel, the general steps outlined in Table 9 can be used as a guide for responding. Once the treatment of contaminated individuals has been addressed, the next priorities are the preven-

Table 9. Guidelines For Decontamination of Persons

1. Protect employees performing decontamination through proper procedures, training, PPE, and equipment.
2. Remove any source of contamination (e.g., clothing); any contaminated material should be placed in a plastic bag and sealed for later decontamination or disposal.
3. Immediately flush the contaminated area with ample flowing water, unless the chemical is water reactive, for at least 15 min. Use soap on skin only if there are no visible burns; if large areas of skin are affected, use an emergency shower to flush entire contaminated area.
4. Consult the MSDS for special treatment or delayed effects; medical attention is always advisable.
5. Seek medical attention if any reddening or pain develops.
6. Chemical splashes in the eyes should be flushed with flowing tepid water for at least 15 min. The eyelids should be held away from the eye, while the eye is being flushed. Do not rub eyes. Immediate medical attention should be sought for all incidents involving the eyes.
7. Report and document all personal exposure to hazardous materials.

Table 10. Generic Spill Response Procedures For Laboratories and Healthcare Facilities

For All Spills or Releases

1. All individuals in the area should be notified of the spill.
2. Access to the area should be restricted to individuals involved in the clean up. Evacuate if conditions warrant it. Post warning signs or guards at points of entry to prevent unauthorized access.
3. The spilled material should be identified, and the scope of the release and the potential for hazard should be assessed; use the MSDS and additional resources if necessary.
4. Determine whether to treat the spill as an incidental release (as defined in the text) or as a large spill requiring emergency responders.
5. Always use appropriate PPE; decontaminate or discard afterwards.
6. Dispose of materials according to applicable local laws.

For Liquid Spills

1. Contain the spill as quickly as possible using appropriate absorbent/neutralizing agent; see Table 6 for list of common absorbents and neutralizing agents for chemical spills.
2. When cleaning up, always start at the outer edges of the spill and work toward the middle of the spill.
3. Once the whole spill has been absorbed/neutralized, pick up the absorbent, the material et al. and place into an appropriate container; securely seal and label the container.
4. Remember, if the material was absorbed only and was not neutralized (or otherwise chemically changed), the waste material can still have the hazardous properties of the original spilled material and so must be handled with caution.

For Solid Spills

1. Low hazard materials, or materials that have been neutralized, can be cautiously swept up.
2. High hazard materials should be cleaned up using HEPA vacuums or charcoal filters or both.

tion of other individuals from becoming contaminated, and, containment of the spill. Table 10 contains a generic spill response and containment model. Subsequent to any chemical release or clean up, as well as a part of normal laboratory operating procedure, proper hygiene techniques should be followed (including washing or showering).

Written Follow-Up and Reporting Requirements

A follow-up written report should be made for all spills. Include information on when and where the spill occurred, what was spilled including approximate amounts, how the spill occurred, the contamination/clean up measures taken, and a list of the individuals involved. A copy of the report should be kept on file in the laboratory and a copy sent to the appropriate individual or department that is responsible for safety for the area, e.g., a Chemical Safety Officer or a Radiation Safety Officer (for a radioactive materials spill).

Most incidental releases of chemicals in the laboratory or healthcare environment will usually not require reporting to outside agencies. However, certain chemical releases may require reporting to federal, state, or local agencies if chemicals are released into the air, water, sewers, etc., in sufficient quantities.

Reportable quantities (RQ) have been defined for certain materials (in the Superfund Amendments Reauthorization Act, more commonly referred to as SARA Title III).[13] For example, if the RQ of chemicals listed in SARA is exceeded in any 24 hour period during a spill or release, then reporting is required by law. While most incidental releases in the laboratory or healthcare workplace will not be large enough to require reporting, the Emergency Plan should be written to take into account what to do if and when reporting of releases to off-site agencies is required. As always, the MSDS and the manufacturer can be consulted as a starting point for identifying any reporting requirements, as can the Environmental Protection Agency.

If reporting is mandated, the exact types of information required can vary but generally will include the name of caller and a call back number, the location and time of the incident, the name of the material released (or any other identifying information), the nature of release including the estimated amount, and its state (i.e., solid, liquid, or gas). Potential adverse health effects, including types of injury or illness, whether they have taken place or are anticipated, should also be communicated.

It should also be noted that SARA additionally requires annual reporting to state and local agencies, as well as to local fire departments, based upon the inventory of certain chemicals (i.e., simply for the presence of, not the release of). These chemicals must be reported annually if present in amounts greater than the Threshold Planning Quantity (TPQ). This is also defined by SARA Title III and elsewhere.[3] EPA publishes and periodically updates the chemicals covered by SARA, including their TPQs and RQs (in pounds).[14]

Chemical Spills

Chemical spills can generally be cleaned up following the general response outlined in Tables 9 and 10. Some common hazardous chemical groups are discussed below along with information and precautions more specific to responding to spills or releases of that type of chemical (specific definitions, e.g., corrosive, are addressed in the Glossary). Each laboratory or healthcare facility is different from the next and so the safest, most reasonable response will differ from place to place, particularly if a spill or release escalates. Specific contingencies should be addressed in the Emergency Plan.

It may be imprudent to generalize the spill response for any material or material type. When in doubt, get specific answers: consult the MSDS, request help from the manufacturer, or consult a properly trained professional such as a CIH or a CSP. The hazards associated with spills or releases of any chemical should be identified while determining the scope of the written Emergency Response Plan.

Flammable Liquids

When a flammable liquid is spilled, make sure all sources of ignition such as non-explosion proof electrical equipment, open flames, and other sources of heat are quickly turned off. Turning off equipment is not necessarily foolproof since turning off equipment can cause sparks also. A fire extinguisher rated for a class B fire should be readily available. When cleaning up spills of flammable materials be aware that the absorbent material used to clean up flammable materials may still give off vapors and should be treated as potentially flammable.

Corrosive Chemicals

Acid and bases are the most common corrosive chemicals found in laboratories. Acid spills should be neutralized using a weak base and base spills can conversely be neutralized using a weak acid. The pH should be checked using a pH indicating paper (litmus paper). If the pH of the spill has been adjusted to between 2 and 12.5 (the characteristic of corrosivity for a hazardous material is defined as an aqueous material that has a pH ≤ 2 or ≥ 12.5), under federal regulations the material can be treated as non-hazardous. However, check local regulations before disposal, especially if material is to be discharged to a water source that might contaminate potable supplies. Additional pH adjustment might be necessary.

Certain acids have additional hazards associated with them. Perchloric acid is a strong oxidizing agent which can react violently with organic materials. Hydrofluoric acid is highly toxic, is bone seeking, and can cause delayed burns several hours after exposure if medical treatment is not received. Such acids are extremely hazardous and the clean up of even small spills should not be attempted by the average laboratory personnel. Restrict access to the area of the spill and contact the appropriate emergency responder as called for in the Emergency Plan.

Toxic Chemicals

In any laboratory or healthcare facility, toxic chemicals, as the phrase is used in this context, applies to a wide range of materials. This includes relatively common chemicals such as formaldehyde (which is an organic gas often found in laboratories in solution form) and mercury (which is a liquid metal element). Some chemicals can usually be characterized as innocuous, but "toxic chemicals" as used here applies to anything chemical, mixture or product which might cause harm during an accident.

Many toxic chemicals will require the use of a respirator for even small spill clean up. Unless the laboratory has a respirator program and respirator-qualified personnel, such releases are automatically beyond the scope of normal operations and the emergency responders should be contacted. However, if the MSDS for a chemical has been consulted and a respirator is not required, the spill can be taken care of in-house providing there is adequate ventilation and that all precautions are taken to limit exposure.

Special training and equipment is needed to handle even small spills of highly toxic chemicals, such as diisopropyl fluorophosphate (DFP) or hydrofluoric acid.

For spills of highly toxic chemicals, block off the area of spill, evacuate the laboratory and contact emergency responders.

Formaldehyde is a suspected human carcinogen. There are many commercially available products for cleaning up formaldehyde solution spills with minimum offgassing. If, after a formaldehyde solution clean up, there is an odor problem, the area should be treated with a dilute solution of sodium bisulfite.

Mercury is a cumulative systemic poison. Small droplets of mercury can be reacted with fine zinc dust to form an amalgam that is much less toxic. Large droplets of mercury can be picked up using an aspirator bulb. There are commercially available vacuum units and vacuum cleaners with special charcoal filters that can be used for mercury cleanups.

Reactive Chemicals

Spills of many reactive chemicals can be handled in-house if cautious but quick action is taken. Spills of water reactive chemicals can be handled in-house provided water is kept away from the spill. Many water reactive solids, such as sodium and potassium, are normally stored in mineral oil or kerosene. If a bottle of a water reactive solid is broken, the solid material should be picked up with dry tongs and placed into a new bottle of mineral oil (or other appropriate material) as quickly as possible. Water reactive liquids, if spilled, should be neutralized or picked up with an appropriate absorbent as quickly as possible and placed into a dry container for disposal.

Air reactive chemicals are harder to handle and can pose special difficulties. Depending on the material, air reactive chemicals are stored under water, in an inert atmosphere, or in a vacuum. Air reactive chemicals that are stored in water, such as phosphorus, should be picked up with tongs and placed back in water before they dry. Air reactive chemicals that are stored in an inert atmosphere or a vacuum may start to react before the required storage conditions can be reproduced. The Emergency Plan should contain contingencies for such possibilities. While cleaning up any reactive material, if any change in the material is noticed (such as the emission of vapor or smoke coming from the material, or if the material is generating heat), and if the clean up cannot be immediately completed, then stop, evacuate the laboratory, and contact an emergency responder.

Leaking Compressed Gas Cylinders

If an unexplained pressure drop is noticed in the gauge readings for a gas cylinder in use, all valves, connections, and tubing should be checked using soapy water or a commercially available leak detection solution. Bubbles will form at the points where leaks are present. Electronic gas detectors, such as photoionization and combustible gas detectors, can also be used for detecting leaks of selective gases. Empirical methods, such as the use of one's nose, a match or other open flame, should not be used for leak detection. If a leak is found outside the tank itself, the valve for the tank should be immediately shut off and the leak repaired.

If a leak is discovered within the tank (including valve threads, emergency pressure release device, valve stem, and valve outlet), the tank should be removed from service and the supplier/manufacturer contacted for repair or removal. A leaking cylinder containing a hazardous gas should be cautiously removed from the laboratory to an isolated area of the building that is well ventilated (such as a fume hood), or better yet, outdoors if possible. Leaks of corrosive, toxic, or any hazardous gases where the cylinder cannot be moved from the laboratory will require the evacuation of the area and contacting emergency responders. If the leak is fast and the tank has to be transported through a populated area of the building, place a plastic bag, rubber shroud, or other similar device over the leak and tape it preferably with duct tape.[15] This should temporarily contain the leak. The tank should be plainly tagged that it is leaking along with a description of the location of the leak. The area where the leaking tank is being stored should be blocked off at a safe distance and warning signs posted, while the final resolution of the problem takes place.

Flammable gas cylinders should be stored away from any ignition source. Extreme care should be taken with leaking hydrogen cylinders. Hydrogen can ignite without normal ignition sources and has a very hot, colorless, almost invisible flame. In order to avoid the possibility of a serious burn, always check a leaking hydrogen cylinder for a flame by passing a straw broom around it before approaching.[16] If a leaking flammable gas cylinder is on fire, evacuate the area and contact the local fire department and emergency responders. Never permit oil, grease, or other combustible substances to come in contact with cylinders, valves, regulators, gases, hoses, and fittings used for oxidizing gases such as oxygen and nitrous oxide, which may combine with these substances with explosive violence.[17]

Cryogenic Spills

Spills of cryogenic materials have the same chemical hazards as a leaking gas cylinder at room temperature with the additional problem of extreme cold. Compound specific information should be obtained for emergency planning purposes.[18] Avoid direct contact with any cryogenic material. The best response to a small cryogenic spill is to block off the area of the spill and allow the chemical to evaporate. For non-flammable, non-oxidizing cryogenics, additional ventilation should be provided if there is a potential for air displacement to occur, that is, if asphyxiation is a possibility. Cryogenics can cause asphyxiation due to displacement of air by escaping liquid (and the rapidly expanding gas). Fire or explosion can result.

All sources of ignition should be shut off during a cryogenic oxygen or inflammable cryogenic gas spill. If liquid oxygen has come into contact with any combustible materials, block off the area for at least 30 min after any frost has disappeared. The chemical properties of substances are severely exaggerated under cryogenic conditions.[19] For example, violent reactions can occur where liquid oxygen has been spilled on surfaces contaminated with combustibles (simply from people walking on asphalt or concrete contaminated with oil).[20]

Biohazard Spills

When a biohazardous material is spilled, especially if it is Biosafety Level 2 or higher (see Chapter 2), the area should be immediately evacuated and all doors closed. Time should be allowed for droplets and aerosols to settle out of the air before the area is re-entered. This will depend on the ventilation for the area, but usually at least 30 min is allowed for settling. Appropriate PPE should be worn when the area is re-entered. The use of disposable gloves, such as latex or vinyl, is required for all biohazardous spill clean ups. Wearing double gloves is recommended if breakage is a potential problem such as when broken glass may be present. Additional protective clothing including splash goggles, face shield, fluid-resistant body protection, and respiratory protection may be required depending on the Biosafety Level of the spilled material and the possibility for droplet and aerosol formation during the clean up.

Spills of solid biohazardous material should be cleaned up using a vacuum with a HEPA filter. Paper towels or other absorbent pads soaked with an appropriate disinfectant should be placed over liquid spills. See Table 8 for a listing of common disinfectants. Bleach diluted 1:10 with water is an effective disinfectant. However, even dilute solutions of bleach can be corrosive to certain materials. There are many effective commercial disinfectants that are far less corrosive than bleach. The selection of a disinfectant will depend on the biohazardous material that is being used. Ethanol (70%) is an effective disinfectant for the human immunodeficiency virus (HIV), but is ineffective against the hepatitis B virus (HBV). A general purpose disinfectant should be tuberculocidal, virucidal, and registered by the EPA as being effective against HIV. The paper towels and spilled material should be picked up, being cautious in case any sharps are involved, and properly disposed of as biohazardous waste (or regulated medical waste in certain states). The area of the spill should be disinfected again and wiped dry. The PPE used should be removed leaving the gloves for last. All personnel involved in the spill and clean up should wash or shower.

If a spill occurs in a biosafety cabinet, the cabinet should be left running and the clean up should start immediately. Other than not waiting for droplets and aerosols to settle, all other procedures for the clean up are the same as above.

Radioactive Spills

Specific radiation spill response procedures for any place of employment using radioactive materials must be included in standard operating procedures (SOPs). These are required as part of the licensing process for use of radioactive materials, unless exempt quantities are in use. These procedures should be posted in the laboratory.[21] The Radiation Safety Officer for the work area should be contacted immediately for information if one is using radioactive material and has not received these SOPs. Major and minor spills for each facility should be defined.

Minor spills will only involve the notification of the people in the area where a spill has occurred. The person cleaning up the spill should wear disposable gloves and use tongs to handle all contaminated materials. Paper towels or other absorbent pads should be used to absorb and pick up the spill. Wet paper towels can be used for solid materials. All contaminated materials should be placed into a plastic bag (minimum 0.006-inch thickness) and disposed of in a radioactive waste container. Remove gloves and dispose of them as radioactive waste. Survey the area with the suitable portable instrument of choice, such as a low-range, thin-window Geiger Muller survey meter. Also take wipe samples from contaminated or potentially contaminated surfaces to be checked using a liquid scintillation counter or other appropriate detection method. Survey the hands and clothing of any individual involved in the spill and/or clean up for contamination. If either the area or the individual show signs of contamination, decontaminate with soap and water and resurvey. Exempt quantity users should follow the above procedure if a spill occurs.

Major spills require the evacuation of the room. However, all individuals who were in the area of the spill and are potentially contaminated should restrict their movements until they can be surveyed and shown to be free of contamination. Before the last person leaves the room, the spill should be covered with paper towels or other absorbent pads and, if practical, shielded. This should be done using PPE, including gloves, disposable coveralls, shoe covers, hoods, and possible lead shielded aprons. When the last person leaves the door should be locked, warning signs posted, and no one allowed to re-enter until the Radiation Safety Officer responds.

If possible, and if no gaseous radioactive materials are involved, the ventilation for the area should be shut down. The Radiation Safety Officer should give direction as to how the clean up should be conducted. All employees found to be contaminated should be decontaminated using procedures outlined for minor spills; medical consultation is always advised. If these procedures do not work, use a commercially available skin decontamination foam. An incident report for all radioactive spills should be filed with the Radiation Safety Officer. This should include information on when and where the spill occurred, how the spill occurred, the names of any people involved in the spill and/or the clean up, the names of anyone found to be contaminated, the types and amounts of radioactive material spilled, measures taken to prevent the spread of contamination, decontamination procedures, and survey results.

Mixed Hazard Spills

When multiple hazards are present in a spill, additional consideration must be given as to how to proceed before starting the clean up. The relative risk of each hazard must be prioritized. The hazard with the highest potential for causing harm to human life and health should be dealt with first. Once the primary hazard is decided, make sure that its clean up method is compatible with the other hazards. Unexpected side reactions can occur that spread existing hazards or create new hazardous byproducts when disinfectants, neutralizing agents, and absorbents for primary hazards remediation are added without considering their compatibilities with other materials already present in a hazardous spill. For example, consider an example of a mixed hazard spill where the primary hazard was biological in nature and the secondary hazard was radioactive iodine. The lab worker wished to use bleach to disinfect the biohazard. Bleach, which would be used to neutralize a biological component, contains sodium hypochlorite which would also serve to volatilize the radioactive iodine and so would spread radioactive contamination.[22] It is better to deal with these contingencies during the preplanning phase than to discover them on the spot.

Once the relative hazards of the spill have been prioritized and all materials to be used have been determined to be compatible, follow the general clean up procedure for the primary hazard. If this procedure reduces the potential for harm from the primary hazard, then the spill procedure for the next hazard with the most potential for human harm should be implemented. Continue addressing each hazard associated with a spill (in order of decreasing potential harm to human health and life) until all hazards have been mitigated and the spill has been cleaned up.

Spills of Unknown Materials

Extreme caution should be taken when dealing with spills of unknown materials or of materials that have been in stock for a long time, either of which are potentially unstable. Some materials which have been in storage for long periods may become explosive. For example, ethers can form explosive organic peroxides over time. In one unfortunate example, a chemist died from injuries resulting from the explosion of a one pint bottle of isopropyl ether. He had been opening the lid when it exploded; investigators theorized that friction caused by the turning lid had set off the sensitive organic peroxides.[23]

Since for most laboratories only hazard communication training is required for the handling of hazardous materials, an in-house clean up should not be attempted for unknowns or old stock since the employee cannot necessarily know the hazards associated with cleaning up of the material. The area of the spill should be blocked off and emergency responders should be contacted. When dealing with an unknown, always assume the worst (be conservative).

LARGE SPILL RESPONSE

In the context of this chapter a large spill refers to a situation where the scale of the release would be greater than an incidental release. This would be beyond the ability of the laboratory staff to safely mitigate, therefore, additional assistance would be required. The clean up of any large spill would have to be conducted by emergency responders trained in accordance with 29 CFR 1910.120. The emergency responders may be from an outside agency such as the local fire department, or they can be an internal group of emergency responders, also referred to as a Hazardous Materials Response Team (HazMat). It is not uncommon for large companies or institutions to have personnel on staff (i.e., a HazMat team) trained to deal with all levels of hazardous material incidents. With the issurance of OSHA's Hazardous Waste Operations and Emergency Response, Final Rule, on March 6, 1989, the requirements, including training, for emergency responders has been standardized (as previously illustrated in Table 2).

Information on contacting emergency responders should be readily available for all laboratory personnel. If the emergency responders are an outside agency, the employer might want to set up a chain of command authorized to make contact with the emergency response team and to relay any required information.[1]

PREVENTION

The best method of dealing with hazardous materials spills is prevention. Product substitution, that is, using a less hazardous material, should always be tried first (on the theory that if a hazardous material is not present, it cannot be spilled). Next, there should be adherence to all safety policies and procedures for handling hazardous materials. Administrative controls such as written policies and procedures should include information on the proper storage, transportation, use, and disposal of all hazardous materials in the laboratory. Training in these policies is key in making sure that all personnel are familiar with them. Enforcement of all such policies and procedures by the supervisory staff has to be done to maintain this familiarity. Engineering controls, such as the proper maintenance and use of fume hoods, will help to minimize releases and their consequences. PPE such as gloves and safety glasses, if used properly and in conjunction with the preventive measures previously discussed, will reduce both workers' exposures and adverse health effects should spills or releases occur (as well as during routine chemical handling). Follow-up by employees and supervisors is important not only after any accident but after any near miss as well (to determine why the near miss occurred and what can be done to prevent it from occurring again) and is a key element in preventing future accidents including spills.

Additional resource materials should always be consulted with regard to work practices, procedures, or specific chemicals.[24-26] Agencies providing such information are shown in Table 11. For any facility, the prevention and mitigation of accidental releases of any size must be dealt with on a case by case basis.

Table 11. Sources of Additional Information

Chemical Spills	Biohazardous Material Spills
American Chemical Society Health and Safety Referral Service Chemical Health and Safety Division 1155 16th St., N.W. Washington, D.C. 20036 (202) 872-4600 (800) 227-5558	Centers for Disease Control Office of Biosafety Atlanta, GA 30333 (404) 639-3883
CHEMTREC (Chemical Transportation Emergency Center, Emergencies Only) (800) 424-9300, 24-hour	National Animal Disease Center U.S. Department of Agriculture Ames, IA 50010 (515) 239-8344
Chemical Manufacturers Association Chemical Referral Center (800) 262-8200, 24-hour	National Institutes of Health Division of Safety Bethesda, MD 20892 (301) 496-1357
Compressed and Cryogenic Gases American Gas Association 1515 Wilson Blvd. Arlington, VA 22209 (703) 841-8400	**Radioactive Material Spills**
Compressed Gas Association 1235 Jefferson Davis Highway Arlington, VA 22202 (703) 979-0900	U.S. Nuclear Regulatory Commission 1717 H St. Bethesda, MD 20814 (301) 492-7000

SUMMARY

Even though many hazardous materials spills can be prevented by following the above suggestions, a hazardous materials spill can occur at any time that chemicals are in use. When a spill does occur make sure that the facility and employees are prepared for it by proper planning and training. Planning and training allow for safe, effective, and quick emergency response action by the employee; this in turn will reduce employee exposure and property damage. Make sure that an Emergency Response Plan exists and that personnel have received training in all its requirements; consider under what circumstance visitors or vendors should receive training. Supplies, including PPE and the necessary tools and materials, should be available in spill response kits wherever a spill can occur. Have a mechanism in place regarding when to call for outside help (a HazMat team or emergency responders) for those circumstances which cannot be termed incidental releases.

REFERENCES

1. Hazardous Waste Operations and Emergency Responses 29 CFR 1910.120, Washington, D.C.: Occupational Safety and Health Administration (1989).
2. National Safety Council, "Job Safety Analysis in an Industrial Laboratory Environment," Data Sheet I-706 Rev. 89 (1989).
3. Brower, J. E. "Worker Right-To-Know," in *The Work Environment, Volume One, Occupational Health Fundamentals,* D. J. Hansen, Ed., Chelsea, MI: Lewis (1991).
4. Burton, D. J. *Laboratory Ventilation Workbook,* Cincinnati, OH: ACGIH (1991).
5. NIOSH. *The Industrial Environment — Its Evaluation and Control,* Washington, D.C.: U.S. Government Printing Office (1973), p. 18.
6. U.S. Department of Labor, 29 CFR 1910.1000, Air Contaminants — Permissible Exposure Limits, OSHA 3112, U.S. Government Printing Office (1989).
7. American Conference of Governmental Industrial Hygienists, Threshold Limit Values and Biological Indices for 1991–92, Cincinnati, OH: ACGIH (1992).
8. American Industrial Hygiene Association. Emergency Response Planning Guidelines, Akron, OH: AIHA (1992).
9. BRIEFS, Emergency Preparedness News (January 6, 1992), p. 7.
10. Forsberg, K. and S. Z. Mansdorf. *Quick Selection Guide to Chemical Protective Clothing,* New York: Van Nostrand Reinhold (1989), p. 60.
11. Keith, L. H. and V. H. Keith. *Instant Chemical Protective Material Selections,* Austin, TX: Instant Reference Sources (1987).
12. National Research Council, *Biosafety in the Laboratory: Prudent Practices for the Handling and Disposal of Infectious Materials,* Washington, D.C.: National Academy Press (1989), p. 40.
13. Superfund Amendments and Reauthorization Act. Title III — Emergency Planning and Community Right-to-Know, 40 CFR Parts 300 and 355, "Extremely Hazardous Substances List and Threshold Planning Quantities; Emergency Planning and Release Requirements; Final Rule," *Federal Register* (April 22, 1987).
14. Environmental Protection Agency, "List of Lists," Office of Toxic Substances, and Office of Solid Waste and Emergency Response, Washington, D.C., EPA 560/4-91-011 (January 1991).
15. National Research Council, *Prudent Practices for Handling Hazardous Chemicals in Laboratories,* (Washington, D.C.: National Academy Press (1981), p. 236.
16. Compressed Gas Association, Inc., *Handbook of Compressed Gases,* 3rd ed., New York: Van Nostrand Reinhold (1990), p. 403.
17. Compressed Gas Association, Inc., *Handbook of Compressed Gases,* 3rd ed., New York: Van Nostrand Reinhold (1990), p. 83.
18. Compressed Gas Association, Inc., *Handbook of Compressed Gases,* 3rd ed., New York: Van Nostrand Reinhold (1990), p. 82.
19. National Safety Council, "Cryogenic Fluids in the Laboratory," Data Sheet I-688-Rev. 86 (1986).
20. Compressed Gas Association, Inc., *Handbook of Compressed Gases,* 3rd ed., New York: Van Nostrand Reinhold (1990), p. 532.
21. *Standards for Protection Against Radiation* (10 CFR 20), Washington, D.C.: United States Nuclear Regulatory Commission (1979).

22. National Research Council, *Biosafety in the Laboratory: Prudent Practices for the Handling and Disposal of Infectious Materials,* Washington, D.C.: National Academy Press (1989), p. 66.
23. Rekus, J. F. "Implementing Chemical-Hygiene Plan Focus of OSHA's New Lab Standard," *Occup. Health Safety* 60(2):52–59 (1991).
24. American Chemical Society, "Safety in Academic Chemistry Laboratories," 3rd ed., Washington, D.C.: ACS (1979).
25. Furr, A. K., Ed. *CRC Handbook of Laboratory Safety,* 3rd ed., Boca Raton, FL: CRC (1990), p. 704.
26. Sax, N. I. and R. J. Lewis. *Dangerous Properties of Industrial Materials,* New York: Van Nostrand Reinhold (1989), p. 5739.

CHAPTER 8

Medical Waste Disposal

John Marchese

INTRODUCTION

Over the past 10 years, medical waste has become a major political, social, and economic issue. The first increases in medical waste occurred in the early 1960s largely as a result of the development of disposable items such as surgical gowns, drapes and diapers. Reusable items require maintenance and always carry the threat of either non-performance or a breakdown in sterility. With the use of disposables, the liabilities of any such occurrence were shifted to the manufacturer (and savings were realized from the decrease in materials processing). However, materials handling and waste costs significantly increased during this transition.

Since 1980, there has been a concern among healthcare facilities with respect to medical waste disposal for occupational health reasons. That concern has spread to communities and to local, state and federal governments due to uncertainty over the spread of human immunodeficiency virus (HIV), and further intensified by medical-type waste washing up on shorelines. These concerns have lead to legislation attempting to control medical waste disposal, by some

0-87371-392-3/93/$0.00 + $.50

states and at the federal level. This has made all aspects of medical waste disposal more restrictive and, therefore, more costly. Regulations are still changing in an attempt to control the disposal of medical waste.

This chapter will review some of the basic issues revolving around medical waste disposal, including definitions, regulations, disposal technologies, and the development of a plan for waste disposal. This chapter will not cover the other types of wastes generated in healthcare facilities (such as hazardous, chemotherapeutic, radioactive, solid, asbestos, and other waste material). It will address waste generated by hospitals and other sources of medical waste (such as healthcare facilities, clinics, laboratories, and long-term care facilities). Though the facilities are different, similar wastes can be generated in each. If a facility generates less than 50 lbs/month of a regulated medical waste, it is exempt from federal requirements (but may be covered under state or local laws). This chapter will address the issue of medical waste in a broad sense, using hospitals as the primary example.

MEDICAL WASTE DEFINITIONS

In the early 1990s, hospitals had little concern for medical waste disposal since most equipment and supplies were reusable. Equipment and supplies were processed on-site and reused until failure. Most of the waste, therefore, was food scraps and office material. As technology progressed, especially in the field of plastics, major advances were also made in the medical field. We as a culture had moved to a largely disposable society. In medical care, disposable equipment was produced quickly and efficiently by manufacturers which enabled the medical community to reduce material processing costs through products having better safety, better quality, and better infection control.

The increase in the use of disposables has had an obvious effect on the amount of waste generated. Since there were no guidelines available, most hospitals used their own ad hoc policies for the disposal of wastes, handling some materials as too dangerous to landfill (preferring instead special handling or incineration). In this way the first medical wastes were somewhat informally defined. In 1976 (and updated in 1980), the Centers for Disease Control (CDC) issued a two page document on the disposal of hospital-type waste.[1] In 1985, the CDC updated its document, defining infective waste in five categories.[2] In 1982, the Environmental Protection Agency (EPA) published a "Draft Manual for Infectious Waste Management", defining infectious waste.[3] The EPA followed that up with a revised document in 1986, "EPA Guide for Infectious Waste Management".[4] In 1988, the Medical Waste Tracking Act (40 CFR 259) defined regulated medical waste in 7 categories. Further, individual states have defined regulated medical waste (also called red-bag waste, potentially infectious waste, infectious waste, infective waste, isolation wastes, and medical waste) with no uniformity through-

Table 1. Overview of "Red Bag" Waste Definitions

Waste Category	EPA	CDC	CA	NY
Contaminated animal carcasses, body parts, and bedding	Yes	Yes	Yes	Yes
Contaminated sharps	Yes	Yes	Yes	Yes
Dialysis wastes	OPT	No	Yes	Yes
Contaminated equipment	OPT	No	Yes	Yes
Human blood and blood products	Yes	Yes	No	Yes
Isolation wastes	Yes	OPT	Yes	Yes
Laboratory wastes	OPT	No	Yes	Yes
Microbiological wastes	Yes	Yes	Yes	Yes
Pathological wastes	Yes	Yes	Yes	Yes
Surgical and autopsy wastes	OPT	No	Yes	Yes
Unused sharps	Yes	No	NA	Yes
Other	No	No	Yes	Yes

OPT — optional.

NA — information not available.

out the U.S. In all but one case, states have been more restrictive than the federal agencies. Table 1 gives examples of the categories of the variety of definitions offered by the CDC, EPA, and selected states.

In 1980, most hospitals were charged low rates (for example $0.25 per bag) for infectious waste disposal. Today, many hospitals are charged that much or more per pound. Based on unpublished data at one hospital in New York, infectious waste disposal costs have risen eightfold since 1984. The volume of infectious waste generated has increased fourfold over the same time period due largely to the increasingly restrictive definitions of infectious waste (regulated medical waste). This anecdotal evidence also suggests that approximately 57% of the increase in volume was associated with the implementation of the Medical Waste Tracking Act of 1988 (MWTA, discussed below).[5]

The EPA defines medical waste as the following:

> Any solid waste which is generated in the diagnosis, treatment, (e.g., provision of medical service) or immunization of human beings or animals in research pertaining thereto or in the production or testing of biologicals. This term does not include any hazardous waste identified or listed under Part 261 of this chapter and household wastes as defined in 261.4(B)(I) of this chapter.[6]

With EPA's definition as a starting point, a more specific definition of Regulated Medical Waste (RMW) was developed. RMW is the waste that is actually subject to the regulations within the MWTA. Table 2 describes the 7 categories of RMW, as defined by the MWTA.

Medical Waste Definitions: How Derived

Understanding how these definitions were derived is interesting. Basically, most waste is not capable of disease transmission. This is the result of several factors. Most organisms found in the waste are already commonly found in the environment. Also, the numbers of organisms which need to be present in the waste to cause disease (infectivity) is insufficient. The time elapsed since the organisms, if present, have been outside of the host body (survival) is not conducive to their survival or propagation and, the proper routes of entry for the organisms into the human body typically are not present.

Regulated medical waste and related definitions have sometimes been derived less from a scientific basis than from a reactionary, non-scientific standpoint. For example, municipal solid waste carries a greater variety and a larger population of bacteria than hospital waste.[7] At a landfill, workers may be more at

Table 2. U.S. Standards For Tracking and Managing Medical Wastes

Cultures and Stocks: Cultures and stocks of infectious agents and associated biologicals include cultures from medical and pathological laboratories, cultures and stocks of infectious agents from research and industrial laboratories; wastes from the production of biologicals; discarded live and attenuated vaccines and culture dishes and devices used to transfer, inoculate, and mix cultures.

Pathological Wastes: Human pathological wastes, including tissues, organs and body parts, and body fluids that are removed during surgery or autopsy or other medical procedures, and specimens of body fluids and their containers.

Human Blood and Blood Products

1. Liquid waste human blood.
2. Products of blood.
3. Items saturated and/or dripping with human blood
4. Items that were saturated and/or dripping with human blood that are now caked with dried human blood including serum, plasma, and other blood components and their containers which were used or intended for use in patient care, testing and laboratory analysis or the development of pharmaceuticals. Intravenous bags are also included in this category.

Sharps: Sharps that have been used in animal or human patient care or treatment or in medical research, or industrial laboratories, including hypodermic needles, syringes (with or without the attached needle). Pasteur pipettes, scalpel blades, blood vials, needles with attached tubing, and culture dishes (regardless of the presence of infectious agents). Also included are other types of broken or unbroken glassware that were in contact with infectious agents, such as used slides and cover slips.

Animal Wastes: Contaminated animal carcasses, body parts and bedding of animals that were known to have been exposed to infectious agents during research (including research in veterinary hospitals), production of biologicals, or testing of pharmaceuticals.

Isolation Wastes: Biological waste and discarded material contaminated with blood, excretion, exudates or secretions from humans who are isolated to protect others from certain highly communicable diseases or isolated animals known to be infected with highly communicable diseases.

Unused Sharps: Unused, discarded sharps (hypodermic needles, suture needles, syringes, and scalpel blades).

From 40 CFR 259 Subpart D, 259.30.

risk handling community waste than hospital waste. This may be because much of the hospital's materials start out in a sterile condition until they are used (whereas household waste does not). It also may be true because most of the infectious individuals are in the community, rather than in the hospital, and therefore waste material from the home may carry more pathogens than hospital waste. In addition, most microorganisms which cause disease are termed opportunistic pathogens. That is, they may always be present in the environment (or even already present in most individuals), resulting in potential exposures every day. Healthy individuals can combat these pathogens, however, those who are in a weakened state may become infected. To this date, there are no reported environmentally transmitted cases of HIV on record, as the Federal Agency for Toxic Substances and Disease Registry (ATSDR) noted in its report to Congress mandated by the MWTA.[8] In fact, HIV is apparently quite fragile outside of the body.

The definition of RMW is primarily a reflection of the public's concern and reaction to hospital wastes and to infectious diseases such as HIV. The definition of RMW is based largely on the aesthetic properties of medical waste and so includes items which may only appear to be infectious or which are simply associated with medical waste. An example of an item which is obviously not infectious, but which is considered RMW (due to esthetics) are intravenous bags/bottles. There is a price associated with these all inclusive definitions of infectious waste; that price is reflected in the increased costs of the removal of infectious waste.

Though the MWTA, states and municipalities may have overdefined infectious wastes, there are categories of infectious waste which must be disposed of properly in order to limit risk of injury and illness to all who handle that material. These categories are covered by the CDC's definition of infective waste. For example, sharps pose a threat of injury and provide the proper route of entry for infection. Blood and blood products, also, can contain organisms which may be infectious, though the appropriate route of entry may or may not be present. However, contamination could occur from splashing into mucous membranes. The category of pathological waste is based on esthetic, legal, and moral questions. Most agree this waste should be considered under this definition, though it has not generally been shown to cause disease. Waste from microbiological laboratories is probably the only waste shown to contain organisms, since that type of laboratory's purpose is to grow infectious organisms. In addition, some isolation waste should be added to this definition for those rare cases of highly communicable diseases as described in the MWTA. This is a workable definition for the healthcare industry which would provide more than adequate protection with a margin of safety for the general public.

THE MEDICAL WASTE TRACKING ACT (MWTA)

The Medical Waste Tracking Act of 1988, developed by the EPA, was a two-year demonstration project. It was enacted, based on public uncertainty over the health and environmental concerns for medical waste (from beach wash-ups of medical waste in some northeastern states) resulting in "beach closures, economic losses for shore communities and public concern over the health hazards associated with medical waste and the general degradation of the shore environment".[9] This waste mismanagement raised public uncertainty of the risks posed by the health and physical hazards associated with the waste material and the environmental harm to a sensitive environment (i.e., beaches). The MWTA was thus promulgated to combat these concerns. It created a tracking system (a cradle-to-grave responsibility) for generators in those areas where waste was mismanaged (much in the same way that hazardous waste is dealt with). The MWTA addressed medical waste mismanagement via several objectives.

1. To control sources contributing to the medical-type waste which had washed up on shores and beaches.
2. To prevent careless management of medical waste by establishing tracking and storage requirements and subjecting violators to a variety of penalties.
3. To assure that medical waste defined under the Act would reach its proper final destination.

The MWTA also established some specific criteria for EPA as well as for generators, transporters, disposal sites et al. to follow in order to attain these objectives. Those criteria are:

1. Proper packaging of waste to assure minimal exposure to waste handlers and the public.
2. Waste identification through a system of labels, tags, and tracking forms; this would reduce the probability of improper disposal and help to limit accidental contamination of the general refuse.
3. Prevention of beach closing in the future.
4. The extent of compliance with this statute.
5. Identification of the quantity and type of debris found on the beaches in 1989 and 1990.
6. Assessment of sources contributing to the medical waste found on beaches and to what extent this regulation can affect it.

Also, the Act required that the EPA prepare three reports on medical waste and the success of the tracking program. It obligated the ATSDR with reporting on the impact of public health risk from medical waste, assessing the infection potential or injury from medical waste, the number of infections related to this waste, and the number of HIV and hepatitis B virus (HBV) infections related to medical waste. Protection from occupational exposure to HIV, HBV, tuberculosis, and others, are discussed in other chapters of this volume.

As the MWTA demonstration project came to an end, many states, including those required to adopt the act (Connecticut, New Jersey, New York, Puerto Rico, and Rhode Island) had already adopted their own (usually more restrictive) requirements for medical waste disposal. Therefore, the end of the demonstration

Table 3. Estimated Annual Range of Injuries*

Employee Group	Sharps Injuries	HBV Infection	HB Cases	HIV Infection	AIDS Cases
Non-Hospital					
Physicians	500–1,700	1–3	<1–2	<1	<1
RNs	17,800–32,500	36–65	18–33	<1	<1
LPNs	10,200–15,400	20–31	10–15	<1	<1
EMs	12,000	24	12	<1	<1
Dentists	100–300	<1	<1	<1	<1
Dental assistants	2,600–3,900	5–8	3–4	<1	<1
Refuse workers	500–7,300	1–15	<1–7	<1	<1
Hospital					
Physician/dentists/interns	100–400	<1	<1	<1	<1
RNs	9,800–17,900	20–36	10–18	<1–1	<1–1
LPNs	2,800–4,300	6–9	3–4	<1	<1
Lab workers	800–7,500	2–15	1–8	<1	<1
Janitorial/laundry workers	11,700–45,300	23–91	12–45	<1–3	<1–3
Hospital engineers	12,000	24	12	<1	<1

* Theoretical Estimated Annual Number of Hepatitis B Virus (HBV) and HIV Infections, and Theoretical Estimate of Annual Number of Hepatitis B (HB) and AIDS Cases in Non-Hospital and Hospital Employees as a Result of Medical Waste Related Injuries from Sharps — U.S., 1990.

From Public Health Implications of Medical Waste: A Report to Congress, USDHH, ATSDR (September 1990), p. 5.18, 5.20, 5.22, and 5.24.

project had little effect on the disposal of medical waste other than to assure that most states did have legislation in place to regulate medical waste disposal. As for the beach wash ups, there is evidence to suggest that the amount of material has decreased since the enactment of the law, but it is unknown if the MWTA was directly responsible for this. Most agree that the sources of the material that did wash up on beaches were from sources other than institutional or clinical offices, mostly from home healthcare and intravenous drug users.[10,11] However, the ATSDR did reach some conclusions as they related to the risks posed to the general public by medical waste. Highlights of the ATSDR report are summarized below:[12-14]

1. The only documented injuries attributed to medical waste have occurred in the occupational setting (processing, transporting, and handling waste at a treatment facility). There is a considerable difference between those injuries which occur occupationally before disposal and those that occur in the environment after disposal. There exists only one documented infection from an occupational injury relating to managing waste and that was a staphylococcal infection from an injury involving a "sharp", an injury not too different than one would receive from stepping on a nail. Table 3 is an estimate of sharps injuries to workers and the likelihood of seroconversions. The numbers are injury statistics of healthcare workers based on the assumption that 50% of those infected with HBV will result in the disease and that all infected with HIV will develop the disease. Outside of the healthcare facility,

needle stick injuries may be a concern, but only from local or systemic secondary infections and the likelihood for HIV or HBV infection is remote. There is no evidence to confirm that HIV or HBV can be communicated to the public through medical waste outside of the healthcare setting.
2. Medical waste does not contain any greater number of microorganisms than residential wastes and viruses present in solid waste tend to absorb into organic matter and deactivate.
3. The public's perception of contracting infections from medical waste is not substantiated by epidemiological evidence.
4. Medical waste from home-bound treatment should not be overlooked as a source of exposure to the general public. These numbers are likely to increase as home healthcare becomes more common, increasing the opportunities for exposure.
5. Another significant source of needles and syringes are the intravenous drug users, which are considered high risk with respect to bloodborne pathogens. Though there is a lack of data regarding contact with the general public, there may be a potential for injury and infection.
6. Medical waste can be effectively treated by several methods, including incineration, steam sterilization, chemical disinfection, and irradiation. With proper worker precautions, even untreated waste can be disposed of in a sanitary landfill.
7. Non-sharp wastes are likely to be less of an infectious hazard than sharps.

The definition of medical waste varies from state to state. This variation affects the volume of waste generated (and, therefore, the cost of disposal) and may indicate that not enough information is available to make sound decisions on what actually constitutes medical waste. Its definition basically relies on the aesthetic value of the waste rather than on scientific data to adequately define it. It is evident that more research is required to further evaluate the actual risks and hazards associated with medical waste. However, as part of the requirements of the MWTA of 1988, the ATSDR was required to submit a report to Congress relative to medical waste disposal.

ATSDR concluded that the actual risks associated with medical waste to the general public, though unknown, are thought to be very low. Also, there are no documented cases of HIV or HBV transmission with respect to the handling or disposal of medical wastes. There are a variety of methods to safely dispose of medical waste, including incineration, steam and chemical sterilization, and irradiation. Also, the landfilling of medical waste is also a viable method for waste disposal, provided worker protection is considered.[15] Currently, there is a bill in the House of Representatives, H.R. 215, which would require the EPA to conduct research in the actual risks associated with medical waste handling and disposal. Other legislation is pending with respect to medical waste disposal. It is safe to say that some type of federal medical waste regulation will be in place in the not so distant future. States will be allowed to be as strict or stricter than the federal regulations, which means there still could be a wide variation in the definition, packaging, transportation, and disposal of medical waste.

Disposal Methods

Infectious wastes can be disposed of in a number of ways, summarized in Table 4 and explained in detail below. Each method offers advantages and disadvantages which must be carefully considered before selection is made.

Incineration

Perhaps the most accepted but still most controversial method for infectious waste disposal is incineration. Incineration is the physical destruction of material by the addition of oxygen to the material being destroyed, releasing heat and products of combustion, basically gaseous compounds and unburned fuel (ash). Because some of these compounds may be toxic, special equipment and/or handling is often required to remove or reduce their toxic effects. A volume reduction of 90% is possible with incineration. Not only will it biologically deactivate the wastes, but incineration will reduce its volume considerably. However, emissions from the incineration of any waste may be of concern to the locality in which the unit is located. Also, ash and its disposal is problematic as well.

There are several types of incinerators. A rotary kiln incinerator has a rotating, cylindrically shaped chamber which increases waste mixing, important for proper

Table 4. Summary of Infectious Waste Disposal Methods

Incineration	Waste burned at high temperatures, ash residue from waste and emission control equipment — ash requires disposal (maybe "special" disposal).
Steam sterilization	High temperature and pressure used to destroy pathogens, waste is basically unchanged — requires further disposal.
Gas sterilization	Sterilant gas used to destroy pathogens, waste is unchanged — requires further disposal.
Chemical sterilization	Liquid chemical used as a bath to destroy pathogens, waste is usually shredded — requires disposal of waste and liquid disinfectant.
Microwave radiation	Waste is heated with microwaves after shredding — requires further disposal.
Ionizing radiation	Waste is bombarded with gamma rays to destroy pathogens, waste is unchanged — requires further disposal.
Pyrolysis	Waste is heated to high temperatures in the absence of oxygen (changes molecular structure of waste) — waste "residue" and emissions may require further treatment.
Encapsulation	Waste is mixed with thick ceramic-like material to form blocks, no further treatment is required.
Ensilation	Using biological agents, non-human pathological waste is converted to pet food, little or no waste remains.
Recycling/reusables	Either remove disposable items from the healthcare setting and convert to reusable (washable) items, or collect and recycle disposable materials — need to be sure material is not biologically active.

Table 5. Range of Legal Emission Values For Hospital Incinerators From Select States

Pollutant	Emission Range
Particulate	0.1–0.015 gr/scf @ 7% O_2
Hydrochloric acid (HCl)	30–50 ppm
	90% removed
	51 lbs/hr
Opacity	5–40%
Carbon monoxide (CO)	50–100 ppm
Sulfur dioxide (SO_2)	30–50 ppm

From Cross, F., *Incineration for Hospital and Medical Wastes for Incinerator Operators,* Matawan, NJ: SciTech (1990), p. 153.

incineration. It can incorporate a number of different feed mechanisms, has continuous ash removal mechanisms, and has no moving internal parts. It has an operating temperature greater than 2500°F and can accommodate many different pollution control configurations. Some of its disadvantages are that it has a high capital cost and is maintenance intensive. Care must be taken when operating the unit to prevent temperature variation from affecting the refractory.

Controlled air incinerators are the most widely used incinerator type over the last two decades. They have relatively low capital costs and less operational and maintenance expenses than other types of units. They are usually available in prefabricated form. Basically, the incinerator is divided into two stages. A measured amount of waste enters into the first chamber through a feeder (e.g., a ram feeder). The material is burned in this, the primary chamber, in an oxygen deficient state. This is necessary to minimize particulate levels while destroying most of the volatile compounds. The material is then sent to the next or secondary chamber, in which oxygen is added to complete the combustion process. The resultant air emissions can be channeled through a variety of air pollution control devices to meet regulatory requirements. Reduced air emissions are also possible from this type of incinerator, thus limiting the need for pollution control devices (depending on local regulations). Some disadvantages are that it can be difficult to control for a variable waste stream and also can lead to incomplete combustion of certain carbon-containing materials.

Several existing incinerators around the country, including those that incinerate medical waste, face closure in 1992 due to new, stricter, emission standards for incinerators as promulgated by the EPA. Some states (e.g., California and New York) have promulgated even stricter regulations. In addition, new incinerators must meet these requirements as well, increasing the costs to a level such that smaller units are less cost effective. Therefore, the retrofitting of existing smaller incinerators or the purchase of smaller incinerators to meet the new requirements will probably not occur. Table 5 shows typical ranges of emissions limits allowed by states. The allowable limits of particulate emissions have been reduced significantly over the last few years and require scrubbers and/or baghouse filters to meet emission standards. Plastics, specifically polyvinyl chloride

(PVC) are a problem because they release hydrogen chloride gas upon incineration and require control devices, such as scrubbers, to remove these contaminants. More recently, a requirement for dioxin scrubbing has been proposed and, in California, implemented. Dioxins are considered by some to be among the most toxic substances known to man. The epidemiological data is inconclusive. However, requirements for scrubbing 99% or more of dioxin emissions or 10 ng or less per kilogram of waste burned must be achieved. Annual stack tests may also be required to assure compliance.[16] In addition, stack monitoring and continuous monitoring devices are usually required by permit conditions. BACT (Best Available Control Technology) is usually applied to these requirements. Also included in these regulations is the requirement of operator training. Incinerator operators must receive pre-use and annual certification on the proper operation of the equipment as well as the environmental ramifications of improper operations.

Perhaps the most important consideration in incineration technology is community response. Location is almost always a problem, with the Not In My Back Yard syndrome (NIMBY) a very big issue. Most communities are reluctant to accept large incinerators into their community, citing toxic gases, ash, and transport of waste through their communities. Most permit applications require public forums or hearings as a final step in the permitting process.

Steam Sterilization

Steam sterilization is a treatment that has been used for several years in hospitals, mostly for the sterilization of medical equipment and, to a lesser extent, sterilization of waste in areas such as the microbiology laboratories (in small amounts). More recently, steam sterilization has moved to the forefront for the treatment of most categories of infectious waste, due to community pressure, increasing regulation, and capital costs as they relate to incineration systems.

A gravity feed steam sterilization unit uses a process whereby steam is injected into a pressurized vessel to destroy any microorganisms present. The temperature required is approximately 250°F at 250 psi for a 60 to 90 min period to allow thorough steam penetration into the waste load. A pre-vacuum system can reduce sterilization time by 30 to 60 min. In this system, a vacuum is drawn at the beginning of the cycle which, in turn, allows for faster and more efficient steam penetration through the chamber. Retort type autoclaves use large volume chambers, designed to withstand high pressure steam injection, further reducing treatment times. Most gravity and pre-vacuum type units have a capacity of between 50 to 200 lbs/cycle. Retort units can process up to 60 tons per day per unit.

Most states require periodic testing of the effectiveness of steam sterilizers. A prepared test strip is used as an indicator of the effectiveness of conditions existing within the autoclave to achieve sterilization. The thermally resistant organism used to conduct this test is bacillus sterathermophillus. In many instances, testing is required for each load of waste sterilized. Careful operating requirements are necessary to assure microbial deactivation, including proper pre/post-sterilization cycles based on the sterilizer manufacturer requirements and the appropriate sterilization contact time, based on temperature and pressure settings. In most cases, waste must be specially packed and/or contained to allow for proper penetration of steam to assure proper sterilization. This usually involves secondary containment or repacking (rehandling) of the waste material into autoclavable bags and containers or metal buckets (in the case of liquids) to assure proper conditions exist with the chamber for optimum steam penetration. Although biologically deactivated, the MWTA (and, therefore, some states) still consider this waste as regulated, needing the appropriate manifests and disposal assurances as untreated regulated medical waste. Once steam sterilized, the waste can be compacted and shipped and properly packaged, to a landfill (if accepted) or incinerator. If shredded on-site, the waste no longer meets the criteria as regulated medical waste and can be disposed of as general refuse (treated regulated medical waste materials must be rendered unrecognizable before disposal as general refuse). The sterilization process does reduce the volume of waste somewhat (approximately 30%). However, it cannot compare with the reduction realized through incineration. Steam sterilization is not an accepted method for the treatment of pathological waste. In all cases of the steam sterilization of infectious waste, there is still a need to further handle the material for ultimate disposal. It is still defined as regulated medical waste and needs to be tracked (manifested).

Gas Sterilization

Many hospitals use ethylene oxide (ETO) to sterilize steam sensitive material. It uses ETO, low heat, and low humidity to biologically deactivate medical supplies; ETO is listed as a suspected carcinogen and regulated by OSHA. In some states, ETO emissions from hospitals may require expensive scrubbing devices to meet local requirements. In addition, the correct penetration of ETO through all the various substances typically found in waste is questionable. This technology may not be a viable alternative at this time.

Chemical Disinfection/Shredding

Chemical disinfection/shredding is designed to treat most categories of infectious waste, producing both a liquid waste which can be discharged to the sanitary sewer and a non-infectious solid waste which can be processed as general refuse. All current chemical disinfection/shredding processes require shredding before the waste contacts the disinfecting solution. Shredding before contact between the waste and the disinfecting solution enhances the effectiveness of the disinfectant. The disinfectant chemical of choice for this process is usually chlorine, though others may be used. As a byproduct of the shredding portion of this process, the waste is rendered unrecognizable and can be disposed of as normal trash. Pathological waste is not normally disposed of using this method.

Problems with chemical disinfection/shredding technology include noise during shredding and excessive chlorine levels in the workplace or surrounding environment. Also, special permits may be required for the discharge of disinfectants through the sewage treatment facility, and discharge may not be allowed at all if discharge is to the ground (septic tank/cesspool).

Ionizing Radiation

The use of radiation to achieve sterilization is not a new technology. Many foods, containers, equipment, and supplies are routinely irradiated to obtain biological deactivation.[17] While not a new technology, its application to infectious waste is new. Cobalt 60 is usually used as the source of the gamma rays which are a highly penetrating form of radiation. This is both an advantage and disadvantage. For example, gamma rays are very penetrating and can reach far into the waste to sterilize. However, that penetrating characteristic also means that massive shielding is required to protect the surrounding work environment from radiation exposure. Also, the cobalt 60 source must be periodically replaced (and itself disposed of) every four years.[18]

Electron beam radiation is also being evaluated as a method for medical waste disposal. An accelerator can focus its electronic beam and scan through the medical waste to obtain biological deactivation. However, with the current technology, it does not have the penetration range comparable to the cobalt 60 source.[18] Waste material therefore must be spread thinly for effective sterilization.

Non-Ionizing Radiation

Laser technology is currently being tested for use in the treatment of municipal solid waste. Lasers can produce a large quantity of energy which can vaporize organic materials and melt inorganic solids. This may be a future technology for waste disposal.

Microwave technology is currently available for the treatment and disposal of most types of infectious waste. In this process, waste material is loaded into a waste feeding hopper. The medical waste is then shredded and steam is injected onto the waste. From here, it is transported to a series of microwave generators. The shredder increases the waste's surface area while the steam moistens the waste to allow for better microwave absorption, and therefore, sterilization. The shredder also renders the waste unrecognizable. After irradiation the waste is held within the unit for a predesignated period of time to allow for proper deactivation. The treated waste can then be handled as general refuse.

Electrothermal radiation is also being studied for use in medical waste disposal. This process takes the form of macrowaves, which is long wave radiation, combined with a shredding process.

Pyrolysis

Pyrolysis (or destructive distillation) is the process of heating material in the absence of oxygen. Typically the temperature of these operations ranges from 1000 to 6000°F. As a result of temperature and oxygen deficiency, the waste undergoes a change in its molecular structure. Organic compounds (i.e., materials largely made of carbon and hydrogen, along with smaller amounts of other elements) break down into their constituent elements. Some advantages of these systems are public acceptance, easier permitability, lower capital costs, and operational efficiencies. Disadvantages include the possibility that further treatment of waste gases and residues from incomplete combustion may be required.

Plasma Sterilization

The plasma sterilization system uses a plasma arc torch, produced from an electrical current flowing through a gas (ionization), to heat a chamber to 3500°F. The waste in the chamber undergoes pyrolysis as described above. The gas is collected and cooled to form a fuel (which potentially could be used to generate electricity). The other non-organic materials, when undergoing this process, form a slag which, when cooled, forms a glass-like pellet which meets the requirements of the EPA's Toxicity Characteristics Leaching Procedure (TCLP). This material can be safely landfilled or used as aggregate in road construction.

Encapsulation

Encapsulation is a very new process and, therefore, not much information is available regarding its development. Waste is mixed with a thick powdered substance (containing sealing and hardening agents) to form a moldable substance called Thomas Ceramic which is reported to be harmless. This may have many applications, such as building and sound attenuating material. It is also reported to be relatively inexpensive.[19]

Another encapsulating method exists whereby solids such as needles and syringes are containerized and a small amount of disinfectant is added. Then, water is added to the container along with an oxidizing agent. This forms a polymer matrix which encapsulates the waste material. For waste liquids, the oxidizing agent is added to the liquid to solidify the fluid. This process may be of some benefit to the small generator, but no reduction in volume is attained by this method; some states accept the material as general refuse.[20]

Ensilation

Researchers at the University of Georgia College of Veterinary Medicine have developed a method for disposing of their research animals. Using a process called ensilation, they recycle their animal carcasses. Corpses, etc., are ground up and treated with active yogurt cultures. The waste is then mixed with a carbohydrate medium (like corn or molasses) and water is added. When the mixture reaches a desired pH, it is fermented for up to five days. From this point, it can be processed as pet food. This process removes all the bacteria from the waste material and, therefore, the former waste material can be kept at room temperature for a long period of time. This process has also been successfully used by the fishing and shrimping industries.[21]

Reusables/Recycling

Though not a disposal technology, these both could have a profound effect on the generation of medical waste and the reduction of medical waste volume and expense. The use of reusable, recyclable items (as opposed to disposables) may have an increasingly more important place in today's healthcare environment. For example, some hospitals (and mothers) have returned to the use of cloth diapers instead of disposables. Though there is controversy over net costs and benefits of recyclable materials (such as the additional water consumption and detergent additives), careful consideration of this option is a definite step in the right direction in the reduction of waste volume. Some items, such as needles and syringes, will not be reused in all cases. The disposable types of these items do have an overwhelming infection control benefit which probably will not be outweighed by disposal costs. Assessment should be careful and case by case. Recycling of medical waste may initially sound odd, but some potentially recyclable items used within the healthcare setting never contact the patient and may have the ability to be recycled (i.e., intravenous bags). With respect to recycling, of equal interest is the character of the waste products produced by many of the disposal technologies previously discussed. For example, the residues from sterilizer/shredders could be thought of as recyclable material because once treated, they can be handled as general refuse. There are some separation problems associated with shredding which need to be addressed (i.e., mixed plastics), but this may be the future of medical waste disposal.

Several technologies are currently available which can be employed to safely dispose of infectious waste, such as incineration, steam sterilization, chemical disinfection, and microwave sterilization. Due to intense interest in the field of solid waste disposal, several promising disposal methods are on the horizon, such as encapsulation, laser and plasma sterilization. These and other ideas can be pursued in order to maximize the efficiency of infectious waste disposal while minimizing costs and the risk of health and environmental risk. In addition, the use of reusable products and recycling practices should also be explored to assist in further reducing medical waste volumes.

MEDICAL WASTE DISPOSAL: ECONOMICS AND OVERVIEW

Table 6 summarizes the different types of medical waste disposal technologies discussed above. Table 7 provides order of magnitude estimates of relative operating and capital costs for these waste disposal methods.

Table 6. Comparison of Medical Waste Treatment Technologies

Factor	Steam Sterilization	Incineration	Hammermill/Chemical	Steam/Compactor	Microwave/Shredding
Operational considerations					
Applicability	Most infectious wastes[a]	Almost all infectious waste	Most infectious wastes[a]	Most infectious wastes[a]	Some infectious wastes[a,b]
Equipment operations	Easy	Complex	Moderately complex	Easy	Moderately complex
Operator requirements	Trained	Highly skilled	Trained	Trained	Trained
Need for waste separation	To eliminate nontreatable wastes	None	To eliminate nontreatable wastes; for proper feeding	To eliminate nontreatable wastes; for proper feeding	To eliminate nontreatable waste, syringes, body and body fluids
Effect of treatment	Appearance of waste unchanged	Waste burned	Waste shredded and ground	Minimum change in appearance	Waste shredded only
Volume reduction	30%	85–90%	Up to 85%	50%	Up to 85%
Occupational hazards	Low	Moderate	Moderate	Low	Low
Treatment validation	Easy, inexpensive[c,d]	Complex, very expensive[d]	Easy, inexpensive[c,d]	Easy, inexpensive[c,d]	Moderate, expensive (culture)
Potential side benefits	None	Energy recovery	None	None	None
Onsite/offsite location	Both	Both	Both	Both	No offsite
Regulatory requirements					
Medical waste tracking regulations	Applicable	Recordkeeping	Not applicable[e]	Applicable	Not applicable[e]
Applicable environmental regulations	Wastewater	Air emissions, ash disposal, wastewater	Wastewater	Wastewater	None

Release to air	Low risk via vent	High risk via emissions	Low risk via HEPA filter	Low risk via vent	Low risk via vent
Release to water	Low risk via drain	Low risk via scrubber water	Moderate risk via wastewater[f]	Low risk via drain	Low risk (evaporation)
Disposal of residue	To sanitary landfill; potential problem with red bags	Ash may be a hazardous waste; if so, to RCRA-permitted landfill	Effluent to sanitary sewer residue to sanitary landfill[g]	To sanitary landfill; potential problem with red bags	Residue to sanitary landfill
Permitting requirements	None[h]	For air emissions[h]	None[h]	None[h]	None[h]
Costs					
Capital costs	Low	High	Moderate	Low	High
Labor costs	Low	High	Moderate	Low	Low
Operating costs	Low	High	Moderate	Low	Moderate
Maintenance costs	Low	High	Moderate	Low	Moderate
Downtime	Low	High	Low to moderate	Low	Moderate to high

[a] Not including chemical, radioactive, or pathological waste.
[b] Not approved for syringes, blood, and body fluids.
[c] Waste treatment efficiency may be tested using a chemical indicator (such as a strip of waste packaging which changes color after treatment); however, chemical indicators are less preferable than biological indicators.
[d] Waste treatment efficiency may be tested by growth or viability of biological indicators, including cultures.
[e] Dependent on local regulations.
[f] This method may discharge disinfectant, etc., via waste stream.
[g] Physical and chemical characterization must meet EPA requirements.
[h] Depending on locality.

Adapted from *Journal of Healthcare Material Management,* Methodist Hospital of Indiana (April 1991).

Table 7. A Comparison of Medical Waste Treatment Technologies

Treatment Technology	RMW Treatable by the Treatment Technology[a]	Volume Reduction (%)	Costs (approximate)[b] Operating Expenses[c] ($/lb/hr)	Capital Costs[d] ($)
Steam autoclaving	All (except pathogens)	0	0.05–0.07	100 K
Autoclaving with compaction	All (except pathogens)	60–80	0.03–0.10	100 K
Mechanical and chemical	All	60–90	0.06	40–350 K
Microwave and shredder	All[e]	60–90	0.07–0.10[f]	500 K
Incinerator	All[g]	90–95	0.07–0.50	1,000 K

[a] The regulated medical waste types which could be treated using the method such that the waste would be rendered non-infectious.
[b] Reliable cost information is difficult to obtain and verify. Further, valid comparisons are difficult to make given the different operational circumstances possible.
[c] Without labor, profit, and depreciation.
[d] Primarily the cost of the equipment and installation into an existing building.
[e] Pathological wastes are usually not treated by microwaves for aesthetic reasons. Cytotoxic or other toxic chemicals cannot be adequately treated to reduce their hazardous nature.
[f] Including an energy cost of $0.07/kwh.
[g] Although separation of non-combustibles and items with problematic constituents improves combustion efficiency.

From Office of Technology Assessment (1990).

Developing a Comprehensive Waste Management Plan

The development of a comprehensive medical waste management plan is an important and necessary component to the overall disposal of medical waste. This plan will have three parts:

1. An extensive review of all factors involved in waste removal.
2. Selection of the best available options for the facility from environmental, social, and economic perspectives.
3. Implementation of the actual mechanisms for waste disposal.

The time and effort taken to thoroughly perform these tasks may help to prevent problems in the future (and help to develop a positive image for the facility within the community).

The first step in the development of the comprehensive waste management plan is to thoroughly assess the regulatory environment. Know what regulations are in place (and what agencies are responsible for enforcing or implementing them) for each phase of the program. This is not a trivial exercise and may include medical waste definitions, storage and transportation requirements, approved disposal methods and their associated requirements, and the permits/licenses necessary to execute the program, among others. In addition, where possible attempt to anticipate any future trends or pending regulations which may affect the program. This may include further definitional requirements or emission standards which will need to be met.

After the regulatory requirements have been assessed, it will be necessary to characterize the waste stream, based on the specific criteria (such as the infectious waste definition) identified in the first step. Once the waste has been defined, a survey should be performed to identify how much waste is generated. Additional information needed to characterize the waste stream includes the areas from which the wastes are generated and the types of wastes generated at those locations. Obtaining present waste volumes and anticipating the growth potential for waste generation, must also be evaluated in order to optimize the effectiveness of the program for the future, particularly if large capital expense is anticipated. Financially, it generally could be less costly to increase the size of the equipment at this stage rather than after the project has been completed. As an aid in planning, total solid waste generated by hospitals has been estimated to be between 17 to 23 lbs/bed/day.[22] Unpublished data suggests that infectious waste could be as high as 11 lbs/bed/day. Obviously, the type of institution and the level of care will influence these factors.

Knowledge of the types of wastes generated, specifically their component parts, is also important when considering the best available method(s) for treatment/disposal. Tables 8 and 9 give some idea of the component parts of the medical waste stream. Of great concern to many, for a variety of reasons, are plastics. Though they have great heat value when incinerated, some of them are associated with the formation of toxic gases/ash components when burned; also, plastics do not break down in the environment, as in the case of landfilling. Because plastics do make up such a large portion of medical waste, how plastics are used needs to be strongly considered when choosing a disposal method.

With the characterization of the facility, and once its waste streams have been determined, the next procedure is the selection of the disposal method.

Table 8. Composition of Medical Waste

	Percent Composition (%)	
Material	**Study 1[a]**	**Study 2[b]**
Plastic	39	71
Paper/fabric	45	26
Metal	0.6	2
Glass	7.2	Minimal
Others	8.2	1

[a] From: HDR Engineering, Inc. (1990).
[b] From International Environmental Health and Safety Report, State University of New York at Stony Brook (1990).

Table 9. Approximate Composition of Medical Waste by Waste Type and Area of Origin

	Percentage						
Type of Operation	**Paper**	**Plastic**	**Pathogenic**	**Food**	**Glass**	**Metal**	**Other**
Administration	100						
Cafeteria	30	30		30	5	5	
Surgery	60	30	10				
Emergency	60	35	5				
Intensive care	60	35	5				
Dialysis	10	85	5				
Lab	35	30	25		10		
Nursery	45	35		5	15		
Pharmacy	50	30			20		
General patient care	60	35		5			
Research	40		30		10		20[a]
Sharps[b]		85			5	10	

[a] Animal bedding.
[b] Sharps are handled separately from other waste regardless of department.

Adapted from Cross, F. L., H. E. Hesketh, and P. K. Rykowski, *Infectious Waste Management,* Lancaster, PA: Technomic (1990), p. 10, and the author's experience.

Essentially, there are two pathways for disposal: on-site and off-site, as discussed below.

On-Site Disposal

On-site disposal requires that an established or emerging technology be employed to manage infectious wastes. Though feasibility must be established, disposing of waste on-site will ensure proper disposal and in most cases, limit liability, handling, transport, and regulatory requirements that would be indicated for off-site disposal. Since the waste stream has been characterized, an evaluation of available alternatives can be completed. Taking into consideration items such as capital costs, waste volumes and types, environmental requirements, volume reductions attained (Tables 10 and 11), political and social concerns, an educated decision can be reached. For a small clinic, an incinerator may not make sense. However, a sterilizer-compactor may be cost effective while satisfying all other requirements. In addition, a combination of methods may be chosen to dispose of the different categories of infectious waste. For example, sharps and pathological waste are not usually steam sterilized. If steam sterilization is the method of choice, then an alternative for these two wastes will need to be developed, whether it be an incinerator or an outside contractor. Table 12 is an example of decision considerations for on-site alternatives.

An important factor for any mechanical device is proper operation and maintenance/repair. Training, which may in fact be required, must be given to operators to not only insure proper destruction of waste, but training also must be given to inform operators of proper operational, health, safety and environmental

Table 10. Disposal Methods and Approximate Weight Reduction

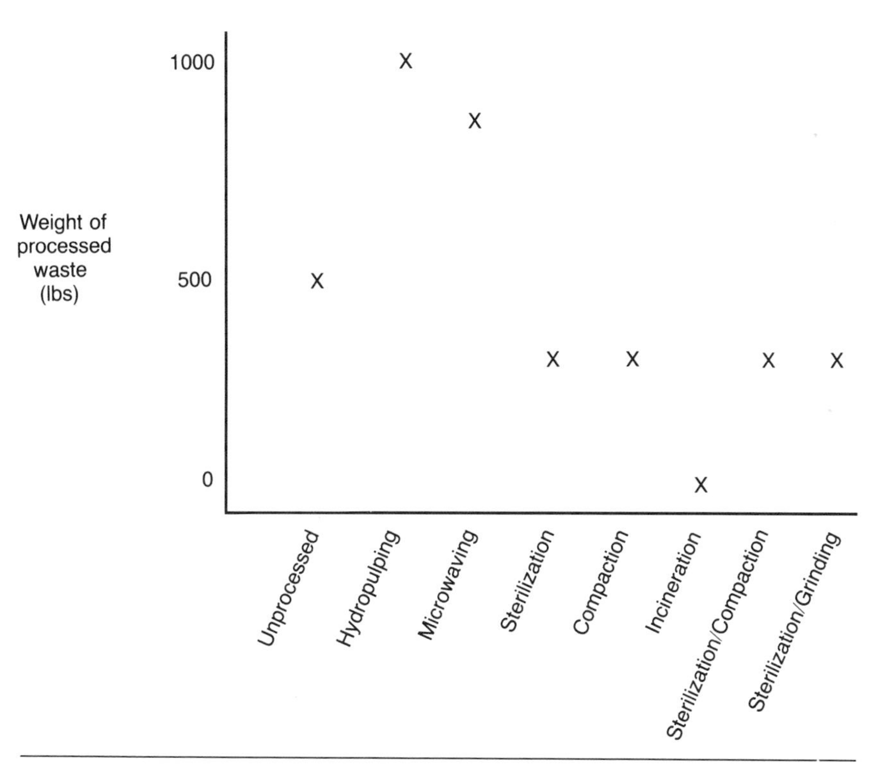

Note: A comparison of the ability of different medical waste disposal methods to reduce the weight of the waste. This example assumes that the starting weight of the medical waste is 500 lbs. Note that the weight of the waste increases for those methods where water must be added (hydropulping and microwaving).

From Cross, F. L., H E. Hesketh, and P. K. Rykowski, *Infectious Waste Management,* Lancaster, PA: Technomic (1990), p. 17.

procedures, and impacts. Additionally, operators must be trained in the recognition and prompt reporting of equipment malfunctions and other emergency conditions, as well as proper maintenance. All equipment requires some type of a preventative maintenance program, often requiring equipment downtime, and will ultimately need repair. It is essential that a program be developed to maintain equipment and to keep it operating in a safe and efficient manner. Included in this should be any required occupational or environmental monitoring, such as stack tests and ash evaluation. Since maintenance and repairs may require the equipment to go off-line, another issue which must be addressed are contingency plans (where does the excess waste get stored during maintenance, etc.). These alternatives are necessary for maintenance and repairs of any type of equipment;

Table 11. Disposal Methods and Approximate Volume Reduction

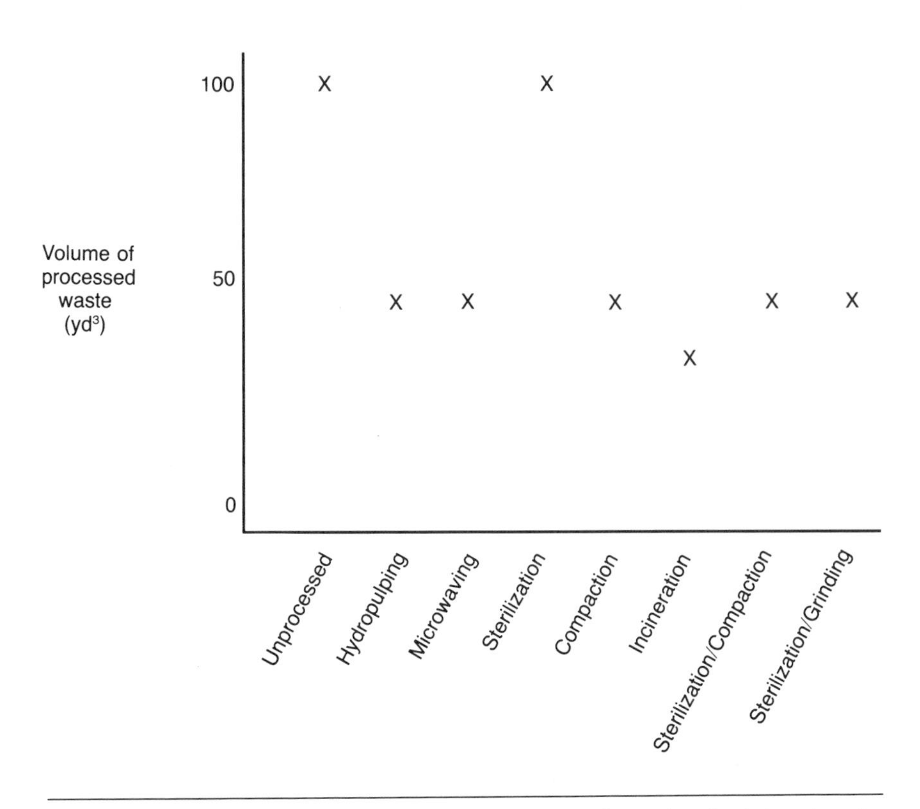

Note: A comparison of the ability of different medical waste disposal methods to reduce the volume of waste. The hypothetical starting volume was 100 cubic yards (100 yd^3).

From Cross, F. L., H. E. Hesketh, and P. K. Rykowski, *Infectious Waste Management,* Lancaster, PA: Technomic (1990), p. 17.

with medical waste, there are additional considerations by virtue of the nature of the commodity being handled. In addition, these issues become increasingly important when developing an operating budget for the program. However, if the contingency plan requires additional waste handling or preparation, be sure that the facilities are present to accommodate the needs of that contingency. If a contract with an outside disposal firm or transporter is used, be sure that the waste can be transported, properly packaged, and delivered to the removal area in a safe, efficient manner.

Each healthcare technology may ultimately have its own optimum mechanism for waste collection. This, too, is an important consideration when choosing an on-site option. If the technology lends itself to single handling mechanisms, it should be pursued. For instance, if the option chosen has an autoloader feature (for example, carts are loaded at the point of waste generation, transported to the disposal area, and can be directly deposited into the unit), waste will not have to be double-handled during disposal processing, which usually will save time and reduce the potential for injury. Carts can be directly loaded with waste, moved through the waste disposal scheme, cleaned and returned for use with the waste being handled only at the generation site. Storage facilities for the waste and for the waste processing products, if any, should also be considered, based on the volume of waste generated, transport (time and distance) to the facility, the amount of waste which is processed in a given time frame, and the amount of space needed to accommodate removal schedules with some contingency. This could vary for each technology studied.

Good practice dictates that records be kept in most healthcare facilities and recordkeeping with respect to medical waste disposal will most likely be required for regulatory compliance almost everywhere. A log of the number of loads, locations, weights, and disposal times and dates should provide regulatory compliance in most cases. A log of maintenance and repairs to equipment should also be kept. If the waste disposal option chosen requires monitoring, whether environmental or biological (i.e., effectiveness of the method chosen), a log should also be kept for these data as well. The entire program should be reviewed periodically by in-house staff to identify problem areas, preventing future problems and to meet changing regulatory compliance to ensure a safe and efficient disposal program. Larger facilities should consider periodic audits by outside parties.

Off-Site Disposal

Off-site disposal of medical waste may be chosen as a primary method for disposal or as a contingency or secondary means of waste disposal. If it is chosen, carefully review any agreement with an off-site disposal firm.

For example, regulatory compliance should be delineated in the agreement which should state all requirements for the proper packaging (containers, bags, and labels), transportation (type of vehicles and vehicle labeling), storage, and disposal (approved) for the waste material. Included in the specification should be all the required permits, licenses, and insurance (amounts and liabilities) for both the transporter (and any sub-transporters) and for the disposal facility (if different). Review them to ensure that all the proper information is present and meets the criteria you have established. Also ask for periodic updates as they become available; this will indicate if the prospective contractor is aware of the pertinent requirements (for example, in your locality as well as other state's requirements if the waste is transported through other states before disposal). If there are no state or local requirements, these requests should be included to limit liability and to ensure proper disposition of the material. Check with local or state authorities on the activities of the prospective contractor; they may have valuable information which may save the facility a costly mistake. In addition,

Table 12. Decision Considerations For On-Site Medical Waste Disposal[a]

Technology	Advantages	Disadvantages	Capacity (lbs/hr)	Volume reduction (%)	Capital (cost/lbs)	Operational (cost/lbs)	Disposal (costs/lbs)
Incineration	Accepted procedure Significant volume reduction (90%) Waste unrecognizable Heat recovery	Operational problems O and M required Temperature variability Licensing and permitting Air emission problems Ash disposal	1200	>90	$2,000,000	$.08–.09	$.02
Mechanical/chemical treatment	Acceptance by landfill operator Waste unrecognizable Waste dilution 1:100 Volume reduction Rapid, on-demand process Negative pressure No odor No air emissions Controllable and flexible Possible recycling of byproduct Local support	Personnel training O and M required Consumable parts Suspended solids Storage of chlorine Some regulatory barriers Sewer permit required Cannot dispose of chemical, radioactive, or pathological waste	1000	80	$350,000	$.044	$.008
Steam/compactions	Accepted procedure General and infectious waste Significant volume reduction Easy operation No double handling No air emissions No licensing/permitting	Landfill acceptance Definitional problem O and M required Long cycles required Validation and monitoring equipment required Odor Cannot dispose of chemical, radioactive, or pathological waste	200	50	$125,000	$.035–.04	$.02

	Possible recycling of byproduct	Located outdoors — freezing potential					
Microwave/shredding	Waste may be unrecognizable Significant volume reduction Reduced liquid effluent No air emissions Easy operation Possible recycling of byproduct	New technology O and M required Consumable parts Waste specifications metallic content: <10% material size: 20″ × 20″ × 20″ moisture: <10% of wt. Destruction quality Temperature hold time Odor Cannot dispose of chemical, radioactive, or pathological waste	500	80	$650,000	$.07–.10	Not determined

[a] Each state residue disposal cost could vary according to solid waste Management rules

From *Journal of Healthcare Material Management,* Methodist Hospital of Indiana (April 1991).

be sure to include a provision in the agreement which will require the successful vendor to meet all future requirements with respect to medical waste disposal (refrigerated storage, transport, etc.).

Insist that all manifests are completed and returned in an appropriate time frame. Manifests are the only documents which can serve as proof that each particular waste load has in fact been handled and disposed of properly. Pay schedules, dependent upon return of the manifest or other documents, can be used to ensure prompt return of the required manifests. Manifests should be maintained on file for future reference or by statute. Even when there are no specific requirements for manifesting the waste, it could be wise to require it in the bid specifications anyway (for obvious liability reasons).

Request the experience and qualifications of the prospective bidders, such as a list of their present clients, their length of time in business, equipment lists, and past or pending legal issues related to the task at hand. Always reserve the right to inspect equipment, storage, and disposal sites. Review the training and education programs as well as written policies and procedures of the perspective bidders, especially with respect to spill response, accidental releases to the environment, and other related emergencies. Also include indemnification assurances (hold-harmless clause) within the specifications.

Other items to consider in off-site infectious waste disposal are:

Contingency plans — Insist that the contractor has (and specifies) an alternate, approved site to dispose of the waste in the event that the primary facility is unable to treat the material. Also, look to see if the permits, licenses, etc. of the alternate site are in order and that the site is listed on the contractor's transportation permit (if required).

Weight — The weight of the waste may be required by statute or dictated by contract terms (i.e., cost per pound). A request for scales would be in order (as would documentation of the weight shipped and the weight received at the disposal site).

Storage — Consider on-site storage requirements. If required, include these facilities as part of the specifications. Also include repairs, cleaning, and removal upon termination of the contract. In addition, materials handling equipment may be required to move material to and from the storage facility; be sure to include this in the specification as well. The proper materials handling equipment, from the standpoint of ergonomics, slips, trips and falls, or general safety, will reduce injuries to those employees who are handling the waste material.

Schedules — Delineate the preferred removal times, locations, and dates. Reserve the right to alter or increase requirements. Also, indicate a time frame for emergency response (emergency removal requirements). Request a contact (an individual and a phone) for problems or concerns.

Termination — In the event of any improper waste disposal activities by the contractor, include an immediate termination clause to release the facility from the contract. Also, it may be a good idea to also include a termination clause with 30 days notice, especially if other methods of waste disposal are being pursued and there is any likelihood that the healthcare facility would wish to switch methods or contractors. For example, if there is a pending alternate waste disposal method (such as one of the previously mentioned disposal technologies), and if the alternate disposal method becomes viable within the contract period, the facility may still be obligated to fulfill the requirements of the contract if no termination clause is present.

THE REGIONAL APPROACH

Many facilities have realized that medical waste disposal was a major expense (particularly for large generators) and a liability. Since on-site treatment may not always be possible, hospitals and other healthcare providers have collaborated to develop plans for infectious waste disposal on a regional level. This has many advantages, such as reduced capital costs for required treatment equipment, reduction of disposal costs, better assurances of proper disposal, and better overall performance over contracted services. In addition, when states have promulgated regulations for existing incineration units which require costly retrofitting to meet the new standards, facilities had to resort to either new technologies or contracted service for disposal. Collaboration allows the costs of retrofits, etc., to be spread around. The major disadvantage of a regional facility, depending on the technology chosen, is location (and NIMBY).

Just as an individual facility must develop a comprehensive plan for the disposal of infectious waste, so too does any organization or confederation considering the regional approach. The considerations for a regional approach are not much different than those for an individual healthcare facility. Important components of such a plan are listed below:

1. Recognition of community concerns (NIMBY).
2. Recognition of what is required for regulatory compliance.
3. Recognition that landfill space is limited.
4. Development of a system which is compatible for all potential generators.
5. Consider the overall characteristics of the regional waste stream; this is necessary in order to apply the appropriate technology for efficient, cost effective, and safe waste disposal.
6. Development of contingency plans to assure proper disposal under any circumstances.
7. Encourage the use of reusables and explore the possibilities of recycling when possible.

In general, there are three approaches which can be employed for medical waste disposal using the regional approach. First, upgrade, retrofit, or develop an existing facility which is approved for medical waste disposal. Second, use a local resource recovery plant that will accept medical waste. Third, design, construct, and operate a new facility.

The first option (upgrade of an existing facility) assumes that there is an existing facility within the region which has the ability to legally process medical waste, and which has the capacity to receive the volume of waste generated in that region. While the existence of such a facility is not likely, if present, its use is feasible and it may become part of a total plan.

The second option (use of a local resource recovery plant) is probably the most logical, since many resource recovery plants are designed such that the temperatures and other parameters meet the requirements for medical waste disposal. If the local communities involved can agree, then this is an excellent potential disposal resource. The use of a local resource recovery plant, even if approval is not likely, could be part of the total regional plan.

The third option (design a new facility from scratch) is extremely costly and again involves local communities, regulatory requirements and is, in some instances, a duplication of the resource recovery operation. Location could also be a problem, particularly for a new facility, and especially if the option is an incinerator. Capital costs can be high, but again, would be spread among several users.

Transportation costs and handling difficulties, packaging requirements, and storage of waste material are common to all three options. For each option there can be an entire set of problems, which may be changeable depending on the option, technology chosen, and regulatory environment.

The Regional Approach: An Example

A hospital council in a northeastern state has developed a regional plan which encompasses all three of these options. Though their intent is ultimately to have a regional facility to manage all of the member hospitals' solid waste, they have developed this plan as an interim step. Almost all solid waste generated by the 24 member council will be incinerated at a resource recovery facility in a local town. This plant had originally been constructed based on expected waste volumes which were not reached. Therefore, the unit had extra capacity and the costs to the community were not as economically advantageous as expected. By adding the medical waste from regional hospitals, not only did the community increase their volume to assist incinerator efficiency, but also assessed a modest fee to the council for incineration of the waste material which decreased the community's responsibility for operating costs of the plant.[23]

The council researched the scope of the problem, realizing the current and future regulatory environment, the processes available, and the lack of locations where material could be processed. They proceeded to collect waste generation statistics and current practices and costs associated with solid waste disposal in order to develop a plan which would not only decrease the current removal costs, but would assure that waste would be handled in an appropriate manner, thereby reducing liability and promising that a location would always be available for solid waste disposal.

General refuse is brought to the resource recovery plant directly. Regulated medical or infectious waste, excluding pathological wastes and sharps, will be treated prior to shipment. The method of treatment chosen in this example is steam sterilization. For most of the member hospitals, a combination of pre-vacuum autoclave and compactor unit will be used. For the larger generators, a separate retort autoclave and compactor will be installed to accommodate the larger volumes generated. These compactors will be properly tagged and labeled in accordance with state and local requirements. Regulatory assistance was sought by the council to help facilitate the process. The pathological waste and sharps will be shipped to a member hospital which already has an incinerator on-site. This incinerator will be retrofitted to meet the new demanding standards associated with infectious waste disposal. As a contingency plan, a central sterilization location, additional storage facilities, and/or a contractor may be employed for removal of waste in the event of equipment failure or other factors which would prevent the proper removal of the waste material. Transportation of the waste will be through a contract issued by the council.[23]

The council's original plan was to have all the waste incinerated at a regional site. However, the location of the plant became an issue. While the council is attempting to develop a site which is both feasible and acceptable to all parties, and until a satisfactory resolution is reached, the plan outlined here will be in effect. It is important to note that steam sterilization of the infectious waste, as well as the requirement of the removal of sharps and pathological waste from the waste stream, was a result of the requirements placed on the council by the resource recovery plant and its community, rather than by regulatory requirements.

Education

Perhaps the most significant factor in any program is education. Waste disposal is no exception. There are two populations which need to be involved at every phase of the program: the community and the facility itself. Including the community in the program is of paramount importance. In some states, it is a required step, depending on the size of the project. However, it is important to educate the community as to the extent of the problems facing the facilities (economic and regulatory) as well as the reasons behind them, which is not an easy task. Many times this is quite an emotional exercise. Despite this, good communication can only help to save future problems associated with a project while developing a positive relationship between the facility and the community. Once the problems are clearly defined for both the facility and community, an acceptable plan to all concerned parties can be developed.

Education within the facility is just as important. Administration must be informed of the problems of solid waste disposal as well as those workers who generate and handle the waste. Having administrative support for the programs is a requirement for success; the plan will surely fail if it is not supported at this level. Employee understanding of the problem, employee input into the development of the program, and employee understanding of what will be required of them must all be clearly defined. It is wise to include members of employee groups directly affected by the plan while it is being developed, such as individual generators (nurses, doctors, laboratory technicians), intermediate waste handlers (housekeeping and maintenance staff), safety committee or staff, disposal equipment operators (incinerator and autoclave operators), administration, and risk management (or equivalent). This integration will address some of the issues during the design phase of the plan and garner cooperation needed for the success of the plan. As with any work environment, health and safety issues need to also be considered as part of the overall education program. With medical waste disposal, education also will include such specific topics as proper source separation, regulatory and institutional requirements, and equipment operation. An educated staff will have less accidents and incidents while assuring that proper procedures are followed for the implementation and subsequent success of the program.

There are many factors which must be addressed in order to determine the best course of action for a generator of infectious waste. Regulatory review, waste characterization, on and off-site disposal options and their respective requirements need to be evaluated by each generator in order to develop a comprehensive waste disposal program which manages waste in a safe and effective manner, while limiting costs, liability, and environmental consequences. Some regions have pooled their resources in order to address the problem in a more efficient manner and created regional plans and facilities for medical waste disposal. In any program, education of the community and staff of a facility and understanding facts surrounding the problems and the solutions are of paramount importance in the development, implementation, and observance of the program.

SUMMARY

Medical waste disposal has increased significantly over the last 10 years as technology has developed newer, safer, cleaner, and less expensive disposable items and equipment for patient care. Along with the concerns about the spread of HIV and beach wash-ups of wastes on some shores, medical waste has become a controversial social, political, and economic problem facing the nation today. As regulations are promulgated, as waste definitions evolve, as disposal techniques become more restrictive, and as technology continues and community concerns are heightened, the problems for the disposal of medical waste will continue to plague the healthcare industry. Regulatory requirements vary from state to state (as do accepted disposal practices, largely based on esthetics and community fears, rather than scientific fact). As the MWTA came to a close, many of these issues were highlighted. In all probability, though, further federal regulations will be forthcoming. Hopefully, these will reflect the facts surrounding these issues as characterized during this demonstration project.

Understanding these problems, industry continues to make advances in the field of medical waste disposal to handle the demand placed upon them. Incineration, steam and chemical sterilization, and other forms of treatment are currently available to generators for the required disposal of this material. Emerging technologies, such as pyrolysis, plasma sterilization, and encapsulation may become viable alternatives to the current accepted practices which may be, among other things, of community concern or economically prohibitive. Emission controls, ash removal, and limited landfill space are also concerns which may limit the disposal methods used for medical waste. Though generators can choose from these technologies, a plan to address the entire program must be developed. By understanding the scope of the problem, particularly including regulatory requirements, and available technologies, combined with waste generation data, economics, and community interaction, a suitable program can be developed.

The three types of medical waste plans discussed in this chapter were on-site, off-site, and regional. Each plan has its own set of advantages and disadvantages which need to be assessed by the individual generator and/or the region as a whole. Education and communication on all levels is required in order to minimize problems and expenses down the road. In addition, the application of reusables and the promotion of recycling, whether before or after treatment, is as important as disposal itself. Minimizing the waste stream will not only decrease all aspects of waste costs, but would set an example for other industries to follow and return the healthcare industry to a less suspicious position within the community.

REFERENCES

1. "Disposal of Solid Wastes from Hospitals," Bacterial Diseases Division, Bureau of Epidemiology, Centers for Disease Control, Atlanta, GA (May 1980).
2. Garner, J. S. and M. S. Favero. "Guideline for Handwashing and Hospital Environmental Control," Section 4: Infective Waste, Hospital Infections Program, Center for Infectious Diseases, Centers for Disease Control, Public Health Service, U.S. Department of Health and Human Services, Atlanta, GA (1985).
3. *Draft Manual for Infectious Waste Management,* U.S. Environmental Protection Agency, Office of Solid Waste, Washington, D.C. (September 1982), pp. 2-5 to 2-15.
4. *EPA Guide for Infectious Waste Management,* U.S. Environmental Protection Agency, Office of Solid Waste, Washington, D.C. (May 1986), pp. 2.1 to 2.5.
5. Greene, W. H. and J. T. Marchese. "Regulated Medical Waste (RMW) Disposal at a University and University Hospital: Future Implications," presented at the *Third International Congress on Nosocomial Infections,* Atlanta, GA (August 1990).

6. *Medical Waste Tracking Act,* Background, 54 FR (March 24, 1989), p. 12328.
7. Kalnowski, G., H. Weigand, and H. Ruden. "On The Microbial Contamination of Hospital Waste," *Zbl. Bakt. Mikrobiol. Hyg.* (1983), p. 1373–1382.
8. *The Public Health Implications of Medical Waste: A Report to Congress,* U.S. Department of Health and Human Services, Public Health Service, Agency for the Toxic Substances and Disease Registry, Atlanta, GA (September 1990), p. 9-4.
9. *Medical Waste Tracking Act,* Background, 54 FR (March 24, 1989), p. 12328.
10. *Medical Waste: Fact or Fiction,* Chicago: American Hospital Association (August 1991).
11. *The Public Health Implications of Medical Waste: A Report to Congress,* U.S. Department of Health and Human Services, Public Health Service, Agency for the Toxic Substances and Disease Registry, Atlanta, GA (September 1990), p. 8-2.
12. *The Public Health Implications of Medical Waste: A Report to Congress,* U.S. Department of Health and Human Services, Public Health Service, Agency for the Toxic Substances and Disease Registry, Atlanta, GA (September 1990), pp. 10-1 to 10-5.
13. "Update: Whither Medical Waste," *Healthcare Hazardous Materials Management,* ECRI, Vol. 4, No. 11 (August 1991), pp. 5–6.
14. Sedor, P. M., "Costs Soar Under the EPA's Waste-Tracking Program," *Health Facilities Management,* (June 1990), p. 62, 64, and 66.
15. *The Public Health Implications of Medical Waste: A Report to Congress,* U.S. Department of Health and Human Services, Public Health Service, Agency for the Toxic Substances and Disease Registry, Atlanta, GA (September 1990), p. E-1.
16. Corbus, D. *Medical Waste Incinerator Emissions and Air Pollution Control,* U.S. EPA Medical and Institutional Waste Incinerator Seminar Workbook, CERT 89-247 (1988–89), p. 3.
17. Cross, F. L., H. E. Hesketh, and P. K. Rykowski. *Infectious Waste Management,* Lancaster, PA: Technomic Publishing (1990), p. 25.
18. Doucet, L. G. *Update of Alternative and Emerging Medical Waste Treatment Technologies* (June 1991), p. 17.
19. "And Now, Building Blocks: New Use Found for Medical Waste," *Infectious Waste News,* Vol. 6, No. 9 (April 29, 1991), p. 1.
20. Spurgin, R. "4 Medical Waste Treatment Technologies," *Medical Waste Management: Recycling and New Technologies; Plant, Technology and Safety Management Series,* The Joint Commission on Accreditation of Healthcare Organizations, No. 2 (1991), p. 16.
21. "University of Georgia Proposing Creative Alternative to Incineration," *Infectious Waste News,* Vol. 6, No. 8 (April 15, 1991), p. 1.
22. Cross, F. L., H. E. Hesketh, and P. K. Rykowski. *Infectious Waste Management,* Lancaster, PA: Technomic Publishing (1990), p. 35.
23. "Long Island Hospitals Band Together To Reduce Incineration, Meets New Regs," *Infectious Waste News,* Vol. 6, No. 22 (October 28, 1991), p. 1.

GLOSSARY

GLOSSARY

Terminology specific to each chapter can also be found explained in the references for that chapter. Gossary terms can have multiple meanings but are defined here only in the context of health and safety in the work environment. Terms listed in **boldface** are defined elsewhere in the glossary.

A

AAIH. American Academy of Industrial Hygiene, a professional society of **CIHs** (Certified Industrial Hygienists; see **ABIH**).

Abatement. The elimination of a potential or actual hazard.

ABIH. American Board of Industrial Hygiene, a professional industrial hygiene accreditation organization, administering the **CIH** and **IHIT** program. ABIH, 4600 W. Saginaw, Suite 101, Lansing, MI, 48917 (517) 321-2638.

Abscesses. A localized collection of pus which is the result of the disintegration or displacement of tissue.

Absorbed Dose. The amount of a substance entering the worker's body after exposure.

Absorption. To take in a substance across the exchange boundaries of an organism (skin, lungs, or gastrointestinal tract).

Accuracy. The measure of the correctness of data, often defined as the difference between the measured value and the ''true'' value.

ACGIH. The American Conference of Governmental Industrial Hygienists. An organization of individuals employed by governmental agencies or educational institutions and engaged in occupational safety and health programs. The ACGIH develops recommended occupational exposure limits for chemical and physical **agents** (see **TLV**). ACGIH, 6500 Glenway Ave., Building D-7, Cincinnati, OH, 45211 (513) 661-7881.

Acid. An inorganic or organic compound that reacts with metals to yield hydrogen, reacts with a base to form a salt, dissociates in water to yield hydrogen or hydronium ions, has a pH of less than 7, and neutralizes bases or alakaline media. Acids turn litmus paper red and are corrosive to human tissue and are to be handled with care. See also **Base, pH**.

Acid-fast bacilli. See **AFB**.

Acquired Immune Deficiency Syndrome. See **AIDS**.

Actinic Burns. Burns caused by ultraviolet light or too much sun.

Action Level. The exposure level at which OSHA regulations to protect employees take effect (29 CFR 1910.1001-1047). Generally, once an action level has been reached, the employer must initiate workplace air or noise analysis, employee training, medical monitoring, and record keeping.

Actuator. A device used to automatically adjust the position of the **damper**; it can be mechanical, electrical, pneumatic, or hydraulic.

Acute. Having a sudden onset, severe symptoms, and/or a short duration.

Acute Exposure. An exposure that occurs over a short period of time and often at high levels.

Acute Health Effect. In this context, an adverse health effect results from one high exposure, with severe symptoms developing rapidly and coming quickly to a crisis. Contrast with **Chronic Health Effect**.

Acute Toxicity. The adverse toxic effects resulting from a single exposure to a material.

Administrative Controls. Approaches for reducing worker exposure other than through the use of engineering controls or personal protective equipment (PPE), for example, training or rotation of workers from job to job. Contrast with **Engineering Controls, Personal Hygiene** and **PPE**.

Adsorption. The attachment of molecules (or atoms if elemental) to the surface of another substance.

Aerosol. A fine suspension in air (or any other gas) of liquid (**mist**, **fog**) or solid (**dust**, **fume**, **smoke**) particles which are sufficiently small in size to be stable and remain airborne (and frequently respirable).

AFB. Acid-fast bacilli. Bacteria which retain certain dyes even when washed in an acid solution. A presumptive diagnosis of tuberculosis can be made on this basis, however, the diagnosis is mot confirmed until a culture is grown and identified as *M. tuberculosis*.

Agent. In this context, any chemical substance, physical force, ionizing or nonionizing radiation, biological organism, or other influence that affects the body.

AHERA. Asbestos Hazard Emergency Response Act; legislation to regulate asbestos in schools.

AIDS. Acquired Immune Deficiency Syndrome. An immunodeficiency syndrome caused by the human immunodeficiency virus (**HIV**). This virus permits opportunistic infections, such as tuberculosis and others.

AIDS-Related Complex. See **ARC**.

AIHA. The American Industrial Hygiene Association, a professional industrial hygiene society of over 7000 persons practicing in the science of industrial hygiene. AIHA, P.O. Box 8390, 345 White Pond Dr., Akron, OH, 44320 (216) 873-2442.

Air Flow Rate. See **Flow Rate**.

Air-Foil. In the context of lab hoods, this is a recommended design for the perimeter surfaces around the face of the hood. In general it consists of tapered, smooth, surfaces leading to the hood face (as opposed to squared) so that turbulence within the hood is reduced.

Air-Line Respirators. A respiratory protection system that supplies fresh air to the user through airlines or hoses which tie into an air compressor located in a clean environment.

Air Sampling. In general, air is pumped through various media (Filter, Sorbent and Impinger are examples) such that the airborne contaminants are captured for chemical analysis. If the volume of air pumped is known, the amount of chemical contaminant captured can be expressed as a concentration (**ppm or mg/m^3**).

ALARA. An acronym for As Low As Reasonably Achievable. Usually used in reference to radiation control, ALARA means that every available method of reducing exposure to radiation is used to minimize exposures and risks.

Alkali. Also see **Base**. In general, any compound having highly basic properties (one that readily ionizes in aqueous solution to yield OH anions, with a pH above 7, and that turns litmus paper blue). Alkalies are caustic. Common commercial alkalies are sodium carbonate (soda ash), NaOH, lye, potash, caustic soda, KOH, and bicarbonate of soda. See also **Acid**; **pH**.

Allergic Reaction. An abnormal physiological response to a foreign agent or substance in an individual who is hypersensitive to that substance as a result of prior exposures.

Alveoli. The small air sacs, which lie at the end or the bronchial tree in the lungs, where gas exchange takes place between blood and air.

Ambient Temperature. The temperature of the air around an individual's body.

American Board of Industrial Hygiene. See **ABIH**.

American Conference of Governmental Industrial Hygienists. See **ACGIH**.

American Industrial Hygiene Association. See **AIHA**.

American Lung Association, Lung Disease Care and Education Program, 1740 Broadway, New York, NY, 10019-4374.

American Public Health Association. See **APHA**.

American Society of Heating, Refrigerating, and Air Conditioning. See **ASHRAE**.

Amps. Amperage is the unit of current flow in an electrical circuit.

Analytical Methods. The chemical or physical analytical techniques used to measure the presence or amount of an **agent** in the workplace, usually in the air. Analytical methods can be qualitative or quantitative.

Anemometer. An instrument used to measure air flow velocity.

Anesthetic. Any substance that causes a loss of sensation.

Anhydrous. Without water. A substance in which no water molecules are present in the form of a hydrate or as water of crystallization.

ANSI. American National Standards Institute. A privately funded, voluntary membership organization that identifies industrial and public need for national consensus standards and coordinates their development. Many ANSI standards relate to safe design/performance of equipment and safe practices or procedures. ANSI, 1430 Broadway, New York, NY, 10018 (212) 642-4900.

Antidote. A substance that can counter the effects of a poison.

APCA. The former Air Pollution Control Association; see **AWMA**.

APHA. American Public Health Association. APHA, 1015 15th St., Washington, D.C., 20005 (202) 789-5600.

Aqueous, aq. Describes a water-based solution or suspension or a gaseous compound dissolved in water. Contrast with **anhydrous**.

ARC. AIDS-related complex. A group of symptoms consisting of progressive generalized disease of the lymph nodes, fever, weight loss, and the presence of antibodies to **HIV**.

Area Sample. A sample collected at a fixed point in the workplace. The data from the area sample may or may not correlate with an individual's personal sample results (due to the often high degree of variability in exposures).

Article. Any manufactured item whose function is dependent on its shape or design (for example, pencils or pens). Unless they give off dust or fumes, articles are usually excluded from hazard laws. An article does not release or result in exposure to a hazardous material in normal use.

ASHRAE. American Society of Heating, Refrigerating and Air Conditioning Engineers, a professional society interested in all aspects of heating and cooling of occupied spaces. ASHRAE, 1791 Tullie Circle NE, Atlanta, GA, 30329 (404) 636-8400.

ASP. Associate Safety Professional, indicating partial certification toward the **CSP**. See **BCSP**.

Asphyxiant. A vapor or gas that can cause unconsciousness or death through suffocation (lack of oxygen). Most simple asphyxiants are harmful to the body only when they become so concentrated that they displace the available oxygen in the air (normally about 21%) to dangerous levels (19.5% or lower). Examples of simple asphyxiants are CO_2, N_2, H_2, and He. Other asphyxiants are chemical asphyxiants like carbon monoxide (CO) or cyanide, which reduce the blood's ability to carry oxygen.

ASSE. American Society of Safety Engineers, a professional society of nearly 23,000 members dedicated to the advancement of the safety profession. ASSE, 1800 E. Oakton St., Des Plaines, IL, 60018-2187 (312) 692-4121.

ASTM. The American Society for Testing and Materials. The ASTM is a voluntary membership organization whose members devise consensus standards for materials characterization and use. ASTM, 1916 Race St., Philadelphia, PA, 19103 (215) 299-5400.

Asymptomatic. The lack of identifiable symptoms.

atm. Atmosphere. 1 atm = 14.7 $lb/in.^2$ 1 atm represents the pressure exerted by the air at sea level that will support a column of mercury 760 mm high (29.92 in. Hg) and expressed as 760 mmHg. 1 torr = 1 mmHg.

Autoignition Temperature. In general, the minimum temperature to which a substance (usually in an enclosure) must be heated, without the application of a flame or spark, which will cause that substance to ignite.

AWMA. Air and Waste Management Association, new name for the **APCA**, a professional society interested in environmental pollution issues. AWMA, P.O. Box 2861, Pittsburgh, PA, 15230 (412) 232-3444.

B

Bactericidal. Capable of killing bacteria.

Barrier Effectiveness. The protection provided from a chemical agent by a PPE as a result of the PPE materials thickness, breakthrough time, permeation rate, and degradation.

Base. A substance that liberates OH anions when dissolved in water, receives a hydrogenation from a strong acid to form a weaker acid, and gives up two electrons to an acid. Bases react with acids to form salts and water. Bases have a pH >7 and turn litmus paper blue. They may be corrosive to human tissue and are to be handled with care. See also **Alkali**, **Acid**, **pH**.

BCG. Bacillus of Calmette and Guerin. Vaccine for tuberculosis, widely used in some parts of the world, but of uncertain efficiency. Seldom used in the U.S.

BCSP. Board of Certified Safety Professionals, a professional safety accreditation organization, administering the **CSP** and **OHST** program. BCSP, 208 Burwash Ave., Savoy, IL, 61874-9510 (217) 359-9263.

BEI, Biological Exposure Indexes. Procedures to determine the amount of material absorbed into the human body by measuring it (or its metabolic products) in tissue, fluid, or exhaled air. See the **ACGIH** publication, "Documentation of the Threshold Limit Values and Biological Exposure Indices," for a full explanation.

Bias. A systematic error inherent in a sampling or analytical method.

Bioassay. A test to determine the potency of a substance at producing some adverse health effect on a biological system.

Biohazard. A contraction of the words "biological" and "hazardous," used to describe occupational hazards resulting from exposure to a **biohazardous agent**.

Biohazardous.

Biohazardous Agent. An agent that is biological in nature, capable of self-replication, and has the capacity to produce deleterious effects upon other biological organisms, particularly humans. Biohazardous agents include microorganisms such as bacteria, viruses, fungi, rickettsiae, and chlamydiae. It also includes blood, blood products, and other biological materials which can harbor these organisms.

Biological Agent. Disease-causing organisms which can cause infection if inhaled, ingested, or absorbed through mucous membranes or impaired skin.

Biological Monitoring. Also see **BEI**. Certain chemicals, if absorbed by the human body, can be detected as metabolites in blood, urine, etc. Periodic examination of body substances can determine the extent of a worker's hazardous material absorption or exposure. Occupational exposures to lead can be monitored by performing blood analysis.

Biological Safety Cabinet. Primary barriers used to protect workers and the environment. Consists of three Classes: Class I, Class II (types A and B1, B2, B3), and Class III. Biological Safety Cabinet Classes should not be confused with Classes of Etiologic Agents.

Biological Time Constant. The time required for a portion of an absorbed dose to undergo metabolic changes in the body, for example, the biological half-life would be the time required for one half of an absorbed dose to be biotransformed.

Biosafety Level. A combination of laboratory practices and techniques, safety equipment, and facility design which are appropriate for the operations performed and the risk posed by a **biohazardous agent** and the laboratory function or activity.

Biotechnology. Any technology that uses living organisms (or parts of organisms) to make or modify products, to improve plants or animals, or to develop microorganisms for specific uses. Biotechnology today consists of recombinant DNA technology, monoclonal antibody technology, and bioprocess technology, all of which require manipulation of various organisms.

Blast Gate. A sliding valve located in **ductwork** to regulate flow.

Bloodborne Pathogen. Pathogenic organisms that are present in human blood, or other potentially infectious body fluids, that can cause disease in humans. These pathogens include, but are not limited to, the hepatitis B virus (**HBV**) and the human immunodeficiency virus (**HIV**).

Body Burden. The total amount of a chemical that a person has ingested or inhaled, from all sources, over time. This includes any source of exposure whether occupational or nonoccupational, and is the net sum of all exposures and excretions, etc.

BOHS. British Occupational Hygiene Society, 1 St. Andrew's Place, Regent's Park, London NW1 4LB, England (01-486-4860).

Boiling Point, BP. The temperature at which the vapor pressure of a liquid is equal to the surrounding atmospheric pressure, i.e., the temperature at which it boils. If a chemical is a mixture of different liquids a range of boiling points can be given.

Bonding. A fire safety practice in which two objects (tanks, cylinders, etc.) are fastened together with clamps and bare wire; electrically connecting two metal containers or surfaces with an electrical strap. This equalizes the electrical potential between the metal containers, etc., and helps prevent static sparks that could ignite flammable materials. See also **Grounding**.

BP. See **Boiling Point**.

Breakthrough. The situation in which contaminated air is passed through a sorbent system, yet a portion of the contamination is not captured and passes out of the sorbent system. When this term is used to refer to chemical cartridge respirators, the implication is that contaminated air is allowed into the facepiece of the respirator. Breakthrough may occur when the sorbent system of the respirator has become depleted, or if the type of sorbent system used is not appropriate for protection against the hazardous **agent** in the workroom **air**.

Breakthrough Time. The length of time it takes a measurable amount of a chemical to permeate a protective material under test conditions.

Breathing Zone. The zone of air from which the worker breathes; not rigidly defined but usually understood to be the volume in front of the worker's face, chest, etc.

British Occupational Hygiene Society. See **BOHS**.

Bronchoscope. An **endoscope** designed to passs through the trachea to allow visual inspection of the tracheobronchial tree.

Btu. British thermal unit. The quantity of heat required to raise the temperature of 1 lb of H_2O 1°F at 39.2°F. See also **Calorie**.

Buffer. A substance that reduces the change in hydrogen ion concentration (i.e., change in **pH**) that otherwise would be produced by adding acids or bases to a solution.

Bypass Vent. The vent positioned above the face of a lab hood; the amount of air that moves through this vent is a function of the height of the hood sash. As the sash is lowered, more of the bypass vent is exposed, which allows air to pass through it rather than through the face of the hood.

C

C. With respect to exposures used by **OSHA** and **ACGIH** to designate ceiling exposure limit. Less commonly, "C" can also mean continuous exposure if used in the expression of toxicological data, for example, "$LD_{50} > 10$ g/kg, 24 H-C" would mean that 10 g/kg was the LD_{50} resulting from continuous exposure for 24 hr. See also **Ceiling Limit**; **TLV**.

°C. Degrees centigrade. Degrees Fahrenheit equals (°C × 1.8) + 32. Degrees centigrade equals (°F = 32) × 5/9. See also **°F**.

CAA. The Clean Air Act (EPA). Public Law PL 91-604, 40 CFR 5080. The CAA sets and monitors airborne pollution hazardous to public health or natural resources. Enforcement and issuance of discharge permits are carried out by the states and are called state implementation plans.

CAER. Community Awareness and Emergency Response program of the **Chemical Manufacturers Association**. See **CMA**.

Calcification. A process in which tissue becomes hardened as a result of the deposition of lime salts in the tissue.

Calorie. A standard unit of heat. A calorie is the amount of heat required to raise 1 g of water 1°C. See also **Btu**.

Capture Velocity. Airflow velocity needed to overcome contaminant movement and cross-drafts in order to draw airborne contaminants into an exhaust hood.

Carcinogen. Used in general to refer to a material that has been found to cause cancer. There are different distinctions in types of carcinogens, for example, a chemical could be an **animal carcinogen**, a **suspect human carcinogen**, or a **human carcinogen**. The designations are a function of the available exposure and laboratory test data. In general, a material is considered to be a carcinogen if it has been evaluated by the International Agency for Research on Cancer (**IARC**) and found to be a carcinogen or potential carcinogen, if it is listed as a carcinogen or potential carcinogen in the "Annual Report on Carcinogens," published by the National Toxicology Program (**NTP**) latest edition, if it is regulated by **OSHA** as a carcinogen, or if one positive study has been published.

Carotid Artery. Principal artery of the neck.

Carpal Tunnel Syndrome. A wrist injury caused by repetitive motion, such as frequent, rapid or forceful movements.

CAS Registration Number, CAS, CAS#. CAS stands for Chemical Abstracts Service. CAS numbers identify "specific" chemicals and are assigned sequentially. Chemicals may have more than one name but only one CAS number. For example, methyl ethyl ketone and 2-butanone are different names for the same chemical (and thus both have a CAS# of 78-93-3). The CAS number is useful because it is a concise, unique means of material identification. Chemical Abstracts Service, Division of the American Chemical Society, Box 3012, Columbus, OH, 43210 (614) 421-3600.

Case. An injured or infected (etc.) worker, test subject, or test animal.

Caseation. The process of converting necrotic tissue into a granular, amorphous mass.

Case-Control Study. A type of epidemiologic study design in which a group of individuals with a disease is identified (**cases**) and an appropriate group of individuals without the disease is identified (**controls**) so that the past exposure experiences of the two groups can be compared.

Case Fatality Rate. A measure of the frequency of deaths, due to a particular cause, among all cases.

Case-History. See **Case-Control**.

Caustic. See Alkali.

cc, cm^3. Cubic centimeter.

CDC. Centers for Disease Control, Atlanta GA, 30333.

CEGL. Continuous Exposure Guidance Levels, established by the Committee on Toxicology of the **National Research Council** for the Department of Defense. CEGLs are intended for normal, long lasting military operations. CEGLs are ceiling concentrations that are designed to avoid adverse health effects, either immediate or delayed, for exposure periods of up to 90 days. See also **EEGL** and **SPEGL**.

Ceiling. See **Ceiling Limit**, also see **C**.

Ceiling Limit, C. The concentration that should not be exceeded during any part of the working exposure. See also **PEL-C** and **TLV-C**.

Centers for Disease Control. See **CDC**.

Centigrade. See **°C**.

CERCLA. The Comprehensive Environmental Response, Compensation, and Liability Act, a.k.a. The Superfund Law, Public Law PL 96-510, 40 CFR 300 (EPA). CERCLA is a major environmental legislative action to control hazardous waste sites. It provides for the identification and the cleanup of the hazardous materials that have been released over the land and into the air, waterways, and groundwater. It covers areas affected by newly released materials and older leaking or abandoned dump sites. CERCLA established the Superfund, a trust fund to help pay for the cleanup and sites where hazardous materials have been released. Report releases of hazardous materials to the National Response Center (800) 424-8802; the EPA Superfund Information Hotline is (800) 535-0202.

cfm. Cubic feet per minute or ft^3/min. The volumetric rate of air flow through a fume hood, duct, etc. Compare with fpm.

CFR. Code of Federal Regulations. Regulations established by law. Contact the agency that issued the regulation for details, interpretations, etc. Copies are sold by the Superintendent of Documents, Government Printing Office, Washington, D.C., 20402 (202) 783-3238.

CGA. The Compressed Gas Association, Inc., 500 Fifth Ave., New York, NY, 10036 (212) 354-1130.

Charcoal Tube. A type of **Solid Sorbent Air Sampling** media.

Chemical Asphyxiant. See **Asphyxiant**.

Chemical Cartridge Respirator. A respirator able to purify inhaled air of certain gases and vapors. Contrast with **HEPA**.

Chemical Family. A group of compounds with a common general name; e. g., acetone, methyl ethyl ketone (MEK), and methyl isobutyl ketone (MIBK) are members of the ketone family.

Chemical Formula. Formula for a molecule, i.e., the number and kind of atoms that comprise a molecule of a material. The chemical formula of water is H_2O (each molecule of water is made up of 2 atoms of hydrogen and 1 of oxygen).

Chemical Hygiene Officer. An employee designated by the employer, and who is qualified by training or experience, to provide technical guidance in the development and implementation of the provisions of the **Chemical Hygiene Plan**.

Chemical Hygiene Plan. A written program developed and implemented by the employer which sets forth procedures, equipment, personal protective equipment and work practices that are capable of protecting employees from the health hazards presented by hazardous chemicals used in that particular workplace.

Chemical Manufacturers Association. See **CMA**.

Chemical Name. The scientific designation of a chemical or a name that will clearly identify the chemical for hazard evaluation purposes. Also see **CAS number**.

Chemical Reactivity. Refers to the ability of a material to chemically change. Undesirable and dangerous effects such as heat, explosions or the production of noxious substances can result.

Chemical Transportation Emergency Center. See **CHEMTREC**.

Chemotherapy. The application of chemical reagents that have a toxic effect uponthe disease causing organism.

CHEMTREC. Chemical Transportation Emergency Center. Established by the Chemical Manufacturers Association (CMA) to provide emergency information on materials involved in transportation accidents. Twenty-four-hour number: (800) 424-9300. In Washington, D.C., Alaska, and Hawaii call (202) 483-7616.

Chronic. In this context, anything of long duration, particularly with respect to exposure. Also usually of frequent occurrence and at low levels.

Chronic Health Effect. An adverse health effect as a result of a long or continuous exposure (usually a low level exposure). Contrast with **Acute Health Effect**.

Chronic Toxicity. Chronic effects result from repeated exposures to a material over a relatively prolonged period of time.

CIH. Certified Industrial Hygienist. An individual who has passed the certification requirements of the American Board of Industrial Hygiene (see **ABIH**).

Class 1. Agents of no or minimal hazard under ordinary conditions of handling.

Class 2. Agents of ordinary potential hazard. This class includes agents which may produce disease of varying degrees of severity from accidental inoculation or injection or other means of cutaneous penetration but which are contained by ordinary laboratory techniques.

Class 3. Agents involving special hazard or agents derived from outside the United States which require as federal permit for importation unless they are specified for a higher classification. This class includes pathogens which require special conditions for containment.

Class 4. Agents that require the most stringent conditions for their containment because they are extremely hazardous to laboratory personnel or may cause serious epidemiologic disease. This class includes Class 3 agents from outside the United States when they are employed in entomological experiments or when other entomological experiments are conducted in the same laboratory.

Class 5. Agents which are excluded from the United States specified by law (for example, virus of foot and mouth disease) or USDA administrative policy (African horse sickness).

Clean Air Act. See **CAA**.

CMA. The Chemical Manufacturers Association; an industry group. Chemical Manufacturers Association, 2501 M St., N.W., Washington, D.C., 20037; (202) 887-1100. See also **CAER**.

CNS. Central Nervous System. "CNS" on an **MSDS** indicates effects on the CNS by the material and can include effects such as headache, tremors, drowsiness, convulsions, hypnosis, anesthesia, nervousness, irritability, narcosis, dizziness, fatigue, lethargy, peripheral myopathy, memory loss, impaired concentration, sleep disturbance, etc.

Code of Federal Regulations. See **CFR**.

Coefficient of Variation, CV. The sample standard deviation divided by the mean, often expressed as a percent.

Coefficient of Water/Oil Distribution. Also called the partition coefficient, this is the ratio of the solubility of a chemical in oil to its solubility in water. The coefficient is used to indicate how easily a material can be absorbed into or stored by the human body.

Cohort. A group of persons selected for a study.

Cohort Study. A type of epidemiologic study design in which disease-free individuals are followed over time and comparisons are made between the exposure experiences of those who subsequently develop disease and those who do not.

Combustible. Used by the **NFPA**, **DOT**, and others to classify on the basis of flash points liquids that will burn. Both NFPA and OSHA define combustible liquids as having a flash point of 100°F (38°C) or higher. Nonliquid materials such as wood and paper are classified as ordinary combustibles by the NFPA. OSHA defines combustible liquid within the Hazard Communication Standard as any liquid having a flash point at or above 100°F (38°C) but below 200°F (93.3°C). Also, any mixture having components with flash points of 200°F (93.3°C) or higher (the total volumes of which make up 99% or more of the total volume of the mixture).

Combustible Liquid. A liquid that has a flash point at or above 140°F.

Common Name. A designation for a material other than its chemical name, such as code name or code number or trade, brand, or generic name.

Communicable Disease. A disease or illness caused by the transmission of an infectious agent, or its toxic products, from a reservoir to a susceptible host.

Compliance Monitoring. Usually, measuring worker exposures for the purpose of evaluating an employer's compliance with governmental standards.

Compressed Gas. A gas (or mixture of gases) in a closed container having an absolute pressure exceeding 40 psi at 70°F (21. 2°C) or having an absolute pressureexceeding 104 psi at 130°F (54.4°C), regardless of the pressure at 70°F (21.2°C); or, a liquid having a vapor pressure exceeding 40 psi or 100°F (38°C) as determined by **ASTM** D-323-72.

Compressed Gas Association. See **CGA**.

conc. Concentration.

Concentration. The quantity of a contaminant per volume of air; see also **ppm** and **mg/m^3**.

Conditioned Air. This is **Replacement Air** which has been heated, cooled, humidified, dehumidified, or otherwise treated in order to make it more comfortable for the occupants (or to meet other needs of processes or systems within the building).

Confidence Interval. For sample data, the range of values about a given number with a probability (for example, 95%) of including the true population value in an underlying distribution.

Confidence Level. The probability that a given confidence interval will include the real population parameter.

Confidence Limits. The upper and lower limits of a confidence interval.

Consumer Product. Any product such as medicines packaged and distributed for retail sale and for consumption by individuals or households. Consumer products are exempt from **HCS** if used in the home or if used in the workplace in the same manner as in normal consumer use; repeated or extensive use would not be exempt.

Consumer Products Safety Commission. See **CPSC**.

Continuous Air Monitors, CAMs. Instrumentation that can be used to continuously measure air concentrations in the workplace (as opposed to area samples and personal samples which are often only taken for a day or part of a day at a time).

Contraindications. Something, such as a symptom or condition, which makes a specific treatment or procedure inadvisable.

Controls. (1) In general, equipment to reduce the amount of contaminant released into the workplace air (or to remove it from the workplace air), etc. See **Administrative Controls** and **Engineering Controls**. (2) The persons or animals (in a comparison group) that differs from the **Cases** with respect to exposure, disease, or experience, etc.

Corrosive. A chemical that causes visible destruction of living tissue or irreversible alterations in living tissue by chemical action at the site of contact. Any liquid that causes a severe corrosion rate in steel. Any waste that exhibits a characteristic of corrosivity (40 CFR 261.22) as defined by **RCRA**, may be regulated by EPA as a hazardous waste.

CPSC. Consumer Products Safety Commission, a federal agency responsible for regulating hazardous materials when they are used in consumer goods (per the Hazardous Substances Act and Poison Prevention Packaging Act of 1970).

Criteria Document. In-depth toxicological review of a hazardous chemical or physical **agent**, published by **NIOSH**.

Critical Items. Medical equipment such as needles, surgical instruments, and cardiac catheters or any appliance which would invade the body. See **semi-critical** and **noncritical**.

Critical Pressure/Critical Temperature. A temperature above which a gas cannot be liquefied by pressure. The critical pressure is that pressure required to liquefy a gas at its critical temperature.

Cross-Sectional Study. A type of epidemiologic study design in which a group of individuals are selected at a single point in time and comparisons are made between the exposure experiences of those who have the disease and those who do not.

Cryogenic. Extremely low temperature (such as for refrigerated gases); gas having a boiling point below -130°F (-90°C).

CSP. A Certified Safety Professional, as certified by the **BCSP**.

Cutaneous. Pertaining to the skin.

CVS. Cardiovascular effects.

Cyclone. An **Air Sampling** device used to collect the **Respirable Fraction** of an airborne particulate.

Cylinder. In this context, a container of various sizes used for the storage of compressed gas. See **CGA**.

D

Damper. An in-line valve (in the ductwork) that controls the flow of air. There are numerous damper designs.

Dangerously Reactive Material. A material that can react by itself or with air or water to produce a hazardous condition. Preventive measures can be taken if the conditions that can cause the dangerous reaction are known.

Dec, Decomp; Decompose, Decomposition. The breakdown of a material (by heat, chemical reaction, electrolysis, decay, or other processes).

Delivered Dose. Refers to the amount of a substance available for interaction in a particular organ or cell.

Demulcent. A material capable of soothing or protecting inflamed, irritated mucous membranes.

Density. Ratio of weight (mass) to volume of a material, usually expressed in grams per cubic centimeter. One cc of H_2O weighs 1 g/cm^3. See also **Specific Gravity**.

Dermal. Pertaining to the skin.

Dermal Toxicity. Adverse effects resulting from the skin's exposure to a material.

Dermatitis. Inflammation of the skin.

Descriptive Statistics. Simple statistical ways to characterize the exposure data obtained in a sample (for example, mean, standard deviation, etc.).

Dewpoint Temperature. The temperature at which water vapor in the air first starts to condense.

Diffusion. Dispersion of sound within a space so that there is uniform energy density throughout the space.

Dilution Ventilation. Ventilation designed to control airborne contaminants by diluting concentrations to acceptable levels. See also **General Ventilation**.

Disease. A deterioration of the normal, healthy health status. Also see **Occupational Disease**.

Disinfection. The destruction of pathogenic microorganisms by chemical or physical methods; a disinfectant is a substance that prevents infection by killing bacteria. Also see **Sterilization**.

Dispensing Drum. Large drum containing chemicals used as a reservoir to fill portable containers. Usually 55 gal drums are fitted with spigots and set on stands horizontally to fill smaller containers.

Dose. The amount of a substance that enters the worker's body; see **absorbed dose** and **delivered dose**. Frequently, **dose** is used to mean the same as absorbed dose.

Dose Rate. The dose per unit time.

Dose-Response. The relationship between dose and health effect (sometimes between exposure and health effect). In general, the curve that results when dose (X-axis) is plotted against response (Y-axis).

Dosimeter. In general, a dosimeter is any individually worn device for measuring exposure to chemical or physical **agents**. An example would be device worn on the person for determining the accumulated sound exposure with regard to level and time.

DOT. U.S. Department of Transportation. Regulates the transportation of materials in order to protect the public as well as fire, law, and other emergency-response personnel. DOT classifications specify appropriate warnings and labels that must be used. DOT Locator, 400 7th St., S.W., Washington, D.C., 20590 (202) 366-4000.

Droplet Nuclei. The microscopic, airborne particles of aerosolized sputum which can carry tubercle bacilli. Droplet nuclei are respirable. See **aerosols**.

Ductwork. A closed pathway for ventilation airflow that connects system components from intake to discharge.

Dust. Solid particles suspended in air (usually produced by some mechanical process such as crushing, grinding, abrading, or blasting). Most dusts are hazardous with respect to inhalation, fire, and dust explosion.

E

Ear Protector. A device worn to reduce the passage of ambient noise into the auditory system. Earplugs are inserted in the external ear canal. Earmuffs fit over the entire ear and snug against the head.

EC_{50}. Median effective concentration. The concentration of a material in water that has been observed to cause a biological effect on 50% of a group of test animals.

EEGL. Emergency Exposure Guidance Level, established by the Committee on Toxicology of the **National Research Council** for the Department of Defense. EEGLs provide guidelines for military personnel operating under emergency conditions whose circumstances are peculiar to military operations. The EEGL is a ceiling guidance level for single emergency exposures usually lasting 1 to 24 h — an occurrence expected to be infrequent in the lifetime of an individual. See also **CEGL** and **SPEGL**.

Effective Exposure. With respect to respirators, the concentration of a contaminant inside the **respirator**.

EHS. Extremely Hazardous Substance. A substance determined by the U.S. EPA to be so potentially hazardous to life and health that local planning committees are required to be notified if an EHS is present in a **facility**. See **SARA**.

Electrolyte. A nonmetallic substance that conducts an electric current in solution by the movement of ions.

Embryotoxin. A material harmful to a developing embryo.

Emergency Response. Mitigation of a spill or leak of a hazardous material.

Emergency Response Planning Guideline. See **ERPG**.

Emetic. An **agent** that induces vomiting.

Endemic. The presence of a disease or infectious agent within a specific geographic area.

Endoscope. A device consisting of a tube and an optical system, used to observe theinside of an organ or a cavity.

Engineering Controls. Environmental, mechanical, or structural factors that serve to encourage, facilitate, or complement safe and healthful behaviors in the worksite. Engineering controls work by removing the worker from the hazard or by removing the hazard from the work environment (using ventilation in particular). Compare with **Administrative Control**, **Personal Hygiene**, **PPE**, **Ventilation**.

EPA, U.S. Environmental Protection Agency A federal agency with environmental protection authority (regulatory and enforcement). EPA, 400 M St., S.W., Washington, D.C., 20460, Public Information (202) 260-2090.

EPCRA. Emergency Planning and Community Right-to-Know Act. The specific title for **SARA Title III**.

Epidemic. An illness or illnesses that occur at a greater rate than normally expected and prevalent in a particular community or region.

Epidemiology. The study of disease occurrence in populations with the aim of relating disease occurrence to characteristics of people and their environment.

Epithelial. Pertaining to the layers of cells forming the epidermis of the skin and the surface layer of mucous and serous membranes.

ERC. Educational Resource Center. ERCs are **NIOSH**-funded interdisciplinary graduate school occupational safety and health training programs.

Ergonomics. The concept that human bodies are components of the workplace, with their own limitations. The study of human characteristics for the appropriate design of living and work environments.

ERPG. Emergency Response Planning Guidelines. An **AIHA** exposure guideline dealing with nonoccupational community exposures to chemicals (i.e., resulting from chemical spills due to accidents, etc.). ERPGs are short-term guidelines for exposure to chemicals during spills or emergencies. ERPGs provide concentration levels at which adverse health effects are expected for the general public. ERPGs are defined as follows:

ERPG-1 is the maximum airborne concentration below which it is believed that nearly all individuals could be exposed for up to 1 h without experiencing other than mild transient health effects or perceiving a clearly defined objectionable odor.

ERPG-2 is the maximum airborne concentration below which it is believed that nearly all individuals could be exposed for up to 1 hr without experiencing or developing irreversible or other serious health effects or symptoms that could impair an individual's ability to take corrective action.

ERPG-3 is the maximum airborne concentration below which it is believed that nearly all individuals could be exposed for up to 1 hr without experiencing or developing life-threatening effects.

Etiologic Agents, Classification of. Human etiologic agents are placed in four classes of increasing hazard. A fifth class is composed of animal agents excluded from the U.S. by law and by U.S. Department of Agriculture administrative policy. For the basis for agent classification, see Class 1, etc.

Etiology. The causes and origins of disease(s).

Evaporation Rate. The rate at which a material will vaporize, volatilize, or evaporate from the liquid or solid state (compared to the rate of vaporization of a known, baseline material). The known material is usually normal butyl acetate, with an evaporization rate of 1.0. Faster evaporization rates are >1 (slower, <1).

Evaporative Heat Loss. The loss of body heat to the environment through the evaporation of sweat.

Excursion Limit. Usually short duration exposure limits (such as **STELs** or **Cs**). The **ACGIH** recommends that excursions in worker exposure levels may exceed 3 times the **TLV-TWA** for no more than a total of 30 minutes during a workday, and under no circumstances should they exceed 5 times the **TLV-TWA**, provided that the **TLV-TWA** is not exceeded.

Explosive. A material that explodes when subjected to abrupt shock, pressure, or high temperature.

Explosive Limits. See **Flammable Limits**.

Exposure. A worker's contact with agents in the workplace. Exposure can be reduced if **engineering controls** are effectively in place and the worker wears the proper **PPE**.

Exposure Assessment. The study and measurement of exposure, usually including its magnitude, frequency, duration, and route.

Exposure Group. See **Homogeneous Exposure Group**.

Exposure Limits. The concentration of a chemical in the workplace or community air that is thought to be safe. See **PEL**, **TLV**, and **ERPG**.

Exposure Route. The manner by which an **agent** enters the body, for example, inhalation, ingestion, or skin absorption.

Extinguishing Media, Agents. The type of fire extinguisher or extinguishing method appropriate for use on a specific material. Some chemicals react violently to the presence of water, so other methods such as the use of foam or CO_2 should be followed.

Extremely Hazardous Substance List. See also **EHS**. Any of approximately 400 chemicals listed under **SARA Title III** as extremely hazardous.

Eye Protection. In general, anything worn to protect the eyes (safety glasses, goggles, face shields, etc.).

F

°F or F. Degrees Fahrenheit. 32°F (0°C) equals the freezing point of water. 212°F (100°C) equals the boiling point of water. Degrees Fahrenheit equals $(°C \times 1.8) + 32$. Degrees centigrade equals $(°F - 32) \times 0.56$.

Face Velocity. The average velocity across an opening (face) of a local exhaust hood; a measurement of the speed at which the air enters the face of the lab hood, usually expressed in feet per minute (fpm) or meters per second (m/s).

Facility. As defined for **SARA Title III**, any building, equipment, structure, and/or other stationary items located on single, adjacent, or contiguous sites.

Fan. A mechanical device that rotates to cause a pressure differential and resulting air movement.

f/cc. Fibers per cubic centimeter of air. A unit of concentrations used to express airborne concentrations of fibers such as asbestos.

Federal Emergency Management Agency. See **FEMA**.

Federal Register. See **FR**, also **CFR**.

FEMA. The Federal Emergency Management Agency, 500 C St., S.W., Washington, D.C., 20472 (202) 646-4600.

Fibrillation. A condition of the heart when the two halves are beating out of synchronization and are not pumping blood. The heart can be shocked into this condition from contact with electrical current.

Fibrosis. The formation of fibrous tissue in excess of amounts normally present in the lung tissue walls. This reduces the oxygen and CO_2 exchange efficiency.

Filter or Filter Collection. In this context, using a filter to collect airborne particulates so that they can be chemically analyzed.

Fire Diamond (NFPA). The fire diamond has four classes of entries by position:

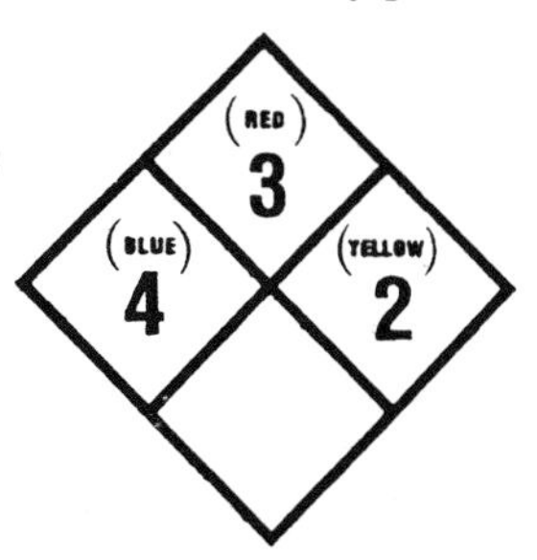

Position A—Health Hazard (Blue):
0 = Ordinary Combustible Hazards in a Fire
1 = Slightly Hazardous
2 = Hazardous
3 = Extreme Danger
4 = Deadly

Position B—Flammability (Red):
0 = Will Not Burn
1 = Will Ignite if Preheated
2 = Will Ignite if Moderately Heated
3 = Will Ignite at Most Ambient Conditions
4 = Burns Readily at Ambient Conditions

Position C—Reactivity (Yellow):
0 = Stable and Not Reactive with Water
1 = Unstable if Heated
2 = Violent Chemical Change
3 = Shock and Heat May Detonate
4 = May Detonate

Position D—Specific Hazard (White):

OXY	= Oxidizer
ACID	= Acid
ALKALI	= Alkali
COR	= Corrosive
~~W~~	= Use No Water
☢	= Radiation Hazard

First Responder Awareness Level. Defined in 29 CFR 1910.120. Individuals who are likely to witness or discover a hazardous substance release and who have been trained to *initiate* an emergency response sequence by *notifying* the proper authorities of the release.

First Responder Operations Level. Individuals who will respond to the hazardous materials release in a defensive fashion from a distance, however, they will *not* be trained to plug or patch or otherwise repair the cause of the release. Defined in 29 CFR 1910.120.

Fit Test. Refers to the testing of respirators (with respect to how a particular respirator fits particular worker's face). See **Qualitative** and **Quantitative Fit Testing**.

Flammable. Describes any solid, liquid, vapor, or gas that will ignite easily and burn rapidly. See also **Combustible**.

Flammable Aerosol. An aerosol is considered to be a flammable aerosol if it is packaged in an aerosol container and can release a flammable material.

Flammable Gas. A gas which, at ambient temperature and pressure, forms a flammable mixture in air at a concentration of 13% by volume or less, or a gas which at ambient temperature and pressure forms a range of flammable mixtures in air greater than 12% by volume, regardless of the lower limit.

Flammable Limits (Flammability Limits). The minimum and maximum concentrations of a flammable gas or vapor at which ignition can occur. Concentrations below the lower flammable limit (LFL) are too lean to burn, while concentrations above the upper flammable limit (UFL) are too rich. All concentrations between LFL and UFL are in the flammable range, and special precautions are needed to prevent ignition or explosion.

Flammable Liquid. A liquid that gives off vapors that can be readily ignited at room temperature. Defined by the **NFPA** and **DOT** as a liquid with a flash point below 100°F (38°C).

Flammable Solid. A solid that will ignite readily and continue to burn.

Flash Back. This takes place when a trail of flammable material is ignited by a distant spark or ignition source (the flame then travels along the trail of the material back to its source).

Flash Point, FP. The lowest temperature at which a flammable liquid gives off sufficient vapor to form an ignitable mixture with air near its surface or within a vessel.

Flow Rate. Generally used to mean the rate of air flow through an air sampling pump; usually expressed in l/min. Also can refer to the volumetric rate of air flow through a ventilation system.

Flux. Rate of transfer of fluid, particles, or energy across a given surface.

Foam. A fire-fighting material consisting of small bubbles of air, water, and concentrating agents. Foam can put out a fire by blanketing it, excluding air and blocking the escape of volatile vapor.

Fog. A visible suspension of fine droplets in a gas; e.g., water in air.

Fomite. An inanimate vector of disease (a comb is a fomite).

Form R. An EPA report required under **SARA Title III** for annual reporting of releases of toxic chemicals from an industrial facility.

Formula Weight. See **Molecular Weight**.

FP. See **Flash Point**.

fpm. Feet per minute. Used to measure and express the velocity of air flow through a fume hood, duct, etc. Contrast with cfm.

FR. The ''Federal Register.'' A daily publication that lists and discusses daily governmental business including regulations of federal agencies. Compare to **CFR**.

Freezing Point. The temperature at which a material changes its physical state from liquid to solid.

Friction. The resistance to movement between two surfaces in contact, for example, resistance to air movement over surfaces (e.g., **ductwork**) which causes pressure loss.

Full Protective Clothing. Fully protective gear that keeps gases, vapor, liquids, and solids (dusts, etc.) away from any contact with skin and prevents them from being inhaled or ingested. Includes **SCBA** (self-contained breathing apparatus).

Fume. Fume is the particulate that is formed by molten metals or molten plastics. It is made up of minute solid particles suspended in air. This heating is often accompanied by a chemical reaction where the particles react with oxygen to form an oxide.

G

g or gm. Gram.

Gas. A formless fluid that occupies the extent of the volume which contains it. It can be changed to its liquid or solid state only by increased pressure and decreased temperature.

General Exhaust. See **General Ventilation**.

General Ventilation. Also known as dilution ventilation. Both terms refer to the removal of contaminated air and its replacement with clean air from the general workplace. Contrast with **local ventilation**.

Generic Name. An appellation such as a code name, code number, trade name, or brand name used to identify a chemical by other than its chemical name.

Granulomatous. Containing granular tumors (i.e., having roughened surfaces) or growths.

GRAS. Generally Recognized As Safe (a phrase applied to food additives approved by the FDA).

Ground Fault Circuit Interrupters (GFCI). An electrical device that opens a circuit when current leakage is sensed by the device, thus preventing electrical shocks to workers.

Grounding. A fire safety practice to conduct any electrical charge to the ground, preventing sparks that could ignite flammable materials. Electrically connecting a metal container to the building ground with an electrical strap. See **Bonding**.

H

Haber's Law. The Haber relationship expresses the constancy of the product of exposure concentration and exposure duration (Ct = K, where C represents exposure concentration, t time, and K is constant). The Haber relationship does not holdover more than small differences in exposure time.

Hand Protection. PPE worn to protect the hands (such as gloves).

Hazard. The probability that an adverse health effect will occur if a substance is used in a specified quantity and under a specified set of conditions. Compare with Safety.

Hazard Assessment (also Hazard Identification). The process of determining what hazards, if any, are associated with exposure to a particular chemical.

Hazard Communication Standard. Requires chemical manufacturers and importers to assess the hazards associated with the materials in their workplace (29 CFR 1910.1200). **Material safety data sheets**, **labeling**, and training are all the result of this law.

Hazardous Decomposition. The breakdown or separation of a substance into its constituent parts (sometimes accompanied by the release of heat, gas, or hazardous materials). Some materials can give off hazardous materials when they burn or decompose.

Hazardous Material. In general, any chemical or mixture having properties capable of producing adverse effects on the health or safety of a human. As defined by **OSHA** with respect to Worker Right-to-Know (Hazard Communication Standard), the term Hazardous Material means a chemical or mixture having one or more of the following characteristics:

- Has a flash point below 140°F, closed cup, or is subject to spontaneous heating
- Has a threshold limit value below 500 ppm for gases and vapors, below 500 mg/m^3 for fumes, and below 25 million particles per cubic foot for dusts
- Has a single dose oral LD_{50} below 50 mg/kg
- Is subject to polymerization with the release of large amounts of energy

- Is a strong oxidizing or reducing **agent**

- Causes first degree burns to skin resulting from a short time exposure, or is systemically toxic by skin contact

- In the course of normal operations, may produce dusts, gases, fumes, vapors, mists, or smokes which have one or more of the above characteristics.

Hazardous Materials Specialists. Individuals trained to act in technical support roles. Defined in 29 CFR 1910.120.

Hazardous Materials Technician. Individuals trained to respond to hazardous materials releases in both a defensive fashion *and* by trying to plug, patch or otherwise repair the cause of the release. Defined in 29 CFR 1910.120.

Hazardous Waste Number. An identification number assigned by the EPA per the **RCRA** law to identify and track wastes.

HazMat. (1) Usually used in reference to any hazardous material. (2) In reference to an emergency response spill control team, program, etc.

HAZWOPER. An term referring to workers regulated under 29 CFR 1910.120 and including emergency responders and workers routinely employed at hazardous waste sites, treatment storage and disposal facilities, among others.

HBV. Hepatitis B. Viral inflammation of the liver.

HCS. See **Hazard Communication Standard**.

Health Education. Any combination of learning methods or activities designed to facilitate voluntary adaptation of behavior conducive to health.

Health Effect. Any effect on human or test animal health as a result of exposure to an **agent**.

Health Promotion. Any combination of health education activities related to organizational, economic, or environmental supports for behavior conducive to health.

Health Risk Appraisal. See **HRA**.

Healthy Worker Effect. A phenomenon in which employed workers appear healthier than the general population due to the fact that ill and disabled individuals are excluded from the workforce. Because of the healthy worker effect, it may be inappropriate in epidemiologic studies to compare the disease experience of workers with that of the general population.

HEG. See **Homogeneous Exposure Group**.

Hematopoietic. Refers to having an effect on the blood cell-forming system of the biological organism.

HEPA. High Efficiency Particulate Air filter. A filter capable of removing very small particles from the airstream. Specifically, it is a filter that will trap particulate material in the size of 0.3 micron (or greater) from the air with a minimum efficiency of 99.97%.

Hepatic. Pertaining to the liver.

Hepatitis B. See **HBV**.

Hepatotoxin. Destructive to the liver.

High Efficiency Particulate Air Filter. See **HEPA**.

Highly Communicable Diseases. Those diseases listed as a Class IV biological agent by the **CDC**.

HIV. Human immunodeficiency virus. A virus which can make the T4 lymphocytes ineffective, leading to opportunistic infections, malignancies, and neurological disease. See **AIDS**.

Homogeneous Exposure Group, HEG. In general, a group of employees who are defined as having the same potential for the same type and degree of exposure. For example, in a factory spray painters may make up one HEG while typists would make up another.

Hood. Inlet to a local exhaust ventilation system specifically designed to capture, enclose, or receive airborne contaminants.

Hood Entry Loss. Loss in static pressure associated with acceleration of airflow and resistance by friction and turbulence.

Household Products Disposal Council, HPDC. 1201 Connecticut Ave., N.W., Washington, D.C., 20036 (202) 659-5535.

HPDC. See **Household Products Disposal Council**.

HRA. Health Risk Appraisal; self-assessment in employee promotion programs.

Humidity. See **Relative Humidity**.

Human Immunodeficiency Virus. See **HIV**.

Hygienic Guides. AIHA Hygienic Guides are publications containing useful industrial hygiene information for many common chemicals.

Hygroscopic. Readily adsorbing moisture from the air.

Hyperbolic. Self-igniting upon contact of its components (without a spark or external aid).

Hyperpyrexia. A body core temperature greater than 104°F (40°C).

Hypothalamus. A part of the brain that activates, controls, and integrates the-peripheral autonomic mechanisms, endocrine activity, and some somatic functions, e.g., regulation of water balance, body temperature, sleep, and food intake, and the development of secondary sex characteristics.

I

I. Intermittent.

IAFF. International Association of Fire Fighters, 1750 New York Ave. N.W., Washington, D.C., 20006 (202) 737-8484.

IARC. International Agency for Research on Cancer. World Health Organization, Geneva, Switzerland; distributed in the U.S. from 49 Sheridan Ave., Albany, NY, 12210 (518) 436-9686.

IDLH. Immediately Dangerous to Life and Health. For certain chemicals, the Standards Completion Program of **NIOSH** has recommended IDLH values. For the chemicals for which they exist, IDLH values are defined only for the purpose of respirator selection and represent the maximum concentration from which, in the event of a respirator failure, one could escape in 30 min without experiencing any escape-impairing symptoms or any irreversible health effects. IDLH values are not to be used for evaluating routine occupational exposures.

IES. Illuminating Engineering Society, 345 E. 47th St., New York, NY, 10017 (212) 705-7926.

Ignition Temperature. The lowest temperature at which a combustible material will catch on fire in air and will continue to burn (independent of the original source of heat).

IHIT. Industrial Hygienist-in-Training, indicating partial certification toward the **CIH**.

Ihl. Inhalation.

Illuminating Engineering Society. See **IES**.

IMDG Code, IMO Classification. The International Maritime Dangerous Goods code classifies materials in shipment (explosives, flammables, oxidizers, poisons, corrosives, and others). International Maritime Organization, #4 Albert Embankment, London, SE 175R, U.K.

Immediately Dangerous to Life and Health. See **IDLH**.

Immunodeficiency. See **HIV**.

Impervious. A material that does not allow another substance to penetrate or pass through it.

Impinger or **Impinger Sampling.** An impinger is a device carrying a liquid and through which the workplace air is bubbled; airborne gases will become dissolved in the liquid if it has been properly selected. Chemical analysis can then be done on the liquid to determine the airborne concentration.

Inches of Water. Units of pressure measurement based upon displacement of a water column exposed to a pressure differential.

Incidence. The number of new cases of a disease that occur in a specified population during a specified time period.

Incidence Rate. The number of new cases of disease occurring during a specified period of time divided by the total number of unaffected individuals who are at risk for acquiring the disease at the beginning of the time interval.

Incident Commanders. Persons trained at this level are capable of assuming overall responsibility for the response to the release. Defined in 29 CFR 1910.120.

Incidental Releases. From 29 CFR 1910.120, "Responses to Incidental releases of hazardous substances where the substance can be absorbed, neutralized, or otherwise controlled at the time of the release by employees in the immediate release area, or by maintenance personnel are not considered to be emergency responses within the scope of this standard."

Incompatible. Materials that could cause dangerous reactions if placed in direct contact with one another are said to be incompatible.

Induration. An area of reddening and swelling around the area of a tuberculosis skin test.

Industrial Hygiene. The profession that devotes itself to the recognition, evaluation, and control of environmental factors arising in or from the workplace which may cause impaired health or significant discomfort among workers or residents of the community.

Inert Ingredients. Anything other than the active ingredient in a product; an ingredient not having active properties. Inert ingredients may still be hazardous.

Infectious. Capable of being transmitted with or without contact.

Infectious Waste. A term used to describe a variable definition of **red bag** waste.

Inflammable. Capable of being easily set on fire and continuing to burn.

Inflammation. A series of reactions produced in tissue. Characterized by redness and swelling caused by an influx of blood and fluids.

Ingestion. The taking in of a substance through the mouth.

Inhalation, ihl. Breathing in a substance (in the form of a gas, vapor, fume, mist, or dust).

Inhibitor. A material that is added to another to prevent an unwanted reaction; e.g., polymerization.

Inorganic Materials. Materials that do not contain carbon bonded to hydrogen, as do hydrocarbons (a.k.a. organics).

insol. Insoluble.

International Association of Fire Fighters. See **IAFF**.

Intubation. Insertion of a tube into any hollow organ (such as an airway into the trachea).

IRR. Irritant effects.

Irreversible Injury. An injury that is not reversible after the exposure has terminated.

Irritant. A material that is noncorrosive and that causes a reversible inflammatory effect on living tissue by chemical action at the site of contact.

Isomers. Chemical compounds that have the same molecular weight and atomic composition but differ in molecular structure.

K

kg, kilogram. 1000 g; 2.2 lb.

Kilocalorie. Amount of heat required to raise the temperature of 1 kg of water 1°C.

L

l. Liter.

Label. Any written sign or symbol displayed on containers of hazardous chemicals. Among other things a label should contain the identity of the hazardous material, appropriate hazard warnings, and the name and address of the chemical manufacturer, importer, or other responsible party.

Lacrimation. Secretion and discharge of tears.

Latency Period. The time that elapses between exposure and the manifestations ofdisease or illness. Latency periods can range from minutes to decades, depending on the hazardous material involved.

LC_{50}. Lethal concentration 50. LC_{50} is the concentration of a material in air that killed 50% of a group of test animals when administered as a single exposure. Other percentages, such as LC_1, are similar in concept. LC_{50} is usually expressed in **ppm** or **mg/m^3**.

LC_{Lo}. Lethal concentration low. LC_{Lo} is the lowest concentration of a substance in air that has been reported to have caused death in test animals.

LD_{50}. Lethal dose 50. The dose of a substance that causes the death of 50% of the animals in a test population. Other lethal-dose percentages such as LD_1, LD_{10}, LD_{30}, and LD_{99} are similar in concept. LD_{50} is usually expressed as milligrams (or grams) of material per kilogram of animal weight.

LD_{Lo}. Lethal dose low. The lowest dose (other than LD_{50}) of a substance reported to have caused death in test animals.

LEL. See **Lower Explosive Limit**.

LEPC. Local Emergency Planning Committee. A committee formed, as required by **SARA Title III**, for the purpose of implementing local planning for chemical emergencies and for implementing requirements of that law.

Leukemia. A progressive, malignant disease of the blood-forming organs.

LFL. Lower Flammable Limit; see **LEL**.

LFM or lfm. Linear feet per minute.

Limit of Detection. The smallest concentration or amount of a substance that an analytical method can distinguish as being different from the background.

Limits of Flammability. See **Flammable Limits**.

Lipophilic. Displaying a preference or affinity for fats in the biologic system.

Local Emergency Planning Committee. See **LEPC**.

Local Exhaust Ventilation. Ventilation designed to capture and remove airborne contaminants near their point(s) of emission.

Local Ventilation. The removal of contaminated air directly from its source. This type of ventilation is recommended for controlling most hazardous airborne materials. Contrast with **General Ventilation**.

LOD. See **Limit of Detection**.

Lower Explosive Limit. The lowest concentration of gas or vapor (% by volume in air) that will burn or explode if an ignition source is present at ambient temperatures.

Lymph Nodes. Small nodules of specialized immune cells, located throughout the body.

M

m^3 or cu m. Cubic meter. m^3 is preferred.

Makeup Air. See **Replacement Air**.

Manifest. A shipping document or tracking document which accompanies a shipment (of medical waste for example) from source of generation to place of disposal. This provides a paper trail to document the proper disposition of the waste product. Copies go to the generator, transporter(s), disposal facility, and then from the disposal or storage facility to the generator.

Manometer. An instrument to measure pressure by displacement of a fluid (e.g., water) column. It usually consists of an inclined or ''U'' shaped tube filled with water or mercury. One end of the tube is attached to a **static pressure tap** while the other is left open to the ambient air. The amount of displacement of the fluid is relative to the to the pressure within the duct.

Mantoux skin test. A skin test which can demonstrate the presence of *M. tuberculosis* bacteria in the absence of active symptoms of the disease.

Material Safety Data Sheet. See **MSDS**.

Mean. The arithmetic average of a set of data.

Measurement Error. The difference between a value obtained using a measuring device and the true value.

Median. In this context, if all exposures in a workplace are ranked from lowest to highest, the **median** is the exposure level which is at the midpoint of the ranking.

Medical Waste. Any solid waste which is generated in the diagnosis, treatment (e.g., provision of medical services), or immunization of human beings or animals, in research pertaining thereto, or in the production or testing of biologicals. It does not include any listed hazardous waste or household waste.

Melting Point. The temperature at which a solid changes to a liquid.

Metabolic Rate. The rate at which energy is made available for use by the body.

Metabolism. The transformation of chemical energy into energy within the body; used for performing work and generating heat.

Metatarsal. The part of the foot between the tarsus (the joining of the leg and foot) and the phalanges (the digital bones).

Meter (m). A measure of length; 100 cm; the equivalent of 39.371 in.

mg. Milligram (1/1000 of a gram).

mg/kg. Milligrams per kilogram. A dosage commonly used in toxicology testing to indicate the dose administered per kilogram of body weight.

mg/m³. Milligrams per cubic meter in air; a unit of airborne concentration. For any substance, the relationship between mg/m^3 and ppm is

$$ppm \times MW = 24.45 \times mg/m^3$$

where ppm = parts per million by volume
MW = molecular weight of the substance in question
24.45 = the number of liters occupied by 1 mole of any gas at **STP.**

Microgram (μg). 1/1,000,000 of a gram.

Micrometer (μm). 1/1,000,000 of a meter, also referred to as a micron.

Micron. See **Micrometer**.

Miliary Tuberculosis. Tuberculosis in which large numbers of tubercle bacilli have been distributed throughout the body by the bloodstream. Miliary refers to the millet seed-like appearance in X-rays.

Millimeter (mm). 1/1000 of a meter.

Mine Safety and Health Administration. See **MSHA**.

Miscible. The extent to which liquids or gases can be mixed or blended.

Mist. Small liquid droplets suspended in the air.

Mixture. A heterogeneous association of materials that cannot be represented by a chemical formula and that does not undergo chemical change as a result of interaction among the mixed materials. The constituent materials may or may not be uniformly dispersed and can usually be separated by mechanical means (as opposed to a chemical reaction). Uniform liquid mixtures are called solutions.

ml. Milliliter. 1/1000 of a liter, equivalent to 1 cubic centimeter (1 cm^3).

MLD. Mild irritation effects.

mmHg. Millimeters of mercury. A measure of pressure. See also **atm**.

MMI. Mucous membrane effects.

MMWR. Morbidity and Mortality Weekly Report, Mailstop C-08, CDC, Atlanta, GA 30333; (404) 332-4555.

MOD. Moderate irritation effects.

Model. A hypothetical representation of a real phenomenon or process.

Mole. The quantity of a chemical substance expressed in g (usually grams) numerically equal to the molecular weight. It is that amount of a material that has 6.023×10^{23} molecules (Avogadro's number).

Molecular Weight. The mass in grams per mole of a substance. See also **Mole**.

Monitoring Protocols. The specific procedures used for different types of monitoring.

Morbidity. The rate of disease or proportion of disease in a specified population.

Morbidity and Mortality Weekly Report. See **MMWR**.

Mortality. A measure of the frequency of deaths within a particular population during a specified time period.

mppcf. An archaic unit of measure of airborne concentration; millions of particles per cubic foot of air. See also **f/cc**.

MSDS. Material Safety Data Sheet. As required by the **Hazard Communication Standard**, an MSDS is supposed to be issued by chemical manufacturers, importers, or distributors for each hazardous material. The MSDS should contain information relating to the various health risks associated with use of the chemical.

MSHA. Mine Safety and Health Administration. A federal agency within the U.S. Department of Labor that devises and promulgates mandatory safety and health in mines. MSHA, Ballston Tower #3, 4015 Wilson Blvd., Arlington, VA, 22203 (703) 235-1452.

MSK. Musculoskeletal effects.

Mucous Membrane. The mucous-secreting membrane lining the hollow organs of the body; i.e., nose, mouth, stomach, intestine, bronchial tubes, and urinary tract.

Muffler. A special duct of pipe that impedes the transmission of sound by reducing the velocity of the air or gas flow.

MUT, Mutagen. A material that induces genetic changes (mutations).

mw or MW. See **Molecular Weight**.

N

n-. Normal; used as a prefix in chemical names to signify a straight-chain structure; i.e, no branches.

NA, ND. Not applicable, not available.

Narcosis. Stupor or unconsciousness produced by narcotics or other materials.

Natality Rate. A measure of the frequency of births in a population during a specified time period.

National Clearinghouse for Alcohol and Drug Information. P. O. Box 2345, Rockville, MD, 20852 (301) 468-2600.

National Fire Protection Association. See **NFPA**.

National Health Information Clearinghouse. P.O. Box 1133, Washington, D.C., 20013 (800) 336-4797, in Maryland (301) 565-4167.

National Injury Information Clearinghouse. 5401 Westbard Ave., Room 625, Washington, D.C., 20207 (301) 492-6424.

National Institute for Occupational Safety and Health. See **NIOSH**.

National Research Council. The NRC develops and publishes emergency exposures limits such as **CEGLs**, **EEGL**, and **SPEGLs**. NRC, National Academy of Sciences, 2101 Constitution Ave., Washington, DC 20418.

National Safety Council. The largest organization in the world dedicating its efforts exclusively to the prevention of accidents. NSC, 444 North Michigan Ave., Chicago, IL, 60611-9811 (800) 621-7619; training information (800) 848-5588 or (415) 341-5649.

National Toxicology Program. See **NTP**.

National Volunteer Fire Council. See **NVFC**.

NCI. National Cancer Institute. A part of the National Institutes of Health.

ND. Not Determined.

Necrosis. Localized death of tissue.

NEO. Neoplastic effects, i.e., the production of tumors.

Neoplasm. A new or abnormal tissue growth that is uncontrollable and progressive.

Nephrotoxic. Poisonous to the kidney.

Neural. Of, relating to, or affecting a nerve or the nervous system.

Neuritis. Inflammation of the nerves.

Neurotoxic. Destructive to nerves or the nervous system.

Neutralize. To render chemically harmless.

NFPA. National Fire Protection Association. An international voluntary membership organization to promote/improve fire protection and prevention and establish safeguards against loss of life and property by fire. NFPA, Battery March Park, Quincy, MA, 02269 (800) 344-3555, (617) 770-3000.

NFPA Fire Diamond. See **Fire Diamond**.

ng. Nanogram. 1/1,000,000,000 or 10^{-9} of 1 g.

NIOSH. National Institute for Occupational Safety and Health. The research and training "arm" of **OSHA**. NIOSH, 4676 Columbia Parkway, Cincinnati, OH, 45226-1998. Technical support (800) 35-NIOSH; the NIOSH Publications Catalog DHHS (NIOSH) Publ. No. 87-115 is free from NIOSH Publications (513) 533-8287.

NOC. Not otherwise classified.

NOEL. No effect level.

Noncritical. Medical equipment which is not invasive and which will touch only the patient's unbroken dermis. See **critical items** and **semi-critical items**.

Nonflammable. Incapable of being easily ignited or burning with extreme rapidity when lighted. Also, a **DOT** hazard class for any compressed gas other than a flammable one.

NOS. Not otherwise specified.

Nosocomial Infection. An infection acquired in a hospital.

NRC. In this context, National Research Council, National Academy of Sciences, 2101 Constitution Ave., Washington, D.C., 20418.

NSC. See **National Safety Council**.

NTP. National Toxicology Program. Overseen by the Department of Health and Human Services (with resources from National Institutes of Health, the Food and Drug Administration, and the Centers for Disease Control). Its goals are to develop tests useful for public health regulations of toxic chemicals, to develop toxicological profiles of materials, to foster testing of materials, and to communicate the results for use by others. NTP Information Office, MD B2-04, Box 12233, Research Triangle Park, NC, 27709 (919) 541-3991.

Nuisance Particulates. Dusts that do not produce significant disease or toxic effects from "reasonable" concentrations and exposures. The **TLV** for nuisance particulates is 10 mg/m^3.

NVFC. The National Volunteer Fire Council, 1325 Pennsylvania Ave., N.W., Ste. 500, Washington, D.C., 20004 (202) 393-3351.

O

OCAW. Oil, Chemical, and Atomic Workers International Union, AFL-CIO, 255 Union Blvd., Lakewood, CO, 80228 (303) 987-2229.

Occupational Disease. A deterioration in health status resulting from circumstances or conditions in the work environment.

Occupational Exposure. See **Action Level**.

Occupational Exposure Work Practice Guideline. AIHA publication, in-depth toxicological and applied industrial hygiene reviews of current interest.

Occupational Safety and Health Act. See **OSHAct**.

Occupational Safety and Health Administration. See **OSHA**.

Odds Ratio. A ratio between two odds as, for example, the odds of exposure among individuals with a disease divided by the odds of exposure among individuals without the disease. The odds ratio is used as a measure of association in **case-control** studies.

Odor Threshold. The lowest concentration of a material's vapor (or a gas) in air that can be detected by odor. Frequently expressed as a percentage of a panel of test individuals.

OEL. Occupational Exposure Limit. See also **Exposure Limits**.

Oil, Chemical, and Atomic Workers Union, AFL-CIO (OCAW). See **OCAW**.

Opaque. Impervious to light rays.

Organic Materials. Compounds composed of carbon, hydrogen, and other elements with chain or ring structures.

ORM. Other Regulated Material. A **DOT** hazard classification for the purpose of labeling hazardous materials in transport. For example, **ORM-A** are materials withan anesthetic, irritating, noxious, toxic, or other property whose leakage can cause extreme discomfort to transportation personnel. **ORM-B** are materials, including solids wet with water, that can cause damage to a vehicle if they leak. **ORM-E** are materials that are not in any other hazard classification but are subject to DOT regulations.

OSHA. The Occupational Safety and Health Administration (Part of the U.S. Department of Labor). The regulatory and enforcement agency for safety and health for most U.S. business and industrial sectors. OSHA Technical Data Center Docket Office, Rm N-3670, 200 Constitution Ave. NW, Washington, D.C., 20210 (202) 523-7894, Public Information (202) 523-8151, Publications Office (202) 523-9667.

OSH Act. The Occupational Safety and Health Act of 1970; Public Law 91-596, effective April 28, 1971. Found at 29 CFR 1910, 1915, 1918, 1926. The regulations intended to ensure the safety and health of workers in firms larger than ten employees by setting standards of safety that will prevent injury and illness among the workers. Regulating employee exposure and informing employees of the dangers of materials are key factors of this act.

OSHA Flammable/Combustible Liquid Classification. (29 CFR 1910.106). Flammable/combustible liquid is a standard classification used to identify the risks of fire or explosion associated with a particular liquid. The distinctions are made on the following basis:

Flammable Liquids (with a flash point below 100°F [38°C]) are class I liquids and are divided into the following classes: class IA—flash point below 73°F (22.8°C), boiling point below 100°F (38°C); class IB—flash point below 73°F (22.8°C), boiling point at or above 100°F (38°C); and class IC—flash point at or above 73°F (22.8°C), boiling point below 100°F (38°C).

Combustible Liquids (with a flash point at or above 100°F) are divided into two classes: class II, with a flash point at or about 100°F (38°C) and below 140°F (60°C), except any mixture having components with flash points of 200°F (93.3°C) or higher, the volume of which makes up 99% or more of the total volume of the mixture; and class III, with flash point at or above 140°F (60°C).

Class III liquids are divided into two subclasses: class IIIA, with a flash point at or above 140°F (60°C) and below 200°F (93.3°C), except any mixture having components with flash points of 200°F (93.3°C) or higher, the volume of which makes up 99% or more of the total volume of the mixture; and class IIIB, with flash point at or above 200°F (93.3°C).

OSHRC. Occupational Safety and Health Review Commission, the review board for appealed **OSHA** decisions.

Overexposure. Any exposure greater than the occupational exposure limit (e.g., TLV-TWA) or any exposure that causes some adverse health effect.

Overhang. In essence, the same as **nosing**, i.e., the projection of the tread past the riser face as measured from the tread surface below.

P

PAH. Polycyclic Aromatic Hydrocarbons. Organic compounds, usually formed from incomplete combustion, that may pose a risk of cancer.

Pandemic. An epidemic occurring over a very wide area and usually affecting a large proportion of the population.

Parenteral. Piercing mucous membranes or the skin barrier through such events as needlesticks, human bites, cuts, and abrasions.

Particulate. Tiny pieces of an airborne material. **Dusts**, **fumes**, **smokes**, **mists**, and **fogs** are examples of particulates.

Partition Coefficient. The quotient of the solubility of compound in oil divided by its solubility in water. See also **Coefficient of Water/Oil Distribution**.

PEL. Permissible Exposure Limit. Exposure limits established by **OSHA**. OSHA PELs have the force of law. Note that **TLVs** (**ACGIH**) and **RELs** (**NIOSH**) are recommendedexposure limits that may or may not be enacted into law by OSHA. PELs may be expressed as

> **PEL-TWA**, a time-weighted average (TWA) limit.
>
> **PEL-C**, a ceiling exposure limit that legally must never be exceeded instantaneously.
>
> **PEL-STEL**, a short term exposure limit (STEL), a 15-min TWA which may not be exceeded.

Percent Volatile. The percentage by volume of a liquid or solid that will evaporate at an ambient temperature of 70°F (20°C).

Permeation Rate. The rate of diffusion of a chemical through a protective material.

Permissible Exposure Limit. See **PEL**.

Personal Dosimeter. With respect to chemical exposures, a personal dosimeter is a small device worn by a worker to collect airborne contaminants and used to measure exposure. It differs from other forms of **Air Sampling** in that a **Sampling Pump** is not used. For noise exposures, a personal noise dosimeter is a mechanical device worn by a worker which will measure and store the sound level experienced by the worker throughout the day. With regard to heat stress, a heat stress dosimeter will monitor the environmental, metabolic, and physiological factors contributing to an individual's **Thermal Stress**.

Personal Hygiene. Precautionary measures taken by the individual worker to maintain good health when exposed to potentially harmful materials. This includes keeping hands, other parts of the body, work clothing, and equipment free of chemicals, dirt, and chemical residue. It includes the practice of not eating, drinking, applying makeup, or using toilet facilities where chemicals are in use.

Personal Measurement. See **Personal Sample**.

Personal Protective Equipment. See **PPE**.

Personal Sample. A measurement collected from an individual's breathing zone.

Personal Sampling Pump. See **Sampling Pump**.

pH. The value that represents the acidity or alkalinity of an aqueous solution. By definition pH is the logarithm to the base 10 of the reciprocal of the hydrogen ion concentration of a solution. The pH scale ranges from 1 to 14; pure water has a pH of 7. The strongest acids have an excess of H^+ ions and a pH of 1 to 3 (HCl, pH = 1). The strongest bases have an excess of OH^- ions and a pH of 11 to 13 (NaOH, pH = 12).

Physical Hazard. A hazard from a chemical that results in potential damage, death, or injury to a person. These hazards include corrosive materials, explosives, reactive substances, compressed gasses, oxidizers, flammables, and organic peroxides.

Physical State. The condition of a material; i.e., solid, liquid, or gas, at room temperature.

Pitot Tube. In industrial ventilation, an instrument to measure velocity pressure as the difference between total and static pressures.

Plano. Nonprescription lenses, as in noncorrective spectacles or goggles.

Plenum. A portion of a ventilation system designed to contain relatively uniform static pressure and usually relatively low airflow velocity.

Pneumatic. Pneumatic systems used compressed air or gases as a source of stored energy, for example, a pressure vessel used to operate a pneumatic press ram.

Pneumoconiosis. Pathological reaction of the organism to the presence of dusts in the lung tissue. Involves activation of the macrophage/phagocyte response system.

PNS. Peripheral nervous system effects.

Poison. A substance that produces injury or death. "*All substances are poisons; there is none which is not a poison. The right dose differentiates between a poison and a remedy.*"—Paracelsus.

Poison, Class A. A **DOT** term for an extremely dangerous poison, for example, a poisonous gas or liquid, such that a very small amount of the gas or vapor of the liquid mixed with air is dangerous to life.

Poison, Class B. A **DOT** term for liquid, solid, paste, or semisolid substances other than class A poisons or irritating materials known (or presumed on the basis of animal tests) to be so toxic to man as to afford a hazard to health during transportation.

Poison Control Center. Provides medical information on a 24-hr basis for accidents involving ingestion of potentially poisonous materials. Call your area's largest hospital to find the one nearest to you.

Polarity. When referring to electrical circuits, one wire delivers the current and the other returns completing the circuit.

Polymerization. A chemical reaction in which many small molecules combine to form large molecules. Sometimes a "hazardous polymerization" can take place at such a rate that large amounts of energy causing fires or explosions or burst containers are released. Materials that can polymerize usually contain inhibitors that can delay the reactions.

Population. All members of a group (for example, all males aged 20 to 30). Contrast with **sample**.

Population Attributable Risk Percent. The proportion of disease in the total population that is attributable to a particular exposure.

Portable Ladder. A ladder that can be easily moved or carried.

Potentiation. See **Synergy**.

Pour Point. The temperature at which a liquid ceases or begins to flow.

ppb. Parts per billion. A volumetric unit of concentration; 1 ppb = 0.001 ppm.

PPD. Purified protein derivative; used in the tuberculin skin test (**Mantoux**).

PPE. Personal Protective Equipment. Equipment (devices or clothing) worn to help isolate a worker from direct exposure to hazardous materials. Examples include gloves and respirators.

ppm. Parts per million by volume; a unit of airborne concentration. For any substance, the relationship between mg/m^3 and ppm is

$$\text{ppm} \times \text{MW} = 24.45 \times \text{mg/m}^3$$

where mg/m^3	= mg of chemical per cubic meter of air
MW	= molecular weight of the substance in question
24.45	= the number of liters occupied by 1 mole of any gas at **STP**

Speaking in very rough terms, one air rifle pellet (a "BB") placed in an empty 10 gal pail is approximately 1 ppm.

ppt. Parts per trillion; 1 part per trillion = 0.000001 ppm.

Precision. A measure of the reproducibility of a measured value under a given set of conditions.

Prevalence Rate. The total number of individuals affected with a disease divided by the total number of individuals who are at risk for having the disease at a particular point in time.

Process Change. In this context, any change in manufacturing process (type of equipment, substitution of different chemicals) that can lead to changes in worker exposure.

Prophylactic. Preventative.

Prospective Study. See also **Cohort Study**. An epidemiological method for identifying the future relationship, if any, between exposure to an **agent** and the increased incidence of some adverse health effect in a population. Contrast with **Retrospective Study**.

Psychotropic, PSY. Acting on the mind.

PUL. A chemical that has pulmonary systems effects.

Pulmonary Fibrosis. A reaction by the biologic system, occurring in the lungs, marked by increased formation of interstitial fibrous tissue.

Pulmonary Tuberculosis. Tuberculosis in the lungs.

Pyrolysis. A chemical decomposition or breaking apart of molecules produced by heating.

Pyrophoric. Describes materials that ignite spontaneously in air below 130°F (54°C). Occasionally friction will ignite them.

Q

Q. See **Ventilation Volume Flow Rate**.

Qualitative Fit Testing. A method of evaluating the fit of a given respirator to the user's face. Irritant smoke is used to indicate whether the seal between the face and the respirator mask is sufficient to prevent the smoke from entering the facepiece.

Quantitative Fit Testing. A method of evaluating the fit of a given respirator to the user's face which measures the concentration of the contaminant inside and outside the mask of the respirator. The method gives a quantitative measure of the percentage of the contaminated air which is allowed to enter the facepiece.

R

Radiograph. An X-ray film.

Random Sampling. Samples collected at random from a population, without bias, such that each sample has an equal probability of being collected. If enough random samples are collected, the data represented in the sample will accurately represent the total population.

Range. The difference between the smallest and largest values in a data set.

RBC. Red blood cell effects.

RCRA. Resource Conservation and Recovery Act; an amendment to the Solid Waste Disposal Act.

Reactive Material. In general, a chemical substance or mixture that will vigorously polymerize, decompose, condense, or become self-reactive due to shock, pressure, or temperature.

Reactivity. Describes a substance's tendency to undergo chemical reaction either by itself or with other materials with the release of energy.

Recommended Exposure Limit. See **REL**.

Red Bag Waste. A generic term for medical or infectious waste.

Regulated Medical Waste. Those items described as coming from cultures and stocks of infectious agents, pathological wastes, human blood and blood products, used and unused sharps, contaminated animal wastes, and highly communicable disease isolation wastes.

REL. Recommended Exposure Limit. An occupational exposure limit recommended by NIOSH as being protective of worker health and safety over a working lifetime; the REL is used in combination with engineering and work practice controls, exposure and medical monitoring, labeling, posting, worker training, and personal protective equipment. The REL is frequently expressed as a **TWA** exposure for up to 10 hr/day during a 40-hr work week. The REL may also be expressed as (1) a short term exposure limit (**STEL**) that should never be exceeded and is to be determined in a specified sampling time (usually 15 min) or (2) a ceiling limit (**C**) that should never be exceeded even instantaneously unless specified over a given time period.

Relative Humidity. The ratio of water vapor present in the ambient air to the water vapor present in saturated air at the same temperature and pressure.

Relative Risk. A ratio between two risks; specifically, the incidence rate among a group of exposed individuals divided by the incidence rate among a group of unexposed individuals.

Release. Any uncontrolled spilling, leaking, emitting, discharging, or disposing.

Repetitive Stress Disorders. Medical conditions of the joints caused by repetitive motion, for example, carpal tunnel syndrome.

Replacement Air. Air brought into a building, by the ventilation system, to replace the air that was removed by the lab hood. Also referred to as **makeup air**.

Reportable Quantity. See **RQ**.

Reproductive Health Hazard. Any chemical or physical agent that has a harmful effect on the adult male or female reproductive system or on the developing fetus or child.

Resource Conservation and Recovery Act. See **RCRA.**

Respirable Fraction. That portion of an airborne particulate that can be inhaled into the lungs; generally accepted to be limited to particles less than 5 **mm**.

Respirator Protection. A device used to provide a source of clean breathing air to the user by either filtering workroom air or by supplying fresh air from another source.

Respirator Protection Factor. A measure of the efficiency of a particular respirator obtained by simultaneously evaluating the concentration of a contaminant inside and outside of the respirator facepiece. The numerical protection factor reflects the ratio of the concentration of the contaminant outside of the facepiece to the concentration of the contaminant inside of the facepiece. See **Fit Test**, **Qualitative Fit Test** and **Quantitative Fit Testing**.

Respiratory System. The breathing system, including the lungs and air passages (trachea or windpipe, larynx, mouth, and nose), as well as the associated system of nerves and circulatory supply.

Retrospective Study. Also, **Case-Control** or **Case History**. A study for identifying the causal relationship between workplace factors (chemical or physical **agents**, temperature, etc.) and the occurrence of adverse health effects in a population. This is done by evaluating disease incidence with past exposures. Contrast with **Prospective Study**.

Reversible Injury. An injury that will cease, or an injury that can be treated, such that the adverse health effect will cease and/or healing can take place.

Risk. The probability of an adverse health effect; risk is often said to be a function of exposure level and toxicity.

Risk Assessment. The evaluation of risk, either qualitatively or quantitatively, that might result from an exposure.

Root-Mean-Square Value (rms). The square root of the arithmetic mean of the squares of a set of values.

Routes of Entry. The method by which a chemical enters the body is called the route of entry. The routes of entry are inhalation, absorption (eye or skin contact), and ingestion.

RQ. Reportable Quantity; used in several pieces of legislation. Briefly, the RQ is the amount of a chemical which, if released into the environment, must be reported to EPA and state agencies (**SARA Title III**).

RTECS. "Registry of Toxic Effects of Chemical Substances," published by **NIOSH**. Presents basic toxicity data on thousands of materials. Its objective is to identify "all known toxic substances" and to reference the original studies.

S

Safety. The practical certainty that an adverse health effect will not occur if an agent is used in a specified quantity and under specified conditions.

Safety Glasses. Spectacle eye protection designed for industrial use with lenses designed for impact (3 mm thick lenses).

Sample. A subset of the population.

Sampling Pump. In general, a calibrated air pump used to perform **Air Sampling**.

SARA. Superfund Amendments and Reauthorization Act. Also referred to as **SARA Title III** and the **Emergency Planning and Community Right-to-Know Act** of 1986. The disaster in Bhopal, India, in 1987 added impetus to the passage of this law. It is intended to support local and state emergency planning efforts. It provides citizens and local governments with information about potential chemical hazards in industries, etc. in their communities. SARA calls for facilities that store hazardous materials to provide officials and citizens with data on the types (flammables, corrosives, etc.), the amounts on hand (daily, yearly), and their specific locations. Facilities are to prepare and submit inventory lists, **MSDSs**, and **Tier I** or **Tier II** inventory forms as well as **Form R** forms as a part of the Toxic Release Inventory reporting. EPA SARA Information Hotline (800) 535-0202.

Sarcoidosis. A disease of unknown etiology, characterized by widespread granulomatous lesions, that may affect any region of the body.

SCBA. Self-Contained Breathing Apparatus. A respirator type in which the user carries a supply of compressed breathing air in a tank on his/her back. Air passes from the tank, through a regulator and is delivered into the facepiece of the respirator. SCBAs offer the highest level of protection of any respirator system and are frequently used for emergency response activities involving highly toxic materials.

Semi-critical items. Medical equipment which comes in contact with mucous membranes of the body. See **critical items** and **noncritical items**.

Sensitization. A state of immune-response reaction in which further exposure elicits an immune or allergic response. A person previously exposed to a certain material is more sensitive when he experiences further contact with it.

Sensitizer. A material which on first exposure causes little or no reaction but which on repeated exposure may cause a marked response in the individual not necessarily limited to the contact site. Skin sensitization is the most common form. Respiratory sensitization to a few chemicals is also known to occur.

SERC. State Emergency Response Commission. The state organization responsible for implementing and enforcing **SARA Title III**.

Serous Membrane. Membrane lining a cavity (such as the pleural, peritoneal, or pericardial cavities).

Sharps. Any object which can penetrate the skin including, but not limited to, needles, scalpels, broken glass, broken capillary tubes, and exposed ends of dental wires.

Sign. Objective evidence of an adverse health effect. Compare with **Symptoms**.

Silica Gel. A type of **Solid Sorbent Air Sampling** media.

Simple Asphyxiant. See **Asphyxiant**.

Skin. In the context of exposure limits such as the **TLV** or the **PEL**, ''skin'' is a notation indicating possible significant contribution to a worker's overall exposure to a material by way of absorption through the skin, mucous membranes, and eyes by either direct or airborne contact.

SKN. Skin effects; e.g., erythema, rash, sensitization of skin.

Slurry. A pourable mixture of solid and liquid.

Smoke. Dry particles and droplets (carbon, soot, PAHs, etc.) generated by incomplete combustion of an organic material combined with and suspended in the gases from combustion.

Solid Sorbent. See **Sorbent**.

Soln. Solution. A uniformly dispersed mixture. Solutions are composed of a solvent (water or another fluid, for example) and a dissolved substance, called the solute.

Solubility in Water. A term used to express the percentage of a material, by weight, that will dissolve in water at ambient temperature; useful in determining cleanup methods for spills and fire-extinguishing methods for a material. Solubility, commonly expressed in g/l, is expressed as:

"negligible," less than 0.1%
"slight," 0.1 to 1.0%
"moderate," 1 to 10 1%
"appreciable," more than 10%
"complete," soluble in all proportions

Solution. See **Soln**.

Solvent. A material that can dissolve other materials to form a uniform mixture. Water is the so-called "universal solvent."

Soot. Particles formed by combustion (complete or incomplete) and consisting mostly of carbon. Soot gives smoke its color.

Sorbent. Also Solid Sorbent; a sampling medium (such as charcoal tubes or silica gel) used to collect airborne chemical contaminants. See **Air Sampling**.

SPEGL. Short-Term Public Emergency Guidance Level, established by the Committee on Toxicology of the **National Research Council** for the Department of Defense. The SPEGL ia defined as a suitable concentration for unpredicted, single, short-term, emergency exposure of the general public. In contrast to the EEGL, the SPEGL takes into account the wide range of susceptibility of the general public. See also **EEGL** and **CEGL**.

Sputum. The substance expelled by coughing or clearing of the throat.

Stability. An expression of the ability of a chemical to remain unchanged under reasonable conditions of storage or use.

Stack. The terminal component of a local exhaust system, i.e., the place at which exhaust air is discharged to the atmosphere.

Standard Deviation, SD. The positive square root of the variance of a distribution; a measure of the representativeness of the mean.

Standard Man. A "representative" human being having a body mass of 70 kg (154 lb) and a body surface area of 1.8 m^2 (19.4 ft^2).

Standardized Mortality Ratio (SMR). The ratio of the observed number of deaths in a population to the number of deaths expected in that population if the population had the same age structure as the standard population.

Static Pressure. Air pressure exerted uniformly in all directions, not affected by air movement. Usually measured in inches of water gauge (in. H_2O) or millimeters of mercury (mmHg).

Static Pressure Traps. This is the name given to many types of hardware used to connect a hose to a hole drilled in the side of ductwork, etc. The hole is the point at which static pressure measurements are taken.

STEL. Short-Term Exposure Limit. The **TWA** exposure for any 15-min period of time cannot exceed the **PEL-STEL (OSHA)** or **TLV-STEL (ACGIH)**.

Sterilization. The process of removing all forms of living organisms on a substance by chemical or physical agents, by radiation, or by filtering through porous materials that remove microorganisms. Also see **Disinfection**.

STP. Standard Temperature and Pressure. Usually refers to 298 K (25°C) and 760 mmHg (1 atm).

Subcutaneous. Beneath the skin.

Sublime. To change from the solid to the vapor phase without passing through the liquid phase, for example, dry ice.

Subpart Z. See **Z list**.

Superfund Amendments and Reauthorization Act. See **SARA**; see also **CERCLA**.

Susceptible. An individual who lacks some or all resistance to a particular disease or environmental **agent**.

Symptom. Subjective evidence of an adverse health effect. Compare with **Sign**.

Synergy. An interaction or potentiation of materials to give a greater result than either material alone. For example, either smoking or exposure to asbestos alone can cause lung cancer. If an individual smokes and is exposed to asbestos, the risk of lung cancer is much greater than the individual risks from smoking of asbestos exposure combined.

Synonyms. Alternative names by which a substance may be known.

SYS. Systemic effects. Effects on the metabolism and excretory functions.

T

Target Organ Effects. Effects, resulting from chemical exposure, to specific organs or organ systems (such as the liver, kidneys, nervous system, lungs, skin, and/or eyes).

TC_{Lo}. Toxic Concentration, low. The lowest airborne concentration of a substance to which humans or animals have been exposed and that has produced a toxic effect.

TDL. Toxic dose level.

TD_{Lo}. The lowest dose of a substance introduced by any route other than inhalation over any given period of time and reported to produce any toxic effect in humans or to produce tumorigenic or reproductive effects in animals or humans.

temp. Temperature.

TER. See **Teratogen**.

Teratogen. Physical defects in a developing embryo caused by the mother's exposure to a chemical **agent** or material.

Threshold Limit Value. See **TLV**.

Threshold Planning Quantity. The amount of an extremely hazardous substance listed in **SARA Title III** that triggers reporting and emergency planning. Usually abbreviated as **TPQ**.

Tier I Report. A hazards inventory report required under SARA Title III that summarizes the amounts of hazardous materials and their storage locations for different hazard categories.

Tier II Report. A chemical inventory report required under **SARA Title III** that summarizes in detail the amounts of specific hazardous chemicals and their storage locations in a facility.

Time-Weighted Average. Also **TWA**. A way of expressing exposure such that the amount of time spent exposed to each different concentration level is weighted by the amount of time the worker was exposed to that level.

$$TWA = \frac{C_1T_1 + C_2T_2 + \cdots C_nT_n}{T_1 + T_2 + \cdots T_n}$$

where

C_1, C_2, C_n, etc. = airborne concentrations of some chemical (expressed in either ppm or mg/m^3)
T_1, T_2, T_n, etc. = the length of time the worker was exposed to C_1, etc.

TLV. Threshold Limit Value. A term used by **ACGIH** to express the airborne concentration of a material to which "nearly" all workers can be exposed day after day without adverse effects. "Workers" means healthy individuals. The young, old, ill, or naturally susceptible will have lower tolerances and need to take additional precautions. The ACGIH expresses TLVs in three ways:

TLV-TWA, the allowable time-weighted average concentration for a normal 8-hr workday or 40-hr week

TLV-STEL, the short-term exposure limit or maximum concentration for a continuous exposure period of 15 min (with a maximum of four such periods per day, with at least 60 min between exposure periods, and provided that the daily TLV-TWA is not exceeded)

TLV-C, the concentration that should not be exceeded at any time

TLV-Skin. See **Skin**.

Torr. 1 mmHg pressure; see **atm**.

Total Pressure. In a ventilation system, Total Pressure is the algebraic sum of **Static Pressure** and **Velocity Pressure**.

Toxic. In general, toxic describes the ability of a material to injure biological tissue. As specifically defined in the **OSHA Hazard Communication Standard**, there are several criteria for a chemical to be considered toxic, for example, having any of the following:

- An LD_{50} of 50 to 500 mg/kg when administered orally to albino rats weighing 200 to 300 g each.

- An LD_{50} of 200 to 1000 mg/kg when administered by continuous contact for 24 hr to the bare skin of albino rabbits weighing 2 to 3 kg each.

- An LC_{50} of 200 to 2000 ppm (gas or vapor) or 2 to 20 mg/L (mist, fume, or dust) when administered by continuous inhalation for 1 hr to albino rats weighing 200 to 300 g each

See also **Acute Toxicity**.

Toxicant. Often used as a synonym for **Poison**.

Toxicity. The degree to which an **agent** can cause harmful effects.

Toxicology. The study of the nature, effects, and detection of poisons in living organisms. Also, substances that are otherwise harmless but prove toxic under particular conditions. The basic assumption of toxicology is that there is a relationship among the dose (amount), the concentration at the affected site, and resulting effects.

Toxic Substance. Any chemical or material that (l) has evidence of an acute or chronic health hazard and (2) is listed in the *NIOSH Registry of Toxic Effects of Chemical Substances* (*RTECS*), provided that:

- The substance causes human toxicity at any dose level
- Causes cancer or reproductive effects in animals at any dose level
- Has a median lethal dose (LD_{50}) of less than 500 mg/kg of body weight when administered orally to rats
- Has a median LD_{50} of less than 1000 mg/kg of body weight when administered by continuous contact to the bare skin of albino rabbits
- Has a median lethal concentration (LC_{50}) in air of less than 2000 ppm by volume of gas or vapor, or less than 20 mg/l of mist, fume, or dust when administered to albino rats.

Toxic Substances Control Act. See **TSCA**.

Toxin. Usually, a poison that is manufactured by some life form (as opposed to manufactured chemicals).

TPQ. See Threshold Planning Quantity. As it relates to **SARA Title III**, for any chemical the TPQ is the quantity above which the chemical's presence at a factory, etc., must be reported to the EPA and the state.

Trade Name. The name given to a chemical product by the manufacturer (or supplier).

Trade Secret. Confidential information (such as formula, recipe, process, etc.) that gives the manufacturer an advantage over competitors, Manufacturers can withhold proprietary information from an **MSDS**. **OSHA** permits this provided that the trade secret claim can be substantiated, the MSDS indicates that data are being withheld and the properties and health effects of the withheld chemicals are included.

Transport Velocity. The minimum velocity needed in ductwork to prevent particulates carried in the air from settling out.

Treated Medical Waste. Waste which has received treatment to significantly reduce or eliminate the potential for disease transmission, but which has not been destroyed or made unrecognizable.

TSCA. Toxic Substances Control Act, Public Law PL 94-469, found in 40 CFR 700-799. Basically, the law says that chemicals are to be evaluated prior to use and can be controlled based on risk. The act provides for a listing of all chemicals that are to be evaluated prior to manufacture or use in the US. EPA, TSCA Office, Office of Pesticides and Toxic Substances, 401 M St., S.W., Washington, D.C., 20460, Information (202) 554-1404.

Tubercle. The characteristic lesion resulting from infection by the tubercle bacilli. Typically it has three parts: a central giant cell, a midzone of epithelioid cells, and a peripheral zone of nonspecific structure.

Tuberculosis. An infectious disease caused by the tubercule bacillus (*Mycobacterium tuberculosis*) which most commoonly affects the respiratory system.

Turbulence. Random movement of air characterized by swirling, changing patterns, causing pressure loss and noise.

TWA. See **Time-Weighted Average**.

U

UEL. See **Upper Explosive Limit**, **Lower Explosive Limit**.

UFL. Upper Flammable Limit; identical to **UEL**.

UL. Underwriters Laboratories, Fire Protection Division, 1285 Walt Whitman Rd., Melville, NY, 11747-3081 (516) 271-6200.

Ultraviolet Light. See **UV**.

Underwriters Laboratories. See **UL**.

Universal Precautions. Handling all blood, and other potentially infectious body fluids or materials, as if they are infected with **HIV**, **HBV**, or other **blood-borne pathogens**.

Unstable. An unstable chemical will vigorously polymerize, decompose, condense, or become self-reactive under conditions of shock, pressure, or temperature.

Upper Explosive Limit. UEL. The highest concentration of a material in air that will produce an explosion in fire or will ignite when it contacts an ignition source (high heat, electric arc, spark, or flame). A higher concentration of the material in a smaller percentage of concentration of air may be too rich to be ignited. See also **Flammable Limits**.

UV. Ultraviolet light. A form of radiation, intermediate between visible light and X-rays, which exists in sunlight or can be generated artificially.

V

Vapor. The gaseous state of a material suspended in air that would be a liquid or solid under ordinary conditions.

Vapor Density. The weight of a vapor or gas as compared to the weight of an equal volume of air; (**MW** of gas)/29 = vapor density, therefore, materials lighter than air have vapor densities of <1.0 and materials heavier than air have vapor densities >1.0. All vapors and gases will mix with air, but the lighter materials will tend to rise and dissipate. Vapors heavier than air are likely to concentrate in low, enclosed places (along or under floors; in sumps, sewers, manholes, trenches, and ditches) creating fire, explosion, or health hazards.

Vapor Pressure. The pressure exerted by a saturated vapor above its own liquid in a closed container. The lower the boiling point of a substance, the higher its vapor pressure. Vapor pressures are useful (with evaporation rates) in estimating how quickly a material becomes airborne within the workplace and thus how quickly a worker can be exposed to it.

Variance. In general, the square of the standard deviation; the mean of the square of the differences between the mean value of an underlying distribution and randomly selected values from the same underlying distribution of a random variable.

Velocity Pressure. Pressure exerted in the direction of airflow, with magnitude increasing with airflow velocity.

Ventilation. The removal and replacement of air to control conditions (e.g., air contaminant concentrations) in the workplace.

Ventilation Volume Flow Rate. The quantity of ventilation air flow measured as air volume per unit of time (e.g., cubic feet per minute, CFM; or cubic meters per second, m^3/s).

Virulence. The ability of a microorganism to produce serious disease; *M. tuberculosis* is virulent.

Vital Statistics. In the context of **epidemiology**, birth death, marriage, divorce, and illness/**morbidity** data.

VOC. Volatile Organic Components, Volatile Organic Compounds, Volatile Organic Chemicals.

Volatility. The measure of a material's tendency to vaporize or evaporate at ambient conditions.

Voltage. Voltage is the potential electrical difference that makes it possible for current to flow in a circuit.

VP. See **Vapor Pressure**.

W

Warning Properties. Warning properties are signs or indications that exposure to a chemical is occurring. Typical warning properties include the presence of an odor and irritation of the mucous membranes, the eyes, the respiratory system, or the skin. Not all chemicals have noticeable warning properties and in general it is not safe to rely upon warning properties as an indicator of exposure.

Water Reactive. Describes a material that reacts with water.

WBC. White blood cell effects.

WEEL. Workplace Environmental Exposure Level; exposure guidelines developed by the American Industrial Hygiene Association (see **AIHA**).

WHF. Workplace Health Fund, 815 16th St., N.W., Washington, D.C., 20006. A nonprofit agency targeting occupational health, (202) 842-7833.

Wipe Testing. The collection of chemicals, etc., by wiping a surface (often with a 100-cm^2 piece of filter paper) and analyzing. This is at best a qualitative method (as opposed to a quantitative method).

Work History. In this context, a worker's history of exposure as a function of task, trade, chemical used, etc. Frequently used for epidemiological studies.

Work Practice Controls. Performing a task or procedure in a way that reduces the chance of exposure to a potentially toxic substance.

Workplace Environmental Exposure Level. See **WEEL**.

Workplace Health Fund. See **WHF**.

WHO. World Health Organization.

Z

Z List. Also known as Z Tables. The list of **OSHA PELs**; OSHA's Toxic and Hazardous Substances Tables Z-1, Z-2, and Z-3 of air contaminants, found in 29 **CFR** 1910.1000.

Zoonosis. Transmission of animal infection to man.

APPENDICES

APPENDIX A

Occupational Exposure to Hazardous Chemicals in Laboratories

PART 1910—OCCUPATIONAL SAFETY AND HEALTH STANDARDS

1. The authority citation for part 1910, subpart Z is amended by adding the following citation at the end. (Citation which precedes asterisk indicates general rulemaking authority.)

Authority: Secs. 6 and 8, Occupational Safety and Health Act, 29 U.S.C. 655, 657; Secretary of Labor's Orders Nos. 12–71 (36 FR 8754), 8–76 (41 FR 25059), or 9–83 (48 FR 35736), as applicable; and 29 CFR part 1911.
* * * Section 1910.1450 is also issued under sec. 6(b), 8(c) and 8(g)(2), Pub. L. 91–596, 84 Stat. 1593, 1599, 1600; 29 U.S.C. 655, 657.

2. Section 1910.1450 is added to subpart Z, part 1910 to read as follows:

§ 191.1450 Occupational exposure to hazardous chemicals in laboratories.

(a) *Scope and application.* (1) This section shall apply to all employers engaged in the laboratory use of hazardous chemicals as defined below.

(2) Where this section applies, it shall supersede, for laboratories, the requirements of all other OSHA health standards in 29 CFR part 1910, subpart Z, except as follows:

(i) For any OSHA health standard, only the requirement to limit employee exposure to the specific permissible exposure limit shall apply for laboratories, unless that particular standard states otherwise or unless the conditions of paragraph (a)(2)(iii) of this section apply.

(ii) Prohibition of eye and skin contact where specified by any OSHA health standard shall be observed.

(iii) Where the action level (or in the absence of an action level, the permissible exposure limit) is routinely exceeded for an OSHA regulated substance with exposure monitoring and medical surveillance requirements, paragraphs (d) and (g)(1)(ii) of this section shall apply.

(3) This section shall not apply to:

(i) Uses of hazardous chemicals which do not meet the definition of laboratory use, and in such cases, the employer shall comply with the relevant standard in 29 CFR part 1910, subpart 2, even if such use occurs in a laboratory.

(ii) Laboratory uses of hazardous chemicals which provide no potential for employee exposure. Examples of such conditions might include:

SOURCE: The contents of Appendix A are reproduced from the *Federal Register*, Vol. 55, No. 21, January 31, 1990, pp. 3327-3335.

(A) Procedures using chemically-impregnated test media such as Dip-and-Read tests where a reagent strip is dipped into the specimen to be tested and the results are interpreted by comparing the color reaction to a color chart supplied by the manufacturer of the test strip; and

(B) Commercially prepared kits such as those used in performing pregnancy tests in which all of the reagents needed to conduct the test are contained in the kit.

(b) *Definitions—*

"*Action level*" means a concentration designated in 29 CFR part 1910 for a specific substance, calculated as an eight (8)-hour time-weighted average, which initiates certain required activities such as exposure monitoring and medical surveillance.

"*Assistant Secretary*" means the Assistant Secretary of Labor for Occupational Safety and Health, U.S. Department of Labor, or designee.

"*Carcinogen*" (see "select carcinogen").

"*Chemical Hygiene Officer*" means an employee who is designated by the employer, and who is qualified by training or experience, to provide technical guidance in the development and implementation of the provisions of the Chemical Hygiene Plan. This definition is not intended to place limitations on the position description or job classification that the designated indvidual shall hold within the employer's organizational structure.

"*Chemical Hygiene Plan*" means a written program developed and implemented by the employer which sets forth procedures, equipment, personal protective equipment and work practices that (i) are capable of protecting employees from the health hazards presented by hazardous chemicals used in that particular workplace and (ii) meets the requirements of paragraph (e) of this section.

"*Combustible liquid*" means any liquid having a flashpoint at or above 100 °F (37.8 °C), but below 200 °F (93.3 °C), except any mixture having components with flashpoints of 200 °F (93.3 °C), or higher, the total volume of which make up 99 percent or more of the total volume of the mixture.

"*Compressed gas*" means:

(i) A gas or mixture of gases having, in a container, an absolute pressure exceeding 40 psi at 70 °F (21.1 °C); or

(ii) A gas or mixture of gases having, in a container, an absolute pressure exceeding 104 psi at 130 °F (54.4 °C) regardless of the pressure at 70 °F (21.1 °C); or

(iii) A liquid having a vapor pressure exceeding 40 psi at 100 °F (37.8 °C) as determined by ASTM D-323-72.

"*Designated area*" means an area which may be used for work with "select carcinogens," reproductive toxins or substances which have a high degree of acute toxicity. A designated area may be the entire laboratory, an area of a laboratory or a device such as a laboratory hood.

"*Emergency*" means any occurrence such as, but not limited to, equipment failure, rupture of containers or failure of control equipment which results in an uncontrolled release of a hazardous chemical into the workplace.

"*Employee*" means an individual employed in a laboratory workplace who may be exposed to hazardous chemicals in the course of his or her assignments.

"*Explosive*" means a chemical that causes a sudden, almost instantaneous release of pressure, gas, and heat when subjected to sudden shock, pressure, or high temperature.

"*Flammable*" means a chemical that falls into one of the following categories:

(i) "*Aerosol, flammable*" means an aerosol that, when tested by the method described in 16 CFR 1500.45, yields a flame protection exceeding 18 inches at full valve opening, or a flashback (a flame extending back to the valve) at any degree of valve opening;

(ii) "*Gas, flammable*" means:

(A) A gas that, at ambient temperature and pressure, forms a flammable mixture with air at a concentration of 13 percent by volume or less; or

(B) A gas that, at ambient temperature and pressure, forms a range of flammable mixtures with air wider than 12 percent by volume, regardless of the lower limit.

(iii) "*Liquid, flammable*" means any liquid having a flashpoint below 100 °F (37.8 °C), except any mixture having components with flashpoints of 100 °F (37.8 °C) or higher, the total of which make up 99 percent or more of the total volume of the mixture.

(iv) "*Solid, flammable*" means a solid, other than a blasting agent or explosive as defined in § 1910.109(a), that is liable to cause fire through friction, absorption of moisture, spontaneous chemical change, or retained heat from manufacturing or processing, or which can be ignited readily and when ignited burns so vigorously and persistently as to create a serious hazard. A chemical shall be considered to be a flammable solid if, when tested by the method described in 16 CFR 1500.44, it ignites and burns with a self-sustained flame at a rate greater than one-tenth of an inch per second along its major axis.

"*Flashpoint*" means the minimum temperature at which a liquid gives off a vapor in sufficient concentration to ignite when tested as follows:

(i) Tagliabue Closed Tester (See American National Standard Method of Test for Flash Point by Tag Closed Tester, Z11.24–1979 (ASTM D 56–79))-for liquids with a viscosity of less than 45 Saybolt Universal Seconds (SUS) at 100 °F (37.8 °C), that do not contain suspended solids and do not have a tendency to form a surface film under test; or

(ii) Pensky-Martens Closed Tester (see American National Standard Method of Test for Flash Point by Pensky-Martens Closed Tester, Z11.7–1979 (ASTM D 93–79))-for liquids with a viscosity equal to or greater than 45 SUS at 100 °F (37.8 °C), or that contain suspended solids, or that have a tendency to form a surface film under test; or

(iii) Setaflash Closed Tester (see American National Standard Method of Test for Flash Point by Setaflash Closed Tester (ASTM D 3278–78)).

Organic peroxides, which undergo autoaccelerating thermal decomposition, are excluded from any of the flashpoint determination methods specified above.

"*Hazardous chemical*" means a chemical for which there is statistically significant evidence based on at least one study conducted in accordance with established scientific principles that acute or chronic health effects may occur in exposed employees. The term "health hazard" includes chemicals which are carcinogens, toxic or highly toxic agents, reproductive toxins, irritants, corrosives, sensitizers, hepatotoxins, nephrotoxins, neurotoxins, agents which act on the hematopoietic systems, and agents which damage the lungs, skin, eyes, or mucous membranes.

Appendices A and B of the Hazard Communication Standard (29 CFR 1910.1200) provide further guidance in defining the scope of health hazards and determining whether or not a chemical is to be considered hazardous for purposes of this standard.

"*Laboratory*" means a facility where the "laboratory use of hazardous chemicals" occurs. It is a workplace where relatively small quantities of hazardous chemicals are used on a non-production basis.

"*Laboratory scale*" means work with substances in which the containers used for reactions, transfers, and other handling of substances are designed to be easily and safely manipulated by one person. "Laboratory scale" excludes those workplaces whose function is to produce commercial quantities of materials.

"*Laboratory-type hood*" means a device located in a laboratory, enclosure on five sides with a moveable sash or fixed partial enclosed on the remaining side; constructed and maintained to draw air from the laboratory and to prevent or minimize the escape of air contaminants into the laboratory; and allows chemical manipulations to be conducted in the enclosure without insertion of any portion of the employee's body other than hands and arms.

Walk-in hoods with adjustable sashes meet the above definition provided that the sashes are adjusted during use so that the airflow and the exhaust of air contaminants are not compromised and employees do not work inside the enclosure during the release of airborne hazardous chemicals.

"*Laboratory use of hazardous chemicals*" means handling or use of such chemicals in which all of the following conditions are met:

(i) Chemical manipulations are carried out on a "laboratory scale;"

(ii) Multiple chemical procedures or chemicals are used;

(iii) The procedures involved are not part of a production process, nor in any way simulate a production process; and

(iv) "Protective laboratory practices and equipment" are available and in common use to minimize the potential for employee exposure to hazardous chemicals.

"Medical consultation" means a consultation which takes place between an employee and a licensed physician for the purpose of determining what medical examinations or procedures, if any, are appropriate in cases where a significant exposure to a hazardous chemical may have taken place.

"Organic peroxide" means an organic compound that contains the bivalent –O–O– structure and which may be considered to be a structural derivative of hydrogen peroxide where one or both of the hydrogen atoms has been replaced by an organic radical.

"Oxidizer" means a chemical other than a blasting agent or explosive as defined in § 1910.109(a), that initiates or promotes combustion in other materials, thereby causing fire either of itself or through the release of oxygen or other gases.

"Physical hazard" means a chemical for which there is scientifically valid evidence that it is a combustible liquid, a compressed gas, explosive, flammable, an organic peroxide, an oxidizer, pyrophoric, unstable (reactive) or water-reactive.

"Protective laboratory practices and equipment" means those laboratory procedures, practices and equipment accepted by laboratory health and safety experts as effective, or that the employer can show to be effective, in minimizing the potential for employee exposure to hazardous chemicals.

"Reproductive toxins" means chemicals which affect the reproductive capabilities including chromosomal damage (mutations) and effects on fetuses (teratogenesis)

"Select carcinogen" means any substance which meets one of the following criteria:

(i) It is regulated by OSHA as a carcinogen; or

(ii) It is listed under the category, "known to be carcinogens," in the Annual Report on Carcinogens published by the National Toxicology Program (NTP) (latest edition); or

(iii) It is listed under Group 1 ("carcinogenic to humans") by the International Agency for Research on Cancer Monographs (IARC) (latest editions); or

(iv) It is listed in either Group 2A or 2B by IARC or under the category, "reasonably anticipated to be carcinogens" by NTP, and causes statistically significant tumor incidence in experimental animals in accordance with any of the following criteria:

(A) After inhalation exposure of 6–7 hours per day, 5 days per week, for a significant portion of a lifetime to dosages of less than 10 mg/m^3;

(B) After repeated skin application of less than 300 (mg/kg of body weight) per week; or

(C) After oral dosages of less than 50 mg/kg of body weight per day.

"Unstable (reactive)" means a chemical which is the pure state, or as produced or transported, will vigorously polymerize, decompose, condense, or will become self-reactive under conditions of shocks, pressure or temperature.

"Water-reactive" means a chemical that reacts with water to release a gas that is either flammable or presents a health hazard.

(c) *Permissible exposure limits.* For laboratory uses of OSHA regulated substances, the employer shall assure that laboratory employees' exposures to such substances do not exceed the permissible exposure limits specified in 29 CFR part 1910, subpart Z.

(d) *Employee exposure determination*—(1) *Initial monitoring.* The employer shall measure the employee's exposure to any substance regulated by a standard which requires monitoring if there is reason to believe that exposure levels for that substance routinely exceed the action level (or in the absence of an action level, the PEL).

(2) *Periodic monitoring.* If the initial monitoring prescribed by paragraph (d)(1) of this section discloses employee exposure over the action level (or in the absence of an action level, the PEL), the employer shall immediately comply with the exposure monitoring provisions of the relevant standard.

(3) *Termination of monitoring.* Monitoring may be terminated in accordance with the relevant standard.

(4) *Employee notification of monitoring results.* The employer shall, within 15 working days after the receipt of any monitoring results, notify the employee of these results in writing either individually or by posting results in an appropriate location that is accessible to employees.

(e) *Chemical hygiene plan—General.* (Appendix A of this section is non-mandatory but provides guidance to assist employers in the development of the Chemical Hygiene Plan.) (1) Where hazardous chemicals as defined by this standard are used in the workplace, the employer shall develop and carry out the provisions of a written Chemical Hygiene Plan which is:

(i) Capable of protecting employees from health hazards associated with hazardous chemicals in that laboratory and

(ii) Capable of keeping exposures below the limits specified in paragraph (c) of this section.

(2) The Chemical Hygiene Plan shall be readily available to employees, employee representatives and, upon request, to the Assistant Secretary.

(3) The Chemical Hygiene Plan shall include each of the following elements and shall indicate specific measures that the employer will take to ensure laboratory employee protection:

(i) Standard operating procedures relevant to safety and health considerations to be followed when laboratory work involves the use of hazardous chemicals;

(ii) Criteria that the employer will use to determine and implement control measures to reduce employee exposure to hazardous chemicals including engineering controls, the use of personal protective equipment and hygiene practices; particular attention shall be given to the selection of control measures for chemicals that are known to be extremely hazardous;

(iii) A requirement that fume hoods and other protective equipment are functioning properly and specific measures that shall be taken to ensure proper and adequate performance of such equipment;

(iv) Provisions for employee information and training as prescribed in paragraph (f) of this section;

(v) The circumstances under which a particular laboratory operation, procedure or activity shall require prior approval from the employer or the employer's designee before implementation;

(vi) Provisions for medical consultation and medical examinations in accordance with paragraph (g) of this section;

(vii) Designation of personnel responsible for implementation of the Chemical Hygiene Plan including the assignment of a Chemical Hygiene Officer and, if appropriate, establishment of a Chemical Hygiene Committee; and

(viii) Provisions for additional employee protection for work with particularly hazardous substances. These include "select carcinogens," reproductive toxins and substances which have a high degree of acute toxicity. Specific consideration shall be given to the following provisions which shall be included where appropriate:

(A) Establishment of a designated area;

(B) Use of containment devices such as fume hoods or glove boxes;

(C) Procedures for safe removal of contaminated waste; and

(D) Decontamination procedures.

(4) The employer shall review and evaluate the effectiveness of the Chemical Hygiene Plan at least annually and update it as necessary.

(f) *Employee information and training.* (1) The employer shall provide employees with information and training to ensure that they are apprised of the hazards of chemicals present in their work area.

(2) Such information shall be provided at the time of an employee's initial assignment to a work area where hazardous chemicals are present and prior to assignments involving new exposure situations. The frequency of refresher information and training shall be determined by the employer.

(3) *Information.* Employees shall be informed of:

(i) The contents of this standard and its appendices which shall be made available to employees;

(ii) The location and availability of the employer's Chemical Hygiene Plan;

(iii) The permissible exposure limits for OSHA regulated substances or recommended exposure limits for other hazardous chemicals where there is no applicable OSHA standard;

(iv) Signs and symptoms associated with exposures to hazardous chemicals used in the laboratory; and

(v) The location and availability of known reference material on the hazards, safe handling, storage and disposal of hazardous chemicals found in the laboratory including, but not limited to, Material Safety Data Sheets received from the chemical supplier.

(4) *Training.* (i) Employee training shall include:

(A) Methods and observations that may be used to detect the presence or release of a hazardous chemical (such as monitoring conducted by the employer, continuous monitoring devices, visual appearance or odor of hazardous chemicals when being released, etc.);

(B) The physical and health hazards of chemicals in the work area; and

(C) The measures employees can take to protect themselves from these hazards, including specific procedures the employer has implemented to protect employees from exposure to hazardous chemicals, such as appropriate work practices, emergency procedures, and personal protective equipment to be used.

(ii) The employee shall be trained on the applicable details of the employer's written Chemical Hygiene Plan.

(g) *Medical consultation and medical examinations.* (1) The employer shall provide all employees who work with hazardous chemicals an opportunity to receive medical attention, including any follow-up examinations which the examining physician determines to be necessary, under the following circumstances:

(i) Whenever an employee develops signs or symptoms associated with a hazardous chemical to which the employee may have been exposed in the laboratory, the employee shall be provided an opportunity to receive an appropriate medical examination.

(ii) Where exposure monitoring reveals an exposure level routinely above the action level (or in the absence of an action level, the PEL) for an OSHA regulated substance for which there are exposure monitoring and medical surveillance requirements, medical surveillance shall be established for the affected employee as prescribed by the particular standard.

(iii) Whenever an event takes place in the work area such as a spill, leak, explosion or other occurrence resulting in the likelihood of a hazardous exposure, the affected employee shall be provided an opportunity for a medical consultation. Such consultation shall be for the purpose of determining the need for a medical examination.

(2) All medical examinations and consultations shall be performed by or under the direct supervision of a licensed physician and shall be provided without cost to the employee, without loss of pay and at a reasonable time and place.

(3) *Information provided to the physician.* The employer shall provide the following information to the physician:

(i) The identity of the hazardous chemical(s) to which the employee may have been exposed;

(ii) A description of the conditions under which the exposure occurred including quantitative exposure data, if available; and

(iii) A description of the signs and symptoms of exposure that the employee is experiencing, if any.

(4) *Physician's written opinion.* (i) For examination or consultation required under this standard, the employer shall obtain a written opinion from the examining physician which shall include the following:

(A) Any recommendation for further medical follow-up;

(B) The results of the medical examination and any associated tests;

(C) Any medical condition which may be revealed in the course of the examination which may place the employee at increased risk as a result of exposure to a hazardous chemical found in the workplace; and

(D) A statement that the employee has been informed by the physician of the results of the consultation or medical examination and any medical condition that may require further examination or treatment.

(ii) The written opinion shall not reveal specific findings of diagnoses unrelated to occupational exposure.

(h) *Hazard identification.* (1) With respect to labels and material safety data sheets:

(i) Employers shall ensure that labels on incoming containers of hazardous chemicals are not removed or defaced.

(ii) Employers shall maintain any material safety data sheets that are received with incoming shipments of hazardous chemicals, and ensure that they are readily accessible to laboratory employees.

(2) The following provisions shall apply to chemical substances developed in the laboratory:

(i) If the composition of the chemical substance which is produced exclusively for the laboratory's use is known, the employer shall determine if it is a hazardous chemical as defined in paragraph (b) of this section. If the chemical is determined to be hazardous, the employer shall provide appropriate training as required under paragraph (f) of this section.

(ii) If the chemical produced is a byproduct whose composition is not known, the employer shall assume that the substance is hazardous and shall implement paragraph (e) of this section.

(iii) If the chemical substance is produced for another user outside of the laboratory, the employer shall comply with the Hazard Communication Standard (29 CFR 1910.1200) including the requirements for preparation of material safety data sheets and labeling.

(i) *Use of respirators.* Where the use of respirators is necessary to maintain exposure below permissible exposure limits, the employer shall provide, at no cost to the employee, the proper respiratory equipment. Respirators shall be selected and used in accordance with the requirements of 29 CFR 1910.134.

(j) *Recordkeeping.* (1) The employer shall establish and maintain for each employee an accurate record of any measurements taken to monitor employee exposures and any medical consultation and examinations including tests or written opinions required by this standard.

(2) The employer shall assure that such records are kept, transferred, and made available in accordance with 29 CFR 1910.20.

(k) *Dates*—(1) *Effective date.* This section shall become effective May 1, 1990.

(2) *Start-up dates.* (i) Employers shall have developed and implemented a written Chemical Hygiene Plan no later than January 31, 1991.

(ii) Paragraph (a)(2) of this section shall not take effect until the employer has developed and implemented a written Chemical Hygiene Plan.

(l) *Appendices.* The information contained in the appendices is not intended, by itself, to create any additional obligations not otherwise imposed or to detract from any existing obligation.

Appendix A to § 1910.1450—National Research Council Recommendations Concerning Chemical Hygiene in Laboratories (Non-Mandatory)

Table of Contents

2. Allergens and Embryotoxins
3. Chemicals of Moderate Chronic or High Acute Toxicity
4. Chemicals of High Chronic Toxicity
5. Animal Work with Chemicals of High Chronic Toxicity

F. Safety Recommendations

G. Material Safety Data Sheets

Foreword

As guidance for each employer's development of an appropriate laboratory Chemical Hygiene Plan, the following non-mandatory recommendations are provided. They were extracted from "Prudent Practices for Handling Hazardous Chemicals in Laboratories" (referred to below as "Prudent Practices"), which was published in 1981 by the National Research Council and is available from the National Academy Press, 2101 Constitution Ave., NW., Washington DC 20418.

"Prudent Practices" is cited because of its wide distribution and acceptance and because of its preparation by members of the laboratory community through the sponsorship of the National Research Council. However, none of the recommendations given here will modify any requirements of the laboratory standard. This Appendix merely presents pertinent recommendations from "Prudent Practices", organized into a form convenient for quick reference during operation of a laboratory facility and during development and application of a Chemical Hygiene Plan. Users of this appendix should consult "Prudent Practices" for a more extended presentation and justification for each recommendation.

"Prudent Practices" deals with both safety and chemical hazards while the laboratory standard is concerned primarily with chemical hazards. Therefore, only those recommendations directed primarily toward control of toxic exposures are cited in this appendix, with the term "chemical hygiene" being substituted for the word "safety". However, since conditions producing or threatening physical injury often pose toxic risks as well, page references concerning major categories of safety hazards in the laboratory are given in section F.

The recommendations from "Prudent Practices" have been paraphrased, combined, or otherwise reorganized, and headings have been added. However, their sense has not been changed.

Corresponding Sections of the Standard and this Appendix

The following table is given for the convenience of those who are developing a Chemical Hygiene Plan which will satisfy the requirements of paragraph (e) of the standard. It indicates those sections of this appendix which are most pertinent to each of the sections of paragraph (e) and related paragraphs.

Paragraph and topic in laboratory standard	Relevant appendix section
(e)(3)(i) Standard operating procedures for handling toxic chemicals.	C, D, E
(e)(3)(ii) Criteria to be used for implementation of measures to reduce exposures.	D
(e)(3)(iii) Fume hood performance..........	C4b
(e)(3)(iv) Employee information and training (including emergency procedures).	D10, D9
(e)(3)(v) Requirements for prior approval of laboratory activities.	E2b, E4b
(e)(3)(vi) Medical consultation and medical examinations.	D5, E4f
(e)(3)(vii) Chemical hygiene responsibilities.	B
(e)(3)(viii) Special precautions for work with particularly hazardous substances.	E2, E3, E4

In this appendix, those recommendations directed primarily at administrators and supervisors are given in sections A–D. Those recommendations of primary concern to employees who are actually handling laboratory chemicals are given in section E. (Reference to page numbers in "Prudent Practices" are given in parentheses.)

A. General Principles for Work with Laboratory Chemicals

In addition to the more detailed recommendations listed below in sections B–E, "Prudent Practices" expresses certain general principles, including the following:

1. *It is prudent to minimize all chemical exposures.* Because few laboratory chemicals are without hazards, general precautions for handling all laboratory chemicals should be adopted, rather than specific guidelines for particular chemicals (2, 10). Skin contact with chemicals should be avoided as a cardinal rule (198).

2. *Avoid underestimation of risk.* Even for substances of no known significant hazard, exposure should be minimized; for work with substances which present special hazards, special precautions should be taken (10, 37, 38). One should assume that any mixture will

be more toxic than its most toxic component (30, 103) and that all substances of unknown toxicity are toxic (3, 34).

3. *Provide adequate ventilation.* The best way to prevent exposure to airborne substances is to prevent their escape into the working atmosphere by use of hoods and other ventilation devices (32, 198).

4. *Institute a chemical hygiene program.* A mandatory chemical hygiene program designed to minimize exposures is needed; it should be a regular, continuing effort, not merely a standby or short-term activity (6, 11). Its recommendations should be followed in academic teaching laboratories as well as by full-time laboratory workers (13).

5. *Observe the PELs, TLVs.* The Permissible Exposure Limits of OSHA and the Threshold Limit Values of the American Conference of Governmental Industrial Hygienists should not be exceeded (13).

B. Chemical Hygiene Responsibilities

Responsibility for chemical hygiene rests at all levels (6, 11, 21) including the:

1. *Chief executive officer,* who has ultimate responsibility for chemical hygiene within the institution and must, with other administrators, provide continuing support for institutional chemical hygiene (7, 11).

2. *Supervisor of the department or other administrative unit,* who is responsible for chemical hygiene in that unit (7).

3. *Chemical hygiene officer(s),* whose appointment is essential (7) and who must:

(a) Work with administrators and other employees to develop and implement appropriate chemical hygiene policies and practices (7);

(b) Monitor procurement, use, and disposal of chemicals used in the lab (8);

(c) See that appropriate audits are maintained (8);

(d) Help project directors develop precautions and adequate facilities (10);

(e) Know the current legal requirements concerning regulated substances (50); and

(f) Seek ways to improve the chemical hygiene program (8, 11).

4. *Laboratory supervisor,* who has overall responsibility for chemical hygiene in the laboratory (21) including responsibility to:

(a) Ensure that workers know and follow the chemical hygiene rules, that protective equipment is available and in working order, and that appropriate training has been provided (21, 22);

(b) Provide regular, formal chemical hygiene and housekeeping inspections including routine inspections of emergency equipment (21, 171);

(c) Know the current legal requirements concerning regulated substances (50, 231);

(d) Determine the required levels of protective apparel and equipment (156, 160, 162); and

(e) Ensure that facilities and training for use of any material being ordered are adequate (215).

5. *Project director or director of other specific operation,* who has primary responsibility for chemical hygiene procedures for that operation (7).

6. *Laboratory worker,* who is responsible for:

(a) Planning and conducting each operation in accordance with the institutional chemical hygiene procedures (7, 21, 22, 230); and

(b) Developing good personal chemical hygiene habits (22).

C. The Laboratory Facility

1. *Design.* The laboratory facility should have:

(a) An appropriate general ventilation system (see C4 below) with air intakes and exhausts located so as to avoid intake of contaminated air (194);

(b) Adequate, well-ventilated stockrooms/storerooms (218, 219);

(c) Laboratory hoods and sinks (12, 162);

(d) Other safety equipment including eyewash fountains and drench showers (162, 169); and

(e) Arrangements for waste disposal (12, 240).

2. *Maintenance.* Chemical-hygiene-related equipment (hoods, incinerator, etc.) should undergo continuing appraisal and be modified if inadequate (11, 12).

3. *Usage.* The work conducted (10) and its scale (12) must be appropriate to the physicial facilities available and, especially, to the quality of ventilation (13).

4. *Ventilation*—(a) *General laboratory ventilation.* This system should: Provide a source of air for breathing and for input to local ventilation devices (199); it should not be relied on for protection from toxic substances released into the laboratory (198); ensure that laboratory air is continually replaced, preventing increase of air concentrations of toxic substances during the working day (194); direct air flow into the laboratory from non-laboratory areas and out to the exterior of the building (194).

(b) *Hoods.* A laboratory hood with 2.5 linear feet of hood space per person should be provided for every 2 workers if they spend most of their time working with chemicals (199); each hood should have a continuous monitoring device to allow convenient confirmation of adequate hood performance before use (200, 209). If this is not possible, work with substances of unknown toxicity should be avoided (13) or other types of local ventilation devices should be provided (199). See pp. 201–206 for a discussion of hood design, construction, and evaluation.

(c) *Other local ventilation devices.* Ventilated storage cabinets, canopy hoods, snorkels, etc. should be provided as needed (199). Each canopy hood and snorkel should have a separate exhaust duct (207).

(d) *Special ventilation areas.* Exhaust air from glove boxes and isolation rooms should be passed through scrubbers or other treatment before release into the regular exhaust system (208). Cold rooms and warm rooms should have provisions for rapid escape and for escape in the event of electrical failure (209).

(e) *Modifications.* Any alteration of the ventilation system should be made only if thorough testing indicates that worker protection from airborne toxic substances will continue to be adequate (12, 193, 204).

(f) *Performance.* Rate: 4–12 room air changes/hour is normally adequate general ventilation if local exhaust systems such as hoods are used as the primary method of control (194).

(g) *Quality.* General air flow should not be turbulent and should be relatively uniform throughout the laboratory, with no high velocity or static areas (194, 195); airflow into and within the hood should not be excessively turbulent (200); hood face velocity should be adequate (typically 60–100 lfm) (200, 204).

(h) *Evaluation.* Quality and quantity of ventilation should be evaluated on installation (202), regularly monitored (at least every 3 months) (6, 12, 14, 195), and reevaluated whenever a change in local ventilation devices is made (12, 195, 207). See pp. 195–198 for methods of evaluation and for calculation of estimated airborne contaminant concentrations.

D. Components of the Chemical Hygiene Plan

1. Basic Rules and Procedures (Recommendations for these are given in section E, below)

2. Chemical Procurement, Distribution, and Storage

(a) *Procurement.* Before a substance is received, information on proper handling, storage, and disposal should be known to those who will be involved (215, 216). No container should be accepted without an adequate identifying label (216). Preferably, all substances should be received in a central location (216).

(b) *Stockrooms/storerooms.* Toxic substances should be segregated in a well-identified area with local exhaust ventilation (221). Chemicals which are highly toxic (227) or other chemicals whose containers have been opened should be in unbreakable secondary containers (219). Stored chemicals should be examined periodically (at least annually) for replacement, deterioration, and container integrity (218–19).

Stockrooms/storerooms should not be used as preparation or repackaging areas, should be open during normal working hours, and should be controlled by one person (219).

(c) *Distribution.* When chemicals are hand carried, the container should be placed in an outside container or bucket. Freight-only elevators should be used if possible (223).

(d) *Laboratory storage.* Amounts permitted should be as small as practical. Storage on bench tops and in hoods is inadvisable. Exposure to heat or direct sunlight should be avoided. Periodic inventories should be conducted, with unneeded items being discarded or returned to the storeroom/stockroom (225–6, 229).

3. Environmental Monitoring

Regular instrumental monitoring of airborne concentrations is not usually justified or practical in laboratories but may be appropriate when testing or redesigning hoods or other ventilation devices (12) or when a highly toxic substance is stored or used regularly (e.g., 3 times/week) (13).

4. Housekeeping, Maintenance, and Inspections

(a) Cleaning. Floors should be cleaned regularly (24).

(b) Inspections. Formal housekeeping and chemical hygiene inspections should be held at least quarterly (6, 21) for units which have frequent pesonnel changes and semiannually for others; informal inspections should be continual (21).

(c) *Maintenance.* Eye wash fountains should be inspected at intervals of not less than 3 months (6). Respirators for routine use should be inspected periodically by the laboratory supervisor (169). Safety showers should be tested routinely (169). Other safety equipment should be inspected regularly. (*e.g.,* every 3–6 months) (6, 24, 171). Procedures to prevent restarting of out-of-service equipment should be established (25).

(d) *Passageways.* Stairways and hallways should not be used as storage areas (24). Access to exits, emergency equipment, and utility controls should never be blocked (24).

5. Medical Program

(a) *Compliance with regulations.* Regular medical surveillance should be established to the extent required by regulations (12).

(b) *Routine surveillance.* Anyone whose work involves regular and frequent handling of toxicologically significant quantities of a chemical should consult a qualified physician to determine on an individual basis whether a regular schedule of medical surveillance is desirable (11, 50).

(c) *First aid.* Personnel trained in first aid should be available during working hours and an emergency room with medical personnel should be nearby (173). See pp. 176–178 for description of some emergency first aid procedures.

6. Protective Apparel and Equipment

These should include for each laboratory:

(a) Protective apparel compatible with the required degree of protection for substances being handled (158–161);

(b) An easily accessible drench-type safety shower (162, 169);

(c) An eyewash fountain (162);

(d) A fire extinguisher (162–164);

(e) Respiratory protection (164–9), fire alarm and telephone for emergency use (162) should be available nearby; and

(f) Other items designated by the laboratory supervisor (156, 160).

7. Records

(a) Accident records should be written and retained (174).

(b) Chemical Hygiene Plan records should document that the facilities and precautions were compatible with current knowledge and regulations (7).

(c) Inventory and usage records for high-risk substances should be kept as specified in sections E3e below.

(d) Medical records should be retained by the institution in accordance with the requirements of state and federal regulations (12).

8. Signs and Labels

Prominent signs and labels of the following types should be posted:

(a) Emergency telephone numbers of emergency personnel/facilities, supervisors, and laboratory workers (28);

(b) Identity labels, showing contents of containers (including waste receptacles) and associated hazards (27, 48);

(c) Location signs for safety showers, eyewash stations, other safety and first aid equipment, exits (27) and areas where food and beverage consumption and storage are permitted (24); and

(d) Warnings at areas or equipment where special or unusual hazards exist (27).

9. Spills and Accidents

(a) A written emergency plan should be established and communicated to all personnel; it should include procedures for ventilation failure (200), evacuation, medical care, reporting, and drills (172).

(b) There should be an alarm system to alert people in all parts of the facility including isolation areas such as cold rooms (172).

(c) A spill control policy should be developed and should include consideration of prevention, containment, cleanup, and reporting (175).

(d) All accidents or near accidents should be carefully analyzed with the results distributed to all who might benefit (8, 28).

10. Information and Training Program

(a) Aim: To assure that all individuals at risk are adequately informed about the work in the laboratory, its risks, and what to do if an accident occurs (5, 15).

(b) Emergency and Personal Protection Training: Every laboratory worker should know the location and proper use of available protective apparel and equipment (154, 169).

Some of the full-time personnel of the laboratory should be trained in the proper use of emergency equipment and procedures (6).

Such training as well as first aid instruction should be available to (154) and encouraged for (176) everyone who might need it.

(c) Receiving and stockroom/storeroom personnel should know about hazards, handling equipment, protective apparel, and relevant regulations (217).

(d) Frequency of Training: The training and education program should be a regular, continuing activity—not simply an annual presentation (15).

(e) Literature/Consultation: Literature and consulting advice concerning chemical hygiene should be readily available to laboratory personnel, who should be encouraged to use these information resources (14).

11. *Waste Disposal Program.*

(a) Aim: To assure that minimal harm to people, other organisms, and the environment will result from the disposal of waste laboratory chemicals (5).

(b) Content (14, 232, 233, 240): The waste disposal program should specify how waste is to be collected, segregated, stored, and transported and include consideration of what materials can be incinerated. Transport from the institution must be in accordance with DOT regulations (244).

(c) Discarding Chemical Stocks: Unlabeled containers of chemicals and solutions should undergo prompt disposal; if partially used, they should not be opened (24, 27).

Before a worker's employment in the laboratory ends, chemicals for which that person was responsible should be discarded or returned to storage (226).

(d) Frequency of Disposal: Waste should be removed from laboratories to a central waste storage area at least once per week and from the central waste storage area at regular intervals (14).

(e) Method of Disposal: Incineration in an environmentally acceptable manner is the most practical disposal method for combustible laboratory waste (14, 238, 241).

Indiscriminate disposal by pouring waste chemicals down the drain (14, 231, 242) or adding them to mixed refuse for landfill burial is unacceptable (14).

Hoods should not be used as a means of disposal for volatile chemicals (40, 200).

Disposal by recycling (233, 243) or chemical decontamination (40, 230) should be used when possible.

E. Basic Rules and Procedures for Working with Chemicals

The Chemical Hygiene Plan should require that laboratory workers know and follow its rules and procedures. In addition to the procedures of the sub programs mentioned above, these should include the rules listed below.

1. General Rules

The following should be used for essentially all laboratory work with chemicals:

(a) *Accidents and spills*—Eye Contact: Promptly flush eyes with water for a prolonged period (15 minutes) and seek medical attention (33, 172).

Ingestion: Encourage the victim to drink large amounts of water (178).

Skin Contact: Promptly flush the affected area with water (33, 172, 178) and remove any contaminated clothing (172, 178). If symptoms persist after washing, seek medical attention (33).

Clean-up. Promptly clean up spills, using appropriate protective apparel and equipment and proper disposal (24 33). See pp. 233–237 for specific clean-up recommendations.

(b) *Avoidance of "routine" exposure:* Develop and encourage safe habits (23); avoid unnecessary exposure to chemicals by any route (23);

Do not smell or taste chemicals (32). Vent apparatus which may discharge toxic chemicals (vacuum pumps, distillation columns, etc.) into local exhaust devices (199).

Inspect gloves (157) and test glove boxes (208) before use.

Do not allow release of toxic substances in cold rooms and warm rooms, since these have contained recirculated atmospheres (209).

(c) *Choice of chemicals:* Use only those chemicals for which the quality of the available ventilation system is appropriate (13).

(d) *Eating, smoking, etc.:* Avoid eating, drinking, smoking, gum chewing, or application of cosmetics in areas where laboratory chemicals are present (22, 24, 32, 40); wash hands before conducting these activities (23, 24).

Avoid storage, handling or consumption of food or beverages in storage areas, refrigerators, glassware or utensils which are also used for laboratory operations (23, 24, 226).

(e) *Equipment and glassware:* Handle and store laboratory glassware with care to avoid damage; do not use damaged glassware (25). Use extra care with Dewar flasks and other evacuated glass apparatus; shield or wrap them to contain chemicals and fragments should implosion occur (25). Use equipment only for its designed purpose (23, 26).

(f) *Exiting:* Wash areas of exposed skin well before leaving the laboratory (23).

(g) *Horseplay:* Avoid practical jokes or other behavior which might confuse, startle or distract another worker (23).

(h) *Mouth suction:* Do not use mouth suction for pipeting or starting a siphon (23, 32).

(i) *Personal apparel:* Confine long hair and loose clothing (23, 158). Wear shoes at all times in the laboratory but do not wear sandals, perforated shoes, or sneakers (158).

(j) *Personal housekeeping:* Keep the work area clean and uncluttered, with chemicals and equipment being properly labeled and stored; clean up the work area on completion of an operation or at the end of each day (24).

(k) *Personal protection:* Assure that appropriate eye protection (154–156) is worn by all persons, including visitors, where chemicals are stored or handled (22, 23, 33, 154).

Wear appropriate gloves when the potential for contact with toxic materials exists (157); inspect the gloves before each use, wash them before removal, and replace them periodically (157). (A table of resistance to chemicals of common glove materials is given p. 159).

Use appropriate (164–168) respiratory equipment when air contaminant concentrations are not sufficiently restricted by engineering controls (164–5), inspecting the respirator before use (169).

Use any other protective and emergency apparel and equipment as appropriate (22, 157–162).

Avoid use of contact lenses in the laboratory unless necessary; if they are used, inform supervisor so special precautions can be taken (155).

Remove laboratory coats immediately on significant contamination (161).

(l) *Planning:* Seek information and advice about hazards (7), plan appropriate protective procedures, and plan positioning of equipment before beginning any new operation (22, 23).

(m) *Unattended operations:* Leave lights on, place an appropriate sign on the door, and provide for containment of toxic substances in the event of failure of a utility service (such as cooling water) to an unattended operation (27, 128).

(n) *Use of hood:* Use the hood for operations which might result in release of toxic chemical vapors or dust (198–9).

As a rule of thumb, use a hood or other local ventilation device when working with any appreciably volatile substance with a TLV of less than 50 ppm (13).

Confirm adequate hood performance before use; keep hood closed at all times except when adjustments within the hood are being made (200); keep materials stored in hoods to a minimum and do not allow them to block vents or air flow (200).

Leave the hood "on" when it is not in active use if toxic substances are stored in it or if it is uncertain whether adequate general laboratory ventilation will be maintained when it is "off" (200).

(o) *Vigilance:* Be alert to unsafe conditions and see that they are corrected when detected (22).

(p) *Waste disposal:* Assure that the plan for each laboratory operation includes plans and training for waste disposal (230).

Deposit chemical waste in appropriately labeled receptacles and follow all other waste disposal procedures of the Chemical Hygiene Plan (22, 24).

Do not discharge to the sewer concentrated acids or bases (231); highly toxic, malodorous, or lachrymatory substances (231); or any substances which might interfere with the biological activity of waste water treatment plants, create fire or explosion hazards, cause structural damage or obstruct flow (242).

(q) *Working alone:* Avoid working alone in a building; do not work alone in a laboratory if the procedures being conducted are hazardous (28).

2. Working with Allergens and Embryotoxins

(a) *Allergens* (examples: diazomethane, isocyanates, bichromates): Wear suitable gloves to prevent hand contact with allergens or substances of unknown allergenic activity (35).

(b) *Embryotoxins* (34–5) (examples: organomercurials, lead compounds, formamide): If you are a woman of childbearing age, handle these substances only in a hood whose satisfactory performance has been confirmed, using appropriate protective apparel (especially gloves) to prevent skin contact.

Review each use of these materials with the research supervisor and review continuing uses annually or whenever a procedural change is made.

Store these substances, properly labeled, in an adequately ventilated area in an unbreakable secondary container.

Notify supervisors of all incidents of exposure or spills; consult a qualified physician when appropriate.

3. Work with Chemicals of Moderate Chronic or High Acute Toxicity

Examples: diisopropylflurophosphate (41), hydrofluoric acid (43), hydrogen cyanide (45).

Supplemental rules to be followed in addition to those mentioned above (Procedure B of "Prudent Practices", pp. 39–41):

(a) *Aim:* To minimize exposure to these toxic substances by any route using all reasonable precautions (39).

(b) *Applicability:* These precautions are appropriate for substances with moderate chronic or high acute toxicity used in significant quantities (39).

(c) *Location:* Use and store these substances only in areas of restricted access with special warning signs (40, 229).

Always use a hood (previously evaluated to confirm adequate performance with a face velocity of at least 60 linear feet per minute) (40) or other containment device for procedures which may result in the generation of aerosols or vapors containing the substance (39); trap released vapors to prevent their discharge with the hood exhaust (40).

(d) *Personal protection:* Always avoid skin contact by use of gloves and long sleeves (and other protective apparel as appropriate) (39). Always wash hands and arms immediately after working with these materials (40).

(e) *Records:* Maintain records of the amounts of these materials on hand, amounts used, and the names of the workers involved (40, 229).

(f) *Prevention of spills and accidents:* Be prepared for accidents and spills (41).

Assure that at least 2 people are present at all times if a compound in use is highly toxic or of unknown toxicity (39).

Store breakable containers of these substances in chemically resistant trays; also work and mount apparatus above such trays or cover work and storage surfaces with removable, absorbent, plastic backed paper (40).

If a major spill occurs outside the hood, evacuate the area; assure that cleanup personnel wear suitable protective apparel and equipment (41).

(g) *Waste:* Thoroughly decontaminate or incinerate contaminated clothing or shoes (41). If possible, chemically decontaminate by chemical conversion (40).

Store contaminated waste in closed, suitably labeled, impervious containers (for liquids, in glass or plastic bottles half-filled with vermiculite) (40).

4. Work with Chemicals of High Chronic Toxicity

(Examples: dimethylmercury and nickel carbonyl (48), benzo-a-pyrene (51), N-nitrosodiethylamine (54), other human carcinogens or substances with high carcinogenic potency in animals (38).)

Further supplemental rules to be followed, in addition to all these mentioned above, for work with substances of known high chronic toxicity (in quantities above a few milligrams to a few grams, depending on the substance) (47). (Procedure A of "Prudent Practices" pp. 47–50).

(a) *Access:* Conduct all transfers and work with these substances in a "controlled area": a restricted access hood, glove box, or portion of a lab, designated for use of highly toxic substances, for which all people with access are aware of the substances being used and necessary precautions (48).

(b) *Approvals:* Prepare a plan for use and disposal of these materials and obtain the approval of the laboratory supervisor (48).

(c) *Non-contamination/Decontamination:* Protect vacuum pumps against contamination by scrubbers or HEPA filters and vent them into the hood (49). Decontaminate vacuum pumps or other contaminated equipment, including glassware, in the hood before removing them from the controlled area (49, 50).

Decontaminate the controlled area before normal work is resumed there (50).

(d) *Exiting:* On leaving a controlled area, remove any protective apparel (placing it in an appropriate, labeled container) and thoroughly wash hands, forearms, face, and neck (49).

(e) *Housekeeping:* Use a wet mop or a vacuum cleaner equipped with a HEPA filter instead of dry sweeping if the toxic substance was a dry powder (50).

(f) *Medical surveillance:* If using toxicologically significant quantities of such a substance on a regular basis (*e.g.*, 3 times per week), consult a qualified physician concerning desirability of regular medical surveillance (50).

(g) *Records:* Keep accurate records of the amounts of these substances stored (229) and used, the dates of use, and names of users (48).

(h) *Signs and labels:* Assure that the controlled area is conspicuously marked with warning and restricted access signs (49) and that all containers of these substances are appropriately labeled with identity and warning labels (48).

(i) *Spills:* Assure that contingency plans, equipment, and materials to minimize exposures of people and property in case of accident are available (233–4).

(j) *Storage:* Store containers of these chemicals only in a ventilated, limited access (48, 227, 229) area in appropriately labeled, unbreakable, chemically resistant, secondary containers (48, 229).

(k) *Glove boxes:* For a negative pressure glove box, ventilation rate must be at least 2 volume changes/hour and pressure at least 0.5 inches of water (48). For a positive pressure glove box, thoroughly check for leaks before each use (49). In either case, trap the exit gases or filter them through a HEPA filter and then release them into the hood (49).

(l) *Waste:* Use chemical decontamination whenever possible; ensure that containers of contaminated waste (including washings from contaminated flasks) are transferred from the controlled area in a secondary container under the supervision of authorized personnel (49, 50, 233).

5. Animal Work with Chemicals of High Chronic Toxicity

(a) *Access:* For large scale studies, special facilities with restricted access are preferable (56).

(b) *Administration of the toxic substance:* When possible, administer the substance by injection or gavage instead of in the diet. If administration is in the diet, use a caging system under negative pressure or under laminar air flow directed toward HEPA filters (56).

(c) *Aerosol suppression:* Devise procedures which minimize formation and dispersal of contaminated aerosols, including those from food, urine, and feces (e.g., use HEPA filtered vacuum equipment for cleaning, moisten contaminated bedding before removal from the cage, mix diets in closed containers in a hood) (55, 56).

(d) *Personal protection:* When working in the animal room, wear plastic or rubber gloves, fully buttoned laboratory coat or jumpsuit and, if needed because of incomplete suppression of aerosols, other apparel and equipment (shoe and head coverings, respirator) (56).

(e) *Waste disposal:* Dispose of contaminated animal tissues and excreta by incineration if the available incinerator can convert the contaminant to non-toxic products (238); otherwise, package the waste appropriately for burial in an EPA-approved site (239).

F. Safety Recommendations

The above recommendations from "Prudent Practices" do not include those which are directed primarily toward prevention of physical injury rather than toxic exposure. However, failure of precautions against injury will often have the secondary effect of causing toxic exposures. Therefore, we list below page references for recommendations concerning some of the major categories of safety hazards which also have implications for chemical hygiene:

1. Corrosive agents: (35–6)
2. Electrically powered laboratory apparatus: (179–92)
3. Fires, explosions: (26, 57–74, 162–4, 174–5, 219–20, 226–7)

4. Low temperature procedures: (26, 88)
5. Pressurized and vacuum operations (including use of compressed gas cylinders): (27, 75–101)

G. Material Safety Data Sheets

Material safety data sheets are presented in "Prudent Practices" for the chemicals listed below. (Asterisks denote that comprehensive material safety data sheets are provided).

*Acetyl peroxide (105)
*Acrolein (106)
*Acrylonilrile (107)
Ammonia (anhydrous) (91)
*Aniline (109)
*Benzene (110)
*Benzo[a]pyrene (112)
*Bis(chloromethyl) ether (113)
Boron trichloride (91)
Boron trifluoride (92)
Bromine (114)
*Tert-butyl hydroperoxide (148)
*Carbon disulfide (116)
Carbon monoxide (92)
*Carbon tetrachloride (118)
*Chlorine (119)
Chlorine trifluoride (94)
*Chloroform (121)
Chloromethane (93)
*Diethyl ether (122)
Diisopropyl fluorophosphate (41)
*Dimethylformamide (123)
*Dimethyl sulfate (125)
*Dioxane (126)
*Ethylene dibromide (128)
*Fluorine (95)
*Formaldehyde (130)
*Hydrazine and salts (132)
Hydrofluoric acid (43)
Hydrogen bromide (98)
Hydrogen chloride (98)
*Hydrogen cyanide (133)
*Hydrogen sulfide (135)
Mercury and compounds (52)
*Methanol (137)
*Morpholine (138)
*Nickel carbonyl (99)
*Nitrobenzene (139)
Nitrogen dioxide (100)
N-nitrosodiethylamine (54)
*Peracetic acid (141)
*Phenol (142)
*Phosgene (143)
*Pyridine (144)
*Sodium azide (145)
*Sodium cyanide (147)
Sulfur dioxide (101)
*Trichloroethylene (149)
*Vinyl chloride (150)

Appendix B to § 1910.1450—References (Non-Mandatory)

The following references are provided to assist the employer in the development of a Chemical Hygiene Plan. The materials listed below are offered as non-mandatory guidance. References listed here do not imply specific endorsement of a book, opinion, technique, policy or a specific solution for a safety or health problem. Other references not listed here may better meet the needs of a specific laboratory. (a) Materials for the development of the Chemical Hygiene Plan:

1. American Chemical Society, Safety in Academic Chemistry Laboratories, 4th edition, 1985.

2. Fawcett, H.H. and W. S. Wood, Safety and Accident Prevention in Chemical Operations, 2nd edition, Wiley-Interscience, New York, 1982.

3. Flury, Patricia A., Environmental Health and Safety in the Hospital Laboratory, Charles C. Thomas Publisher, Springfield IL, 1978.

3. Green, Michael E. and Turk, Amos, Safety in Working with Chemicals, Macmillan Publishing Co., NY, 1978.

5. Kaufman, James A., Laboratory Safety Guidelines, Dow Chemical Co., Box 1713, Midland, MI 48640, 1977.

6. National Institutes of Health, NIH Guidelines for the Laboratory use of Chemical Carcinogens, NIH Pub. No. 81–2385, GPO, Washington, DC 20402, 1981.

7. National Research Council, Prudent Practices for Disposal of Chemicals from Laboratories, National Academy Press, Washington, DC, 1983.

8. National Research Council, Prudent Practices for Handling Hazardous Chemicals in Laboratories, National Academy Press, Washington, DC, 1981.

9. Renfrew, Malcolm, Ed., Safety in the Chemical Laboratory, Vol. IV, *J. Chem. Ed.*, American Chemical Society, Easlon, PA, 1981.

10. Steere, Norman V., Ed., Safety in the Chemical Laboratory, *J. Chem. Ed.* American Chemical Society, Easlon, PA, 18042, Vol. I, 1967, Vol. II, 1971, Vol. III 1974.

11. Steere, Norman V., Handbook of Laboratory Safety, the Chemical Rubber Company Cleveland, OH, 1971.

12. Young, Jay A., Ed., Improving Safety in the Chemical Laboratory, John Wiley & Sons, Inc. New York, 1987.

(b) Hazardous Substances Information:

1. American Conference of Governmental Industrial Hygienists, Threshold Limit Values for Chemical Substances and Physical Agents in the Workroom Environment with Intended Changes, P.O. Box 1937 Cincinnati, OH 45201 (latest edition).

2. Annual Report on Carcinogens, National Toxicology Program U.S. Department of Health and Human Services, Public Health Service, U.S. Government Printing Office, Washington, DC, (latest edition).

3. Best Company, Best Safety Directory, Vols. I and II, Oldwick, N.J., 1981.

4. Bretherick, L., Handbook of Reactive Chemical Hazards, 2nd edition, Butterworths, London, 1979.

5. Bretherick, L., Hazards in the Chemical Laboratory, 3rd edition, Royal Society of Chemistry, London, 1986.

6. Code of Federal Regulations, 29 CFR part 1910 subpart Z. U.S. Govt. Printing Office, Washington, DC 20402 (latest edition).

7. IARC Monographs on the Evaluation of the Carcinogenic Risk of Chemicals to Man, World Health Organization Publications Center, 49 Sheridan Avenue, Albany, New York 12210 (latest editions).

8. NIOSH/OSHA Pocket Guide to Chemical Hazards. NIOSH Pub. No. 85–114, U.S. Government Printing Office, Washington, DC, 1985 (or latest edition).

9. Occupational Health Guidelines, NIOSH/OSHA NIOSH Pub. No. 81–123 U.S. Government Printing Office, Washington, DC, 1981.

10. Patty, F.A., Industrial Hygiene and Toxicology, John Wiley & Sons, Inc., New York, NY (Five Volumes).

11. Registry of Toxic Effects of Chemical Substances, U.S. Department of Health and Human Services, Public Health Service, Centers for Disease Control, National Institute for Occupational Safety and Health, Revised Annually, for sale from Superintendent of Documents U.S. Govt. Printing Office, Washington, DC 20402.

12. The Merck Index: An Encyclopedia of Chemicals and Drugs. Merck and Company Inc. Rahway, N.J., 1976 (or latest edition).

13. Sax, N.I. Dangerous Properties of Industrial Materials, 5th edition, Van Nostrand Reinhold, NY., 1979.

14. Sittig, Marshall, Handbook of Toxic and Hazardous Chemicals, Noyes Publications, Park Ridge, NJ, 1981.

(c) Information on Ventilation:

1. American Conference of Governmental Industrial Hygienists Industrial Ventilation, 16th edition Lansing, MI, 1980.

2. American National Standards Institute, Inc. American National Standards Fundamentals Governing the Design and Operation of Local Exhaust Systems ANSI Z 9.2–1979 American National Standards Institute, N.Y. 1979.

3. Imad, A.P. and Watson, C.L. Ventilation Index: An Easy Way to Decide about Hazardous Liquids, Professional Safety pp 15–18, April 1980.

4. National Fire Protection Association, Fire Protection for Laboratories Using Chemicals NFPA–45, 1982.

Safety Standard for Laboratories in Health Related Institutions, NFPA, 56c, 1980.

Fire Protection Guide on Hazardous Materials, 7th edition, 1978.

National Fire Protection Association, Batterymarch Park, Quincy, MA 02269.

5. Scientific Apparatus Makers Association (SAMA), Standard for Laboratory Fume Hoods, SAMA LF7–1980, 1101 16th Street, NW., Washington, DC 20036.

(d) Information on Availability of Referenced Material:

1. American National Standards Institute (ANSI), 1430 Broadway, New York, NY 10018.

2. American Society for Testing and Materials (ASTM), 1916 Race Street, Philadelphia, PA 19103.

(Approved by the Office of Management and Budget under control number 1218–0131)

[FR Doc. 90–1717 Filed 1–30–90; 8:45 am]

APPENDIX B

Recommendations for Prevention of HIV Transmission in Health-Care Settings

Introduction

Human immunodeficiency virus (HIV), the virus that causes acquired immunodeficiency syndrome (AIDS), is transmitted through sexual contact and exposure to infected blood or blood components and perinatally from mother to neonate. HIV has been isolated from blood, semen, vaginal secretions, saliva, tears, breast milk, cerebrospinal fluid, amniotic fluid, and urine and is likely to be isolated from other body fluids, secretions, and excretions. However, epidemiologic evidence has implicated only blood, semen, vaginal secretions, and possibly breast milk in transmission.

The increasing prevalence of HIV increases the risk that health-care workers will be exposed to blood from patients infected with HIV, especially when blood and body-fluid precautions are not followed for all patients. Thus, this document emphasizes the need for health-care workers to consider **all** patients as potentially infected with HIV and/or other blood-borne pathogens and to adhere rigorously to infection-control precautions for minimizing the risk of exposure to blood and body fluids of all patients.

The recommendations contained in this document consolidate and update CDC recommendations published earlier for preventing HIV transmission in health-care settings: precautions for clinical and laboratory staffs (*1*) and precautions for health-care workers and allied professionals (*2*); recommendations for preventing

---

SOURCE: The contents of Appendix B are reproduced from *Morbidity and Mortality Weekly Report,* Vol. 36, No. 2S, August 21, 1987, pp. 3S-18S. Published by Centers for Disease Control, U.S. Department of Health and Human Services, Atlanta, GA.

HIV transmission in the workplace (*3*) and during invasive procedures (*4*); recommendations for preventing possible transmission of HIV from tears (*5*); and recommendations for providing dialysis treatment for HIV-infected patients (*6*). These recommendations also update portions of the "Guideline for Isolation Precautions in Hospitals" (*7*) and reemphasize some of the recommendations contained in "Infection Control Practices for Dentistry" (*8*). The recommendations contained in this document have been developed for use in health-care settings and emphasize the need to treat blood and other body fluids from **all** patients as potentially infective. These same prudent precautions also should be taken in other settings in which persons may be exposed to blood or other body fluids.

Definition of Health-Care Workers

Health-care workers are defined as persons, including students and trainees, whose activities involve contact with patients or with blood or other body fluids from patients in a health-care setting.

Health-Care Workers with AIDS

As of July 10, 1987, a total of 1,875 (5.8%) of 32,395 adults with AIDS, who had been reported to the CDC national surveillance system and for whom occupational information was available, reported being employed in a health-care or clinical laboratory setting. In comparison, 6.8 million persons—representing 5.6% of the U.S. labor force—were employed in health services. Of the health-care workers with AIDS, 95% have been reported to exhibit high-risk behavior; for the remaining 5%, the means of HIV acquisition was undetermined. Health-care workers with AIDS were significantly more likely than other workers to have an undetermined risk (5% versus 3%, respectively). For both health-care workers and non-health-care workers with AIDS, the proportion with an undetermined risk has not increased since 1982.

AIDS patients initially reported as not belonging to recognized risk groups are investigated by state and local health departments to determine whether possible risk factors exist. Of all health-care workers with AIDS reported to CDC who were initially characterized as not having an identified risk and for whom follow-up information was available, 66% have been reclassified because risk factors were identified or because the patient was found not to meet the surveillance case definition for AIDS. Of the 87 health-care workers currently categorized as having no identifiable risk, information is incomplete on 16 (18%) because of death or refusal to be interviewed; 38 (44%) are still being investigated. The remaining 33 (38%) health-care workers were interviewed or had other follow-up information available. The occupations of these 33 were as follows: five physicians (15%), three of whom were surgeons; one dentist (3%); three nurses (9%); nine nursing assistants (27%); seven housekeeping or maintenance workers (21%); three clinical laboratory technicians (9%); one therapist (3%); and four others who did not have contact with patients (12%). Although 15 of these 33 health-care workers reported parenteral and/or other non-needlestick exposure to blood or body fluids from patients in the 10 years preceding their diagnosis of AIDS, none of these exposures involved a patient with AIDS or known HIV infection.

Risk to Health-Care Workers of Acquiring HIV in Health-Care Settings

Health-care workers with documented percutaneous or mucous-membrane exposures to blood or body fluids of HIV-infected patients have been prospectively evaluated to determine the risk of infection after such exposures. As of June 30, 1987, 883 health-care workers have been tested for antibody to HIV in an ongoing surveillance project conducted by CDC (*9*). Of these, 708 (80%) had percutaneous exposures to blood, and 175 (20%) had a mucous membrane or an open wound contaminated by blood or body fluid. Of 396 health-care workers, each of whom had only a convalescent-phase serum sample obtained and tested ≥90 days post-exposure, one—for whom heterosexual transmission could not be ruled out—was seropositive for HIV antibody. For 425 additional health-care workers, both acute- and convalescent-phase serum samples were obtained and tested; none of 74 health-care workers with nonpercutaneous exposures seroconverted, and three (0.9%) of 351 with percutaneous exposures seroconverted. None of these three health-care workers had other documented risk factors for infection.

Two other prospective studies to assess the risk of nosocomial acquisition of HIV infection for health-care workers are ongoing in the United States. As of April 30, 1987, 332 health-care workers with a total of 453 needlestick or mucous-membrane exposures to the blood or other body fluids of HIV-infected patients were tested for HIV antibody at the National Institutes of Health (*10*). These exposed workers included 103 with needlestick injuries and 229 with mucous-membrane exposures; none had seroconverted. A similar study at the University of California of 129 health-care workers with documented needlestick injuries or mucous-membrane exposures to blood or other body fluids from patients with HIV infection has not identified any seroconversions (*11*). Results of a prospective study in the United Kingdom identified no evidence of transmission among 150 health-care workers with parenteral or mucous-membrane exposures to blood or other body fluids, secretions, or excretions from patients with HIV infection (*12*).

In addition to health-care workers enrolled in prospective studies, eight persons who provided care to infected patients and denied other risk factors have been reported to have acquired HIV infection. Three of these health-care workers had needlestick exposures to blood from infected patients (*13-15*). Two were persons who provided nursing care to infected persons; although neither sustained a needlestick, both had extensive contact with blood or other body fluids, and neither observed recommended barrier precautions (*16,17*). The other three were health-care workers with non-needlestick exposures to blood from infected patients (*18*). Although the exact route of transmission for these last three infections is not known, all three persons had direct contact of their skin with blood from infected patients, all had skin lesions that may have been contaminated by blood, and one also had a mucous-membrane exposure.

A total of 1,231 dentists and hygienists, many of whom practiced in areas with many AIDS cases, participated in a study to determine the prevalence of antibody to HIV; one dentist (0.1%) had HIV antibody. Although no exposure to a known HIV-infected person could be documented, epidemiologic investigation did not identify any other risk factor for infection. The infected dentist, who also had a history of sustaining needlestick injuries and trauma to his hands, did not routinely wear gloves when providing dental care (*19*).

Precautions To Prevent Transmission of HIV

Universal Precautions

Since medical history and examination cannot reliably identify all patients infected with HIV or other blood-borne pathogens, blood and body-fluid precautions should be consistently used for **all** patients. This approach, previously recommended by CDC (*3,4*), and referred to as "universal blood and body-fluid precautions" or "universal precautions," should be used in the care of **all** patients, especially including those in emergency-care settings in which the risk of blood exposure is increased and the infection status of the patient is usually unknown (*20*).

1. All health-care workers should routinely use appropriate barrier precautions to prevent skin and mucous-membrane exposure when contact with blood or other body fluids of any patient is anticipated. Gloves should be worn for touching blood and body fluids, mucous membranes, or non-intact skin of all patients, for handling items or surfaces soiled with blood or body fluids, and for performing venipuncture and other vascular access procedures. Gloves should be changed after contact with each patient. Masks and protective eyewear or face shields should be worn during procedures that are likely to generate droplets of blood or other body fluids to prevent exposure of mucous membranes of the mouth, nose, and eyes. Gowns or aprons should be worn during procedures that are likely to generate splashes of blood or other body fluids.
2. Hands and other skin surfaces should be washed immediately and thoroughly if contaminated with blood or other body fluids. Hands should be washed immediately after gloves are removed.
3. All health-care workers should take precautions to prevent injuries caused by needles, scalpels, and other sharp instruments or devices during procedures; when cleaning used instruments; during disposal of used needles; and when handling sharp instruments after procedures. To prevent needlestick injuries, needles should not be recapped, purposely bent or broken by hand, removed from disposable syringes, or otherwise manipulated by hand. After they are used, disposable syringes and needles, scalpel blades, and other sharp items should be placed in puncture-resistant containers for disposal; the puncture-resistant containers should be located as close as practical to the use area. Large-bore reusable needles should be placed in a puncture-resistant container for transport to the reprocessing area.
4. Although saliva has not been implicated in HIV transmission, to minimize the need for emergency mouth-to-mouth resuscitation, mouthpieces, resuscitation bags, or other ventilation devices should be available for use in areas in which the need for resuscitation is predictable.
5. Health-care workers who have exudative lesions or weeping dermatitis should refrain from all direct patient care and from handling patient-care equipment until the condition resolves.
6. Pregnant health-care workers are not known to be at greater risk of contracting HIV infection than health-care workers who are not pregnant; however, if a health-care worker develops HIV infection during pregnancy, the infant is at risk of infection resulting from perinatal transmission. Because of this risk, pregnant health-care workers should be especially familiar with and strictly adhere to precautions to minimize the risk of HIV transmission.

Implementation of universal blood and body-fluid precautions for **all** patients eliminates the need for use of the isolation category of "Blood and Body Fluid Precautions" previously recommended by CDC (*7*) for patients known or suspected to be infected with blood-borne pathogens. Isolation precautions (e.g., enteric, "AFB" [*7*]) should be used as necessary if associated conditions, such as infectious diarrhea or tuberculosis, are diagnosed or suspected.

Precautions for Invasive Procedures

In this document, an invasive procedure is defined as surgical entry into tissues, cavities, or organs or repair of major traumatic injuries 1) in an operating or delivery room, emergency department, or outpatient setting, including both physicians' and dentists' offices; 2) cardiac catheterization and angiographic procedures; 3) a vaginal or cesarean delivery or other invasive obstetric procedure during which bleeding may occur; or 4) the manipulation, cutting, or removal of any oral or perioral tissues, including tooth structure, during which bleeding occurs or the potential for bleeding exists. The universal blood and body-fluid precautions listed above, combined with the precautions listed below, should be the minimum precautions for **all** such invasive procedures.

1. All health-care workers who participate in invasive procedures must routinely use appropriate barrier precautions to prevent skin and mucous-membrane contact with blood and other body fluids of all patients. Gloves and surgical masks must be worn for all invasive procedures. Protective eyewear or face shields should be worn for procedures that commonly result in the generation of droplets, splashing of blood or other body fluids, or the generation of bone chips. Gowns or aprons made of materials that provide an effective barrier should be worn during invasive procedures that are likely to result in the splashing of blood or other body fluids. All health-care workers who perform or assist in vaginal or cesarean deliveries should wear gloves and gowns when handling the placenta or the infant until blood and amniotic fluid have been removed from the infant's skin and should wear gloves during post-delivery care of the umbilical cord.
2. If a glove is torn or a needlestick or other injury occurs, the glove should be removed and a new glove used as promptly as patient safety permits; the needle or instrument involved in the incident should also be removed from the sterile field.

Precautions for Dentistry*

Blood, saliva, and gingival fluid from **all** dental patients should be considered infective. Special emphasis should be placed on the following precautions for preventing transmission of blood-borne pathogens in dental practice in both institutional and non-institutional settings.

1. In addition to wearing gloves for contact with oral mucous membranes of all patients, all dental workers should wear surgical masks and protective eyewear or chin-length plastic face shields during dental procedures in which splashing or spattering of blood, saliva, or gingival fluids is likely. Rubber dams, high-speed evacuation, and proper patient positioning, when appropriate, should be utilized to minimize generation of droplets and spatter.
2. Handpieces should be sterilized after use with each patient, since blood, saliva, or gingival fluid of patients may be aspirated into the handpiece or waterline. Handpieces that cannot be sterilized should at least be flushed, the outside surface cleaned and wiped with a suitable chemical germicide, and then rinsed. Handpieces should be flushed at the beginning of the day and after use with each patient. Manufacturers' recommendations should be followed for use and maintenance of waterlines and check valves and for flushing of handpieces. The same precautions should be used for ultrasonic scalers and air/water syringes.

*General infection-control precautions are more specifically addressed in previous recommendations for infection-control practices for dentistry (*8*).

3. Blood and saliva should be thoroughly and carefully cleaned from material that has been used in the mouth (e.g., impression materials, bite registration), especially before polishing and grinding intra-oral devices. Contaminated materials, impressions, and intra-oral devices should also be cleaned and disinfected before being handled in the dental laboratory and before they are placed in the patient's mouth. Because of the increasing variety of dental materials used intra-orally, dental workers should consult with manufacturers as to the stability of specific materials when using disinfection procedures.
4. Dental equipment and surfaces that are difficult to disinfect (e.g., light handles or X-ray-unit heads) and that may become contaminated should be wrapped with impervious-backed paper, aluminum foil, or clear plastic wrap. The coverings should be removed and discarded, and clean coverings should be put in place after use with each patient.

Precautions for Autopsies or Morticians' Services

In addition to the universal blood and body-fluid precautions listed above, the following precautions should be used by persons performing postmortem procedures:

1. All persons performing or assisting in postmortem procedures should wear gloves, masks, protective eyewear, gowns, and waterproof aprons.
2. Instruments and surfaces contaminated during postmortem procedures should be decontaminated with an appropriate chemical germicide.

Precautions for Dialysis

Patients with end-stage renal disease who are undergoing maintenance dialysis and who have HIV infection can be dialyzed in hospital-based or free-standing dialysis units using conventional infection-control precautions (*21*). Universal blood and body-fluid precautions should be used when dialyzing **all** patients.

Strategies for disinfecting the dialysis fluid pathways of the hemodialysis machine are targeted to control bacterial contamination and generally consist of using 500-750 parts per million (ppm) of sodium hypochlorite (household bleach) for 30-40 minutes or 1.5%-2.0% formaldehyde overnight. In addition, several chemical germicides formulated to disinfect dialysis machines are commercially available. None of these protocols or procedures need to be changed for dialyzing patients infected with HIV.

Patients infected with HIV can be dialyzed by either hemodialysis or peritoneal dialysis and do not need to be isolated from other patients. The type of dialysis treatment (i.e., hemodialysis or peritoneal dialysis) should be based on the needs of the patient. The dialyzer may be discarded after each use. Alternatively, centers that reuse dialyzers—i.e., a specific single-use dialyzer is issued to a specific patient, removed, cleaned, disinfected, and reused several times on the same patient only—may include HIV-infected patients in the dialyzer-reuse program. An individual dialyzer must never be used on more than one patient.

Precautions for Laboratories[†]

Blood and other body fluids from **all** patients should be considered infective. To supplement the universal blood and body-fluid precautions listed above, the following precautions are recommended for health-care workers in clinical laboratories.

[†]Additional precautions for research and industrial laboratories are addressed elsewhere (*22,23*).

1. All specimens of blood and body fluids should be put in a well-constructed container with a secure lid to prevent leaking during transport. Care should be taken when collecting each specimen to avoid contaminating the outside of the container and of the laboratory form accompanying the specimen.
2. All persons processing blood and body-fluid specimens (e.g., removing tops from vacuum tubes) should wear gloves. Masks and protective eyewear should be worn if mucous-membrane contact with blood or body fluids is anticipated. Gloves should be changed and hands washed after completion of specimen processing.
3. For routine procedures, such as histologic and pathologic studies or microbiologic culturing, a biological safety cabinet is not necessary. However, biological safety cabinets (Class I or II) should be used whenever procedures are conducted that have a high potential for generating droplets. These include activities such as blending, sonicating, and vigorous mixing.
4. Mechanical pipetting devices should be used for manipulating all liquids in the laboratory. Mouth pipetting must not be done.
5. Use of needles and syringes should be limited to situations in which there is no alternative, and the recommendations for preventing injuries with needles outlined under universal precautions should be followed.
6. Laboratory work surfaces should be decontaminated with an appropriate chemical germicide after a spill of blood or other body fluids and when work activities are completed.
7. Contaminated materials used in laboratory tests should be decontaminated before reprocessing or be placed in bags and disposed of in accordance with institutional policies for disposal of infective waste (*24*).
8. Scientific equipment that has been contaminated with blood or other body fluids should be decontaminated and cleaned before being repaired in the laboratory or transported to the manufacturer.
9. All persons should wash their hands after completing laboratory activities and should remove protective clothing before leaving the laboratory.

Implementation of universal blood and body-fluid precautions for **all** patients eliminates the need for warning labels on specimens since blood and other body fluids from all patients should be considered infective.

Environmental Considerations for HIV Transmission

No environmentally mediated mode of HIV transmission has been documented. Nevertheless, the precautions described below should be taken routinely in the care of **all** patients.

Sterilization and Disinfection

Standard sterilization and disinfection procedures for patient-care equipment currently recommended for use (*25,26*) in a variety of health-care settings—including hospitals, medical and dental clinics and offices, hemodialysis centers, emergency-care facilities, and long-term nursing-care facilities—are adequate to sterilize or disinfect instruments, devices, or other items contaminated with blood or other body fluids from persons infected with blood-borne pathogens including HIV (*21,23*).

Instruments or devices that enter sterile tissue or the vascular system of any patient or through which blood flows should be sterilized before reuse. Devices or items that contact intact mucous membranes should be sterilized or receive high-level disinfection, a procedure that kills vegetative organisms and viruses but not necessarily large numbers of bacterial spores. Chemical germicides that are registered with the U.S. Environmental Protection Agency (EPA) as "sterilants" may be used either for sterilization or for high-level disinfection depending on contact time.

Contact lenses used in trial fittings should be disinfected after each fitting by using a hydrogen peroxide contact lens disinfecting system or, if compatible, with heat (78 C-80 C [172.4 F-176.0 F]) for 10 minutes.

Medical devices or instruments that require sterilization or disinfection should be thoroughly cleaned before being exposed to the germicide, and the manufacturer's instructions for the use of the germicide should be followed. Further, it is important that the manufacturer's specifications for compatibility of the medical device with chemical germicides be closely followed. Information on specific label claims of commercial germicides can be obtained by writing to the Disinfectants Branch, Office of Pesticides, Environmental Protection Agency, 401 M Street, SW, Washington, D.C. 20460.

Studies have shown that HIV is inactivated rapidly after being exposed to commonly used chemical germicides at concentrations that are much lower than used in practice (*27-30*). Embalming fluids are similar to the types of chemical germicides that have been tested and found to completely inactivate HIV. In addition to commercially available chemical germicides, a solution of sodium hypochlorite (household bleach) prepared daily is an inexpensive and effective germicide. Concentrations ranging from approximately 500 ppm (1:100 dilution of household bleach) sodium hypochlorite to 5,000 ppm (1:10 dilution of household bleach) are effective depending on the amount of organic material (e.g., blood, mucus) present on the surface to be cleaned and disinfected. Commercially available chemical germicides may be more compatible with certain medical devices that might be corroded by repeated exposure to sodium hypochlorite, especially to the 1:10 dilution.

Survival of HIV in the Environment

The most extensive study on the survival of HIV after drying involved greatly concentrated HIV samples, i.e., 10 million tissue-culture infectious doses per milliliter (*31*). This concentration is at least 100,000 times greater than that typically found in the blood or serum of patients with HIV infection. HIV was detectable by tissue-culture techniques 1-3 days after drying, but the rate of inactivation was rapid. Studies performed at CDC have also shown that drying HIV causes a rapid (within several hours) 1-2 log (90%-99%) reduction in HIV concentration. In tissue-culture fluid, cell-free HIV could be detected up to 15 days at room temperature, up to 11 days at 37 C (98.6 F), and up to 1 day if the HIV was cell-associated.

When considered in the context of environmental conditions in health-care facilities, these results do not require any changes in currently recommended sterilization, disinfection, or housekeeping strategies. When medical devices are contaminated with blood or other body fluids, existing recommendations include the cleaning of these instruments, followed by disinfection or sterilization, depending on the type of medical device. These protocols assume "worst-case" conditions of extreme virologic and microbiologic contamination, and whether viruses have been inactivated after drying plays no role in formulating these strategies. Consequently, no changes in published procedures for cleaning, disinfecting, or sterilizing need to be made.

Housekeeping

Environmental surfaces such as walls, floors, and other surfaces are not associated with transmission of infections to patients or health-care workers. Therefore, extraordinary attempts to disinfect or sterilize these environmental surfaces are not necessary. However, cleaning and removal of soil should be done routinely.

Cleaning schedules and methods vary according to the area of the hospital or institution, type of surface to be cleaned, and the amount and type of soil present. Horizontal surfaces (e.g., bedside tables and hard-surfaced flooring) in patient-care areas are usually cleaned on a regular basis, when soiling or spills occur, and when a patient is discharged. Cleaning of walls, blinds, and curtains is recommended only if they are visibly soiled. Disinfectant fogging is an unsatisfactory method of decontaminating air and surfaces and is not recommended.

Disinfectant-detergent formulations registered by EPA can be used for cleaning environmental surfaces, but the actual physical removal of microorganisms by scrubbing is probably at least as important as any antimicrobial effect of the cleaning agent used. Therefore, cost, safety, and acceptability by housekeepers can be the main criteria for selecting any such registered agent. The manufacturers' instructions for appropriate use should be followed.

Cleaning and Decontaminating Spills of Blood or Other Body Fluids

Chemical germicides that are approved for use as "hospital disinfectants" and are tuberculocidal when used at recommended dilutions can be used to decontaminate spills of blood and other body fluids. Strategies for decontaminating spills of blood and other body fluids in a patient-care setting are different than for spills of cultures or other materials in clinical, public health, or research laboratories. In patient-care areas, visible material should first be removed and then the area should be decontaminated. With large spills of cultured or concentrated infectious agents in the laboratory, the contaminated area should be flooded with a liquid germicide before cleaning, then decontaminated with fresh germicidal chemical. In both settings, gloves should be worn during the cleaning and decontaminating procedures.

Laundry

Although soiled linen has been identified as a source of large numbers of certain pathogenic microorganisms, the risk of actual disease transmission is negligible. Rather than rigid procedures and specifications, hygienic and common-sense storage and processing of clean and soiled linen are recommended (*26*). Soiled linen should be handled as little as possible and with minimum agitation to prevent gross microbial contamination of the air and of persons handling the linen. All soiled linen should be bagged at the location where it was used; it should not be sorted or rinsed in patient-care areas. Linen soiled with blood or body fluids should be placed and transported in bags that prevent leakage. If hot water is used, linen should be washed with detergent in water at least 71 C (160 F) for 25 minutes. If low-temperature($\leq$70 C [158 F]) laundry cycles are used, chemicals suitable for low-temperature washing at proper use concentration should be used.

Infective Waste

There is no epidemiologic evidence to suggest that most hospital waste is any more infective than residential waste. Moreover, there is no epidemiologic evidence that hospital waste has caused disease in the community as a result of improper disposal. Therefore, identifying wastes for which special precautions are indicated is largely a matter of judgment about the relative risk of disease transmission. The most practical approach to the management of infective waste is to identify those wastes with the potential for causing infection during handling and disposal and for which some special precautions appear prudent. Hospital wastes for which special precautions appear prudent include microbiology laboratory waste, pathology waste, and blood specimens or blood products. While any item that has had contact with blood, exudates, or secretions may be potentially infective, it is not usually considered practical or necessary to treat all such waste as infective (*23,26*). Infective waste, in general, should either be incinerated or should be autoclaved before disposal in a sanitary landfill. Bulk blood, suctioned fluids, excretions, and secretions may be carefully poured down a drain connected to a sanitary sewer. Sanitary sewers may also be used to dispose of other infectious wastes capable of being ground and flushed into the sewer.

Implementation of Recommended Precautions

Employers of health-care workers should ensure that policies exist for:

1. Initial orientation and continuing education and training of all health-care workers—including students and trainees—on the epidemiology, modes of transmission, and prevention of HIV and other blood-borne infections and the need for routine use of universal blood and body-fluid precautions for **all** patients.
2. Provision of equipment and supplies necessary to minimize the risk of infection with HIV and other blood-borne pathogens.
3. Monitoring adherence to recommended protective measures. When monitoring reveals a failure to follow recommended precautions, counseling, education, and/or re-training should be provided, and, if necessary, appropriate disciplinary action should be considered.

Professional associations and labor organizations, through continuing education efforts, should emphasize the need for health-care workers to follow recommended precautions.

Serologic Testing for HIV Infection

Background

A person is identified as infected with HIV when a sequence of tests, starting with repeated enzyme immunoassays (EIA) and including a Western blot or similar, more specific assay, are repeatedly reactive. Persons infected with HIV usually develop antibody against the virus within 6-12 weeks after infection.

The sensitivity of the currently licensed EIA tests is at least 99% when they are performed under optimal laboratory conditions on serum specimens from persons infected for ≥12 weeks. Optimal laboratory conditions include the use of reliable reagents, provision of continuing education of personnel, quality control of procedures, and participation in performance-evaluation programs. Given this performance, the probability of a false-negative test is remote except during the first several weeks after infection, before detectable antibody is present. The proportion of infected persons with a false-negative test attributed to absence of antibody in the early stages of infection is dependent on both the incidence and prevalence of HIV infection in a population (Table 1).

The specificity of the currently licensed EIA tests is approximately 99% when repeatedly reactive tests are considered. Repeat testing of initially reactive specimens by EIA is required to reduce the likelihood of laboratory error. To increase further the specificity of serologic tests, laboratories must use a supplemental test, most often

TABLE 1. Estimated annual number of patients infected with HIV not detected by HIV-antibody testing in a hypothetical hospital with 10,000 admissions/year*

Beginning prevalence of HIV infection	Annual incidence of HIV infection	Approximate number of HIV-infected patients	Approximate number of HIV-infected patients not detected
5.0%	1.0%	550	17-18
5.0%	0.5%	525	11-12
1.0%	0.2%	110	3-4
1.0%	0.1%	105	2-3
0.1%	0.02%	11	0-1
0.1%	0.01%	11	0-1

*The estimates are based on the following assumptions: 1) the sensitivity of the screening test is 99% (i.e., 99% of HIV-infected persons with antibody will be detected); 2) persons infected with HIV will not develop detectable antibody (seroconvert) until 6 weeks (1.5 months) after infection; 3) new infections occur at an equal rate throughout the year; 4) calculations of the number of HIV-infected persons in the patient population are based on the mid-year prevalence, which is the beginning prevalence plus half the annual incidence of infections.

the Western blot, to validate repeatedly reactive EIA results. Under optimal laboratory conditions, the sensitivity of the Western blot test is comparable to or greater than that of a repeatedly reactive EIA, and the Western blot is highly specific when strict criteria are used to interpret the test results. The testing sequence of a repeatedly reactive EIA and a positive Western blot test is highly predictive of HIV infection, even in a population with a low prevalence of infection (Table 2). If the Western blot test result is indeterminant, the testing sequence is considered equivocal for HIV infection. When this occurs, the Western blot test should be repeated on the same serum sample, and, if still indeterminant, the testing sequence should be repeated on a sample collected 3-6 months later. Use of other supplemental tests may aid in interpreting of results on samples that are persistently indeterminant by Western blot.

Testing of Patients

Previous CDC recommendations have emphasized the value of HIV serologic testing of patients for: 1) management of parenteral or mucous-membrane exposures

TABLE 2. Predictive value of positive HIV-antibody tests in hypothetical populations with different prevalences of infection

	Prevalence of infection	Predictive value of positive test*
Repeatedly reactive enzyme immunoassay (EIA)†	0.2%	28.41%
	2.0%	80.16%
	20.0%	98.02%
Repeatedly reactive EIA followed by positive Western blot (WB)§	0.2%	99.75%
	2.0%	99.97%
	20.0%	99.99%

*Proportion of persons with positive test results who are actually infected with HIV.
†Assumes EIA sensitivity of 99.0% and specificity of 99.5%.
§Assumes WB sensitivity of 99.0% and specificity of 99.9%.

of health-care workers, 2) patient diagnosis and management, and 3) counseling and serologic testing to prevent and control HIV transmission in the community. In addition, more recent recommendations have stated that hospitals, in conjunction with state and local health departments, should periodically determine the prevalence of HIV infection among patients from age groups at highest risk of infection (*32*).

Adherence to universal blood and body-fluid precautions recommended for the care of all patients will minimize the risk of transmission of HIV and other blood-borne pathogens from patients to health-care workers. The utility of routine HIV serologic testing of patients as an adjunct to universal precautions is unknown. Results of such testing may not be available in emergency or outpatient settings. In addition, some recently infected patients will not have detectable antibody to HIV (Table 1).

Personnel in some hospitals have advocated serologic testing of patients in settings in which exposure of health-care workers to large amounts of patients' blood may be anticipated. Specific patients for whom serologic testing has been advocated include those undergoing major operative procedures and those undergoing treatment in critical-care units, especially if they have conditions involving uncontrolled bleeding. Decisions regarding the need to establish testing programs for patients should be made by physicians or individual institutions. In addition, when deemed appropriate, testing of individual patients may be performed on agreement between the patient and the physician providing care.

In addition to the universal precautions recommended for all patients, certain additional precautions for the care of HIV-infected patients undergoing major surgical operations have been proposed by personnel in some hospitals. For example, surgical procedures on an HIV-infected patient might be altered so that hand-to-hand passing of sharp instruments would be eliminated; stapling instruments rather than hand-suturing equipment might be used to perform tissue approximation; electrocautery devices rather than scalpels might be used as cutting instruments; and, even though uncomfortable, gowns that totally prevent seepage of blood onto the skin of members of the operative team might be worn. While such modifications might further minimize the risk of HIV infection for members of the operative team, some of these techniques could result in prolongation of operative time and could potentially have an adverse effect on the patient.

Testing programs, if developed, should include the following principles:

- Obtaining consent for testing.
- Informing patients of test results, and providing counseling for seropositive patients by properly trained persons.
- Assuring that confidentiality safeguards are in place to limit knowledge of test results to those directly involved in the care of infected patients or as required by law.
- Assuring that identification of infected patients will not result in denial of needed care or provision of suboptimal care.
- Evaluating prospectively 1) the efficacy of the program in reducing the incidence of parenteral, mucous-membrane, or significant cutaneous exposures of health-care workers to the blood or other body fluids of HIV-infected patients and 2) the effect of modified procedures on patients.

Testing of Health-Care Workers

Although transmission of HIV from infected health-care workers to patients has not been reported, transmission during invasive procedures remains a possibility. Transmission of hepatitis B virus (HBV)—a blood-borne agent with a considerably greater potential for nosocomial spread—from health-care workers to patients has been documented. Such transmission has occurred in situations (e.g., oral and gynecologic surgery) in which health-care workers, when tested, had very high concentrations of HBV in their blood (at least 100 million infectious virus particles per milliliter, a concentration much higher than occurs with HIV infection), and the health-care workers sustained a puncture wound while performing invasive procedures or had exudative or weeping lesions or microlacerations that allowed virus to contaminate instruments or open wounds of patients (*33,34*).

The hepatitis B experience indicates that only those health-care workers who perform certain types of invasive procedures have transmitted HBV to patients. Adherence to recommendations in this document will minimize the risk of transmission of HIV and other blood-borne pathogens from health-care workers to patients during invasive procedures. Since transmission of HIV from infected health-care workers performing invasive procedures to their patients has not been reported and would be expected to occur only very rarely, if at all, the utility of routine testing of such health-care workers to prevent transmission of HIV cannot be assessed. If consideration is given to developing a serologic testing program for health-care workers who perform invasive procedures, the frequency of testing, as well as the issues of consent, confidentiality, and consequences of test results—as previously outlined for testing programs for patients—must be addressed.

Management of Infected Health-Care Workers

Health-care workers with impaired immune systems resulting from HIV infection or other causes are at increased risk of acquiring or experiencing serious complications of infectious disease. Of particular concern is the risk of severe infection following exposure to patients with infectious diseases that are easily transmitted if appropriate precautions are not taken (e.g., measles, varicella). Any health-care worker with an impaired immune system should be counseled about the potential risk associated with taking care of patients with any transmissible infection and should continue to follow existing recommendations for infection control to minimize risk of exposure to other infectious agents (*7,35*). Recommendations of the Immunization Practices Advisory Committee (ACIP) and institutional policies concerning requirements for vaccinating health-care workers with live-virus vaccines (e.g., measles, rubella) should also be considered.

The question of whether workers infected with HIV—especially those who perform invasive procedures—can adequately and safely be allowed to perform patient-care duties or whether their work assignments should be changed must be determined on an individual basis. These decisions should be made by the health-care worker's personal physician(s) in conjunction with the medical directors and personnel health service staff of the employing institution or hospital.

Management of Exposures

If a health-care worker has a parenteral (e.g., needlestick or cut) or mucous-membrane (e.g., splash to the eye or mouth) exposure to blood or other body fluids or has a cutaneous exposure involving large amounts of blood or prolonged contact with blood—especially when the exposed skin is chapped, abraded, or afflicted with dermatitis—the source patient should be informed of the incident and tested for serologic evidence of HIV infection after consent is obtained. Policies should be developed for testing source patients in situations in which consent cannot be obtained (e.g., an unconscious patient).

If the source patient has AIDS, is positive for HIV antibody, or refuses the test, the health-care worker should be counseled regarding the risk of infection and evaluated clinically and serologically for evidence of HIV infection as soon as possible after the exposure. The health-care worker should be advised to report and seek medical evaluation for any acute febrile illness that occurs within 12 weeks after the exposure. Such an illness—particularly one characterized by fever, rash, or lymphadenopathy—may be indicative of recent HIV infection. Seronegative health-care workers should be retested 6 weeks post-exposure and on a periodic basis thereafter (e.g., 12 weeks and 6 months after exposure) to determine whether transmission has occurred. During this follow-up period—especially the first 6-12 weeks after exposure, when most infected persons are expected to seroconvert—exposed health-care workers should follow U.S. Public Health Service (PHS) recommendations for preventing transmission of HIV (*36,37*).

No further follow-up of a health-care worker exposed to infection as described above is necessary if the source patient is seronegative unless the source patient is at high risk of HIV infection. In the latter case, a subsequent specimen (e.g., 12 weeks following exposure) may be obtained from the health-care worker for antibody testing. If the source patient cannot be identified, decisions regarding appropriate follow-up should be individualized. Serologic testing should be available to all health-care workers who are concerned that they may have been infected with HIV.

If a patient has a parenteral or mucous-membrane exposure to blood or other body fluid of a health-care worker, the patient should be informed of the incident, and the same procedure outlined above for management of exposures should be followed for both the source health-care worker and the exposed patient.

References

1. CDC. Acquired immunodeficiency syndrome (AIDS): Precautions for clinical and laboratory staffs. MMWR 1982;31:577-80.
2. CDC. Acquired immunodeficiency syndrome (AIDS): Precautions for health-care workers and allied professionals. MMWR 1983;32:450-1.
3. CDC. Recommendations for preventing transmission of infection with human T-lymphotropic virus type III/lymphadenopathy-associated virus in the workplace. MMWR 1985;34:681-6, 691-5.
4. CDC. Recommendations for preventing transmission of infection with human T-lymphotropic virus type III/lymphadenopathy-associated virus during invasive procedures. MMWR 1986;35:221-3.
5. CDC. Recommendations for preventing possible transmission of human T-lymphotropic virus type III/lymphadenopathy-associated virus from tears. MMWR 1985;34:533-4.
6. CDC. Recommendations for providing dialysis treatment to patients infected with human T-lymphotropic virus type III/lymphadenopathy-associated virus infection. MMWR 1986;35:376-8, 383.
7. Garner JS, Simmons BP. Guideline for isolation precautions in hospitals. Infect Control 1983;4 (suppl) :245-325.
8. CDC. Recommended infection control practices for dentistry. MMWR 1986;35:237-42.
9. McCray E, The Cooperative Needlestick Surveillance Group. Occupational risk of the acquired immunodeficiency syndrome among health care workers. N Engl J Med 1986;314:1127-32.
10. Henderson DK, Saah AJ, Zak BJ, et al. Risk of nosocomial infection with human T-cell lymphotropic virus type III/lymphadenopathy-associated virus in a large cohort of intensively exposed health care workers. Ann Intern Med 1986;104:644-7.
11. Gerberding JL, Bryant-LeBlanc CE, Nelson K, et al. Risk of transmitting the human immunodeficiency virus, cytomegalovirus, and hepatitis B virus to health care workers exposed to patients with AIDS and AIDS-related conditions. J Infect Dis 1987;156:1-8.

12. McEvoy M, Porter K, Mortimer P, Simmons N, Shanson D. Prospective study of clinical, laboratory, and ancillary staff with accidental exposures to blood or other body fluids from patients infected with HIV. Br Med J 1987;294:1595-7.
13. Anonymous. Needlestick transmission of HTLV-III from a patient infected in Africa. Lancet 1984;2:1376-7.
14. Oksenhendler E, Harzic M, Le Roux JM, Rabian C, Clauvel JP. HIV infection with seroconversion after a superficial needlestick injury to the finger. N Engl J Med 1986;315:582.
15. Neisson-Vernant C, Arfi S, Mathez D, Leibowitch J, Monplaisir N. Needlestick HIV seroconversion in a nurse. Lancet 1986;2:814.
16. Grint P, McEvoy M. Two associated cases of the acquired immune deficiency syndrome (AIDS). PHLS Commun Dis Rep 1985;42:4.
17. CDC. Apparent transmission of human T-lymphotropic virus type III/lymphadenopathy-associated virus from a child to a mother providing health care. MMWR 1986;35:76-9.
18. CDC. Update: Human immunodeficiency virus infections in health-care workers exposed to blood of infected patients. MMWR 1987;36:285-9.
19. Kline RS, Phelan J, Friedland GH, et al. Low occupational risk for HIV infection for dental professionals [Abstract]. In: Abstracts from the III International Conference on AIDS, 1-5 June 1985. Washington, DC: 155.
20. Baker JL, Kelen GD, Sivertson KT, Quinn TC. Unsuspected human immunodeficiency virus in critically ill emergency patients. JAMA 1987;257:2609-11.
21. Favero MS. Dialysis-associated diseases and their control. In: Bennett JV, Brachman PS, eds. Hospital infections. Boston: Little, Brown and Company, 1985:267-84.
22. Richardson JH, Barkley WE, eds. Biosafety in microbiological and biomedical laboratories, 1984. Washington, DC : US Department of Health and Human Services, Public Health Service. HHS publication no. (CDC) 84-8395.
23. CDC. Human T-lymphotropic virus type III/lymphadenopathy-associated virus: Agent summary statement. MMWR 1986;35:540-2, 547-9.
24. Environmental Protection Agency. EPA guide for infectious waste management. Washington, DC :U.S. Environmental Protection Agency, May 1986 (Publication no. EPA/530-SW-86-014).
25. Favero MS. Sterilization, disinfection, and antisepsis in the hospital. In: Manual of clinical microbiology. 4th ed. Washington, DC: American Society for Microbiology, 1985;129-37.
26. Garner JS, Favero MS. Guideline for handwashing and hospital environmental control, 1985. Atlanta: Public Health Service, Centers for Disease Control, 1985. HHS publication no. 99-1117.
27. Spire B, Montagnier L, Barré-Sinoussi F, Chermann JC. Inactivation of lymphadenopathy associated virus by chemical disinfectants. Lancet 1984;2:899-901.
28. Martin LS, McDougal JS, Loskoski SL. Disinfection and inactivation of the human T lymphotropic virus type III/lymphadenopathy-associated virus. J Infect Dis 1985; 152:400-3.
29. McDougal JS, Martin LS, Cort SP, et al. Thermal inactivation of the acquired immunodeficiency syndrome virus-III/lymphadenopathy-associated virus, with special reference to antihemophilic factor. J Clin Invest 1985;76:875-7.
30. Spire B, Barré-Sinoussi F, Dormont D, Montagnier L, Chermann JC. Inactivation of lymphadenopathy-associated virus by heat, gamma rays, and ultraviolet light. Lancet 1985;1:188-9.
31. Resnik L, Veren K, Salahuddin SZ, Tondreau S, Markham PD. Stability and inactivation of HTLV-III/LAV under clinical and laboratory environments. JAMA 1986;255:1887-91.
32. CDC. Public Health Service (PHS) guidelines for counseling and antibody testing to prevent HIV infection and AIDS. MMWR 1987;3:509-15..
33. Kane MA, Lettau LA. Transmission of HBV from dental personnel to patients. J Am Dent Assoc 1985;110:634-6.
34. Lettau LA, Smith JD, Williams D, et. al. Transmission of hepatitis B with resultant restriction of surgical practice. JAMA 1986;255:934-7.
35. Williams WW. Guideline for infection control in hospital personnel. Infect Control 1983;4 (suppl) :326-49.
36. CDC. Prevention of acquired immune deficiency syndrome (AIDS): Report of inter-agency recommendations. MMWR 1983;32:101-3.
37. CDC. Provisional Public Health Service inter-agency recommendations for screening donated blood and plasma for antibody to the virus causing acquired immunodeficiency syndrome. MMWR 1985;34:1-5.

APPENDIX C

Recommendations for Preventing Transmission of Human Immunodeficiency Virus and Hepatitis B Virus to Patients During Exposure-Prone Invasive Procedures

This document has been developed by the Centers for Disease Control (CDC) to update recommendations for prevention of transmission of human immunodeficiency virus (HIV) and hepatitis B virus (HBV) in the health-care setting. Current data suggest that the risk for such transmission from a health-care worker (HCW) to a patient during an invasive procedure is small; a precise assessment of the risk is not yet available. This document contains recommendations to provide guidance for prevention of HIV and HBV transmission during those invasive procedures that are considered exposure-prone.

INTRODUCTION

Recommendations have been made by the Centers for Disease Control (CDC) for the prevention of transmission of the human immunodeficiency virus (HIV) and the hepatitis B virus (HBV) in health-care settings (*1-6*). These recommendations empha-

---

SOURCE: The contents of Appendix C are reproduced from *Morbidity and Mortality Weekly Report,* Vol. 40, No. RR-8, July 21, 1991, pp. 1-9. Published by Centers for Disease Control, U.S. Department of Health and Human Services, Atlanta, GA.

size adherence to universal precautions that require that blood and other specified body fluids of **all** patients be handled as if they contain blood-borne pathogens (*1,2*).

Previous guidelines contained precautions to be used during invasive procedures (defined in Appendix) and recommendations for the management of HIV- and HBV-infected health-care workers (HCWs) (*1*). These guidelines did not include specific recommendations on testing HCWs for HIV or HBV infection, and they did not provide guidance on which invasive procedures may represent increased risk to the patient.

The recommendations outlined in this document are based on the following considerations:

- Infected HCWs who adhere to universal precautions and who do not perform invasive procedures pose no risk for transmitting HIV or HBV to patients.
- Infected HCWs who adhere to universal precautions and who perform certain exposure-prone procedures (see page 4) pose a small risk for transmitting HBV to patients.
- HIV is transmitted much less readily than HBV.

In the interim, until further data are available, additional precautions are prudent to prevent HIV and HBV transmission during procedures that have been linked to HCW-to-patient HBV transmission or that are considered exposure-prone.

BACKGROUND

Infection-Control Practices

Previous recommendations have specified that infection-control programs should incorporate principles of universal precautions (i.e., appropriate use of hand washing, protective barriers, and care in the use and disposal of needles and other sharp instruments) and should maintain these precautions rigorously in all health-care settings (*1,2,5*). Proper application of these principles will assist in minimizing the risk of transmission of HIV or HBV from patient to HCW, HCW to patient, or patient to patient.

As part of standard infection-control practice, instruments and other reusable equipment used in performing invasive procedures should be appropriately disinfected and sterilized as follows (*7*):

- Equipment and devices that enter the patient's vascular system or other normally sterile areas of the body should be sterilized before being used for each patient.
- Equipment and devices that touch intact mucous membranes but do not penetrate the patient's body surfaces should be sterilized when possible or undergo high-level disinfection if they cannot be sterilized before being used for each patient.
- Equipment and devices that do not touch the patient or that only touch intact skin of the patient need only be cleaned with a detergent or as indicated by the manufacturer.

Compliance with universal precautions and recommendations for disinfection and sterilization of medical devices should be scrupulously monitored in all health-care settings (*1, 7, 8*). Training of HCWs in proper infection-control technique should begin in professional and vocational schools and continue as an ongoing process. Institutions should provide all HCWs with appropriate inservice education regarding infection control and safety and should establish procedures for monitoring compliance with infection-control policies.

All HCWs who might be exposed to blood in an occupational setting should receive hepatitis B vaccine, preferably during their period of professional training and before any occupational exposures could occur (*8, 9*).

Transmission of HBV During Invasive Procedures

Since the introduction of serologic testing for HBV infection in the early 1970s, there have been published reports of 20 clusters in which a total of over 300 patients were infected with HBV in association with treatment by an HBV-infected HCW. In 12 of these clusters, the implicated HCW did not routinely wear gloves; several HCWs also had skin lesions that may have facilitated HBV transmission (*10-22*). These 12 clusters included nine linked to dentists or oral surgeons and one cluster each linked to a general practitioner, an inhalation therapist, and a cardiopulmonary-bypass-pump technician. The clusters associated with the inhalation therapist and the cardiopulmonary-bypass-pump technician—and some of the other 10 clusters—could possibly have been prevented if current recommendations on universal precautions, including glove use, had been in effect. In the remaining eight clusters, transmission occurred despite glove use by the HCWs; five clusters were linked to obstetricians or gynecologists, and three were linked to cardiovascular surgeons (*6, 22-28*). In addition, recent unpublished reports strongly suggest HBV transmission from three surgeons to patients in 1989 and 1990 during colorectal (CDC, unpublished data), abdominal, and cardiothoracic surgery (*29*).

Seven of the HCWs who were linked to published clusters in the United States were allowed to perform invasive procedures following modification of invasive techniques (e.g., double gloving and restriction of certain high-risk procedures) (*6,11-13,15,16, 24*). For five HCWs, no further transmission to patients was observed. In two instances involving an obstetrician/gynecologist and an oral surgeon, HBV was transmitted to patients after techniques were modified (*6, 12*).

Review of the 20 published studies indicates that a combination of risk factors accounted for transmission of HBV from HCWs to patients. Of the HCWs whose hepatitis B e antigen (HBeAg) status was determined (17 of 20), all were HBeAg positive. The presence of HBeAg in serum is associated with higher levels of circulating virus and therefore with greater infectivity of hepatitis-B-surface-antigen (HBsAg)-positive individuals; the risk of HBV transmission to an HCW after a percutaneous exposure to HBeAg-positive blood is approximately 30% (*30-32*). In addition, each report indicated that the potential existed for contamination of surgical wounds or traumatized tissue, either from a major break in standard infection-control practices (e.g., not wearing gloves during invasive procedures) or from unintentional injury to the infected HCW during invasive procedures (e.g., needle sticks incurred while manipulating needles without being able to see them during suturing).

Most reported clusters in the United States occurred before awareness increased of the risks of transmission of blood-borne pathogens in health-care settings and before emphasis was placed on the use of universal precautions and hepatitis B vaccine among HCWs. The limited number of reports of HBV transmission from HCWs to patients in recent years may reflect the adoption of universal precautions and increased use of HBV vaccine. However, the limited number of recent reports does not preclude the occurrence of undetected or unreported small clusters or individual instances of transmission; routine use of gloves does not prevent most injuries caused by sharp instruments and does not eliminate the potential for exposure of a patient to an HCW's blood and transmission of HBV (*6, 22-29*).

Transmission of HIV During Invasive Procedures

The risk of HIV transmission to an HCW after percutaneous exposure to HIV-infected blood is considerably lower than the risk of HBV transmission after percutaneous exposure to HBeAg-positive blood (0.3% versus approximately 30%) (*33-35*). Thus, the risk of transmission of HIV from an infected HCW to a patient during an invasive procedure is likely to be proportionately lower than the risk of HBV transmission from an HBeAg-positive HCW to a patient during the same procedure. As with HBV, the relative infectivity of HIV probably varies among individuals and over time for a single individual. Unlike HBV infection, however, there is currently no readily available laboratory test for increased HIV infectivity.

Investigation of a cluster of HIV infections among patients in the practice of one dentist with acquired immunodeficiency syndrome (AIDS) strongly suggested that HIV was transmitted to five of the approximately 850 patients evaluated through June 1991 (*36-38*). The investigation indicates that HIV transmission occurred during dental care, although the precise mechanisms of transmission have not been determined. In two other studies, when patients cared for by a general surgeon and a surgical resident who had AIDS were tested, all patients tested, 75 and 62, respectively, were negative for HIV infection (*39, 40*). In a fourth study, 143 patients who had been treated by a dental student with HIV infection and were later tested were all negative for HIV infection (*41*). In another investigation, HIV antibody testing was offered to all patients whose surgical procedures had been performed by a general surgeon within 7 years before the surgeon's diagnosis of AIDS; the date at which the surgeon became infected with HIV is unknown (*42*). Of 1,340 surgical patients contacted, 616 (46%) were tested for HIV. One patient, a known intravenous drug user, was HIV positive when tested but may already have been infected at the time of surgery. HIV test results for the 615 other surgical patients were negative (95% confidence interval for risk of transmission per operation = 0.0%-0.5%).

The limited number of participants and the differences in procedures associated with these five investigations limit the ability to generalize from them and to define precisely the risk of HIV transmission from HIV-infected HCWs to patients. A precise estimate of the risk of HIV transmission from infected HCWs to patients can be determined only after careful evaluation of a substantially larger number of patients whose exposure-prone procedures have been performed by HIV-infected HCWs.

Exposure-Prone Procedures

Despite adherence to the principles of universal precautions, certain invasive surgical and dental procedures have been implicated in the transmission of HBV from infected HCWs to patients, and should be considered exposure-prone. Reported examples include certain oral, cardiothoracic, colorectal (CDC, unpublished data), and obstetric/gynecologic procedures (*6, 12, 22-29*).

Certain other invasive procedures should also be considered exposure-prone. In a prospective study CDC conducted in four hospitals, one or more percutaneous injuries occurred among surgical personnel during 96 (6.9%) of 1,382 operative procedures on the general surgery, gynecology, orthopedic, cardiac, and trauma services (*43*). Percutaneous exposure of the patient to the HCW's blood may have occurred when the sharp object causing the injury recontacted the patient's open wound in 28 (32%) of the 88 observed injuries to surgeons (range among surgical specialties = 8%-57%; range among hospitals = 24%-42%).

Characteristics of exposure-prone procedures include digital palpation of a needle tip in a body cavity or the simultaneous presence of the HCW's fingers and a needle or other sharp instrument or object in a poorly

visualized or highly confined anatomic site. Performance of exposure-prone procedures presents a recognized risk of percutaneous injury to the HCW, and—if such an injury occurs—the HCW's blood is likely to contact the patient's body cavity, subcutaneous tissues, and/or mucous membranes.

Experience with HBV indicates that invasive procedures that do not have the above characteristics would be expected to pose substantially lower risk, if any, of transmission of HIV and other blood-borne pathogens from an infected HCW to patients.

RECOMMENDATIONS

Investigations of HIV and HBV transmission from HCWs to patients indicate that, when HCWs adhere to recommended infection-control procedures, the risk of transmitting HBV from an infected HCW to a patient is small, and the risk of transmitting HIV is likely to be even smaller. However, the likelihood of exposure of the patient to an HCW's blood is greater for certain procedures designated as exposure-prone. To minimize the risk of HIV or HBV transmission, the following measures are recommended:

- **All HCWs should adhere to universal precautions, including the appropriate use of hand washing, protective barriers, and care in the use and disposal of needles and other sharp instruments. HCWs who have exudative lesions or weeping dermatitis should refrain from all direct patient care and from handling patient-care equipment and devices used in performing invasive procedures until the condition resolves. HCWs should also comply with current guidelines for disinfection and sterilization of reusable devices used in invasive procedures.**
- **Currently available data provide no basis for recommendations to restrict the practice of HCWs infected with HIV or HBV who perform invasive procedures not identified as exposure-prone, provided the infected HCWs practice recommended surgical or dental technique and comply with universal precautions and current recommendations for sterilization/disinfection.**
- **Exposure-prone procedures should be identified by medical/surgical/dental organizations and institutions at which the procedures are performed.**
- **HCWs who perform exposure-prone procedures should know their HIV antibody status. HCWs who perform exposure-prone procedures and who do not have serologic evidence of immunity to HBV from vaccination or from previous infection should know their HBsAg status and, if that is positive, should also know their HBeAg status.**
- **HCWs who are infected with HIV or HBV (and are HBeAg positive) should not perform exposure-prone procedures unless they have sought counsel from an expert review panel and been advised under what circumstances, if any, they may continue to perform these procedures.* Such circumstances would include notifying prospective patients of the HCW's seropositivity before they undergo exposure-prone invasive procedures.**

---

*The review panel should include experts who represent a balanced perspective. Such experts might include all of the following: a) the HCW's personal physician(s), b) an infectious disease specialist with expertise in the epidemiology of HIV and HBV transmission, c) a health professional with expertise in the procedures performed by the HCW, and d) state or local public health official(s). If the HCW's practice is institutionally based, the expert review panel might also include a member of the infection-control committee, preferably a hospital epidemiologist. HCWs who perform exposure-prone procedures outside the hospital/institutional setting should seek advice from appropriate state and local public health officials regarding the review process. Panels must recognize the importance of confidentiality and the privacy rights of infected HCWs.

- **Mandatory testing of HCWs for HIV antibody, HBsAg, or HBeAg is not recommended. The current assessment of the risk that infected HCWs will transmit HIV or HBV to patients during exposure-prone procedures does not support the diversion of resources that would be required to implement mandatory testing programs. Compliance by HCWs with recommendations can be increased through education, training, and appropriate confidentiality safeguards.**

HCWS WHOSE PRACTICES ARE MODIFIED BECAUSE OF HIV OR HBV STATUS

HCWs whose practices are modified because of their HIV or HBV infection status should, whenever possible, be provided opportunities to continue appropriate patient-care activities. Career counseling and job retraining should be encouraged to promote the continued use of the HCW's talents, knowledge, and skills. HCWs whose practices are modified because of HBV infection should be reevaluated periodically to determine whether their HBeAg status changes due to resolution of infection or as a result of treatment (*44*).

NOTIFICATION OF PATIENTS AND FOLLOW-UP STUDIES

The public health benefit of notification of patients who have had exposure-prone procedures performed by HCWs infected with HIV or positive for HBeAg should be considered on a case-by-case basis, taking into consideration an assessment of specific risks, confidentiality issues, and available resources. Carefully designed and implemented follow-up studies are necessary to determine more precisely the risk of transmission during such procedures. Decisions regarding notification and follow-up studies should be made in consultation with state and local public health officials.

ADDITIONAL NEEDS

- Clearer definition of the nature, frequency, and circumstances of blood contact between patients and HCWs during invasive procedures.
- Development and evaluation of new devices, protective barriers, and techniques that may prevent such blood contact without adversely affecting the quality of patient care.
- More information on the potential for HIV and HBV transmission through contaminated instruments.
- Improvements in sterilization and disinfection techniques for certain reusable equipment and devices.
- Identification of factors that may influence the likelihood of HIV or HBV transmission after exposure to HIV- or HBV-infected blood.

References

1. CDC. Recommendations for prevention of HIV transmission in health-care settings. MMWR 1987;36(suppl. no. 2S):1-18S.
2. CDC. Update: Universal precautions for prevention of transmission of human immunodeficiency virus, hepatitis B virus, and other bloodborne pathogens in health-care settings. MMWR 1988;37:377-82,387-8.
3. CDC. Hepatitis Surveillance Report No. 48. Atlanta: U.S. Department of Health and Human Services, Public Health Service, 1982:2-3.
4. CDC. CDC Guideline for Infection Control in Hospital Personnel, Atlanta, Georgia: Public Health Service, 1983. 24 pages. (GPO# 6AR031488305).

5. CDC. Guidelines for prevention of transmission of human immunodeficiency virus and hepatitis B virus to health-care and public-safety workers. MMWR 1989;38;(suppl. no. S-6):1-37.
6. Lettau LA, Smith JD, Williams D, et al. Transmission of hepatitis B with resultant restriction of surgical practice. JAMA 1986;255:934-7.
7. CDC. Guidelines for the prevention and control of nosocomial infections: guideline for handwashing and hospital environmental control. Atlanta, Georgia: Public Health Service, 1985. 20 pages. (GPO# 544-436/24441).
8. Department of Labor, Occupational Safety and Health Administration. Occupational exposure to bloodborne pathogens: proposed rule and notice of hearing. Federal Register 1989;54:23042-139.
9. CDC. Protection against viral hepatitis: recommendations of the immunization practices advisory committee (ACIP). MMWR 1990;39:(no. RR-2).
10. Levin ML, Maddrey WC, Wands JR, Mendeloff AI. Hepatitis B transmission by dentists. JAMA 1974; 228:1139-40.
11. Rimland D, Parkin WE, Miller GB, Schrack WD. Hepatitis B outbreak traced to an oral surgeon. N Engl J Med 1977;296:953-8.
12. Goodwin D, Fannin SL, McCracken BB. An oral-surgeon related hepatitis-B outbreak. California Morbidity 1976;14.
13. Hadler SC, Sorley DL, Acree KH, et al. An outbreak of hepatitis B in a dental practice. Ann Intern Med 1981;95:133-8.
14. Reingold AL, Kane MA, Murphy BL, Checko P, Francis DP, Maynard JE. Transmission of hepatitis B by an oral surgeon. J Infect Dis 1982;145:262-8.
15. Goodman RA, Ahtone JL, Finton RJ. Hepatitis B transmission from dental personnel to patients: unfinished business [Editorial]. Ann Intern Med 1982;96:119.
16. Ahtone J, Goodman RA. Hepatitis B and dental personnel: transmission to patients and prevention issues. J Am Dent Assoc 1983;106:219-22.
17. Shaw FE, Jr, Barrett CL, Hamm R, et al. Lethal outbreak of hepatitis B in a dental practice. JAMA 1986;255:3260-4.
18. CDC. Outbreak of hepatitis B associated with an oral surgeon, New Hampshire. MMWR 1987;36:132-3.
19. Grob PJ, Moeschlin P. Risk to contacts of a medical practitioner carrying HBsAg. [Letter]. N Engl J Med 1975;293:197.
20. Grob PJ, Bischof B, Naeff F. Cluster of hepatitis B transmitted by a physician. Lancet 1981;2:1218-20.
21. Snydman DR, Hindman SH, Wineland MD, Bryan JA, Maynard JE. Nosocomial viral hepatitis B. A cluster among staff with subsequent transmission to patients. Ann Intern Med 1976;85:573-7.
22. Coutinho RA, Albrecht-van Lent P, Stoutjesdijk L, et al. Hepatitis B from doctors [Letter]. Lancet 1982;1:345-6.
23. Anonymous. Acute hepatitis B associated with gynaecological surgery. Lancet 1980;1:1-6.
24. Carl M, Blakey DL, Francis DP, Maynard JE. Interruption of hepatitis B transmission by modification of a gynaecologist's surgical technique. Lancet 1982;1:731-3.
25. Anonymous. Acute hepatitis B following gynaecological surgery. J Hosp Infect 1987;9:34-8.
26. Welch J, Webster M, Tilzey AJ, Noah ND, Banatvala JE. Hepatitis B infections after gynaecological surgery. Lancet 1989;1:205-7.
27. Haeram JW, Siebke JC, Ulstrup J, Geiram D, Helle I. HBsAg transmission from a cardiac surgeon incubating hepatitis B resulting in chronic antigenemia in four patients. Acta Med Scand 1981;210:389-92.
28. Flower AJE, Prentice M, Morgan G, et al. Hepatitis B infection following cardiothoracic surgery [Abstract]. 1990 International Symposium on Viral Hepatitis and Liver Diseases, Houston. 1990;94.
29. Heptonstall J. Outbreaks of hepatitis B virus infection associated with infected surgical staff in the United Kingdom. Communicable Disease Reports 1991 (in press).
30. Alter HJ, Seef LB, Kaplan PM, et al. Type B hepatitis: the infectivity of blood positive for e antigen and DNA polymerase after accidental needlestick exposure. N Engl J Med 1976; 295:909-13.
31. Seeff LB, Wright EC, Zimmerman HJ, et al. Type B hepatitis after needlestick exposure: prevention with hepatitis B immunoglobulin: final report of the Veterans Administration Cooperative Study. Ann Intern Med 1978;88:285-93.
32. Grady GF, Lee VA, Prince AM, et al. Hepatitis B immune globulin for accidental exposures among medical personnel: final report of a multicenter controlled trial. J Infect Dis 1978;138:625-38.

33. Henderson DK, Fahey BJ, Willy M, et al. Risk for occupational transmission of human immunodeficiency virus type 1 (HIV-1) associated with clinical exposures: a prospective evaluation. Ann Intern Med 1990;113:740-6.
34. Marcus R, CDC Cooperative Needlestick Study Group. Surveillance of health-care workers exposed to blood from patients infected with the human immunodeficiency virus. N Engl J Med 1988;319:1118-23.
35. Gerberding JL, Bryant-LeBlanc CE, Nelson K, et al. Risk of transmitting the human immunodeficiency virus, cytomegalovirus, and hepatitis B virus to health-care workers exposed to patients with AIDS and AIDS-related conditions. J Infect Dis 1987;156:1-8.
36. CDC. Possible transmission of human immunodeficiency virus to a patient during an invasive dental procedure. MMWR 1990;39:489-93.
37. CDC. Update: transmission of HIV infection during an invasive dental procedure - Florida. MMWR 1991;40:21-27,33.
38. CDC. Update: transmission of HIV infection during invasive dental procedures - Florida. MMWR 1991;40:377-81.
39. Porter JD, Cruikshank JG, Gentle PH, Robinson RG, Gill ON. Management of patients treated by a surgeon with HIV infection. [Letter] Lancet 1990;335:113-4.
40. Armstrong FP, Miner JC, Wolfe WH. Investigation of a health-care worker with symptomatic human immunodeficiency virus infection: an epidemiologic approach. Milit Med 1987; 152:414-8.
41. Comer RW, Myers DR, Steadman CD, Carter MJ, Rissing JP, Tedesco FJ. Management considerations for an HIV positive dental student. J Dent Educ 1991;55:187-91.
42. Mishu B, Schaffner W, Horan JM, Wood LH, Hutcheson R, McNabb P. A surgeon with AIDS: lack of evidence of transmission to patients. JAMA 1990;264:467-70.
43. Tokars J, Bell D, Marcus R, et al. Percutaneous injuries during surgical procedures [Abstract]. VII International Conference on AIDS. Vol 2. Florence, Italy, June 16-21, 1991:83.
44. Perrillo RP, Schiff ER, Davis GL, et al. A randomized, controlled trial of interferon alfa-2b alone and after prednisone withdrawal for the treatment of chronic hepatitis B. N Engl J Med 1990;323:295-301.

APPENDIX

Definition of Invasive Procedure

An invasive procedure is defined as "surgical entry into tissues, cavities, or organs or repair of major traumatic injuries" associated with any of the following: "1) an operating or delivery room, emergency department, or outpatient setting, including both physicians' and dentists' offices; 2) cardiac catheterization and angiographic procedures; 3) a vaginal or cesarean delivery or other invasive obstetric procedure during which bleeding may occur; or 4) the manipulation, cutting, or removal of any oral or perioral tissues, including tooth structure, during which bleeding occurs or the potential for bleeding exists."

Reprinted from: Centers for Disease Control. Recommendation for prevention of HIV transmission in health-care settings. *MMWR* 1987;36 (suppl. no. 2S):6S-7S.

APPENDIX D

Bloodborne Pathogens

XI. The Standard

General Industry

Part 1910 of title 29 of the Code of Federal Regulations is amended as follows:

PART 1910—[AMENDED]

Subpart Z—[Amended]

1. The general authority citation for subpart Z of 29 CFR part 1910 continues to read as follows and a new citation for § 1910.1030 is added:

Authority: Secs. 6 and 8, Occupational Safety and Health Act, 29 U.S.C. 655, 657, Secretary of Labor's Orders Nos. 12–71 (36 FR 8754), 8–76 (41 FR 25059), or 9–83 (48 FR 35736), as applicable; and 29 CFR part 1911.

* * * * *

Section 1910.1030 also issued under 29 U.S.C. 653.

* * * * *

2. Section 1910.1030 is added to read as follows:

§ 1910.1030 Bloodborne Pathogens.

(a) *Scope and Application.* This section applies to all occupational exposure to blood or other potentially infectious materials as defined by paragraph (b) of this section.

(b) *Definitions.* For purposes of this section, the following shall apply:

Assistant Secretary means the Assistant Secretary of Labor for Occupational Safety and Health, or designated representative.

Blood means human blood, human blood components, and products made from human blood.

Bloodborne Pathogens means pathogenic microorganisms that are present in human blood and can cause disease in humans. These pathogens include, but are not limited to, hepatitis B virus (HBV) and human immunodeficiency virus (HIV).

Clinical Laboratory means a workplace where diagnostic or other screening procedures are performed on blood or other potentially infectious materials.

Contaminated means the presence or the reasonably anticipated presence of blood or other potentially infectious materials on an item or surface.

Contaminated Laundry means

SOURCE: The contents of Appendix D are reproduced from the *Federal Register,* Vol. 56, No. 235, December 6, 1991, pp. 64175-64182.

laundry which has been soiled with blood or other potentially infectious materials or may contain sharps.

Contaminated Sharps means any contaminated object that can penetrate the skin including, but not limited to, needles, scalpels, broken glass, broken capillary tubes, and exposed ends of dental wires.

Decontamination means the use of physical or chemical means to remove, inactivate, or destroy bloodborne pathogens on a surface or item to the point where they are no longer capable of transmitting infectious particles and the surface or item is rendered safe for handling, use, or disposal.

Director means the Director of the National Institute for Occupational Safety and Health, U.S. Department of Health and Human Services, or designated representative.

Engineering Controls means controls (e.g., sharps disposal containers, self-sheathing needles) that isolate or remove the bloodborne pathogens hazard from the workplace.

Exposure Incident means a specific eye, mouth, other mucous membrane, non-intact skin, or parenteral contact with blood or other potentially infectious materials that results from the performance of an employee's duties.

Handwashing Facilities means a facility providing an adequate supply of running potable water, soap and single use towels or hot air drying machines.

Licensed Healthcare Professional is a person whose legally permitted scope of practice allows him or her to independently perform the activities required by paragraph (f) Hepatitis B Vaccination and Post-exposure Evaluation and Follow-up.

HBV means hepatitis B virus.

HIV means human immunodeficiency virus.

Occupational Exposure means reasonably anticipated skin, eye, mucous membrane, or parenteral contact with blood or other potentially infectious materials that may result from the performance of an employee's duties.

Other Potentially Infectious Materials means

(1) The following human body fluids: semen, vaginal secretions, cerebrospinal fluid, synovial fluid, pleural fluid, pericardial fluid, peritoneal fluid, amniotic fluid, saliva in dental procedures, any body fluid that is visibly contaminated with blood, and all body fluids in situations where it is difficult or impossible to differentiate between body fluids;

(2) Any unfixed tissue or organ (other than intact skin) from a human (living or dead); and

(3) HIV-containing cell or tissue cultures, organ cultures, and HIV- or HBV-containing culture medium or other solutions; and blood, organs, or other tissues from experimental animals infected with HIV or HBV.

Parenteral means piercing mucous membranes or the skin barrier through such events as needlesticks, human bites, cuts, and abrasions.

Personal Protective Equipment is specialized clothing or equipment worn by an employee for protection against a hazard. General work clothes (e.g., uniforms, pants, shirts or blouses) not intended to function as protection against a hazard are not considered to be personal protective equipment.

Production Facility means a facility engaged in industrial-scale, large-volume or high concentration production of HIV or HBV.

Regulated Waste means liquid or semi-liquid blood or other potentially infectious materials; contaminated items that would release blood or other potentially infectious materials in a liquid or semi-liquid state if compressed; items that are caked with dried blood or other potentially infectious materials and are capable of releasing these materials during handling; contaminated sharps; and pathological and microbiological wastes containing blood or other potentially infectious materials.

Research Laboratory means a laboratory producing or using research-laboratory-scale amounts of HIV or HBV. Research laboratories may produce high concentrations of HIV or HBV but not in the volume found in production facilities.

Source Individual means any individual, living or dead, whose blood or other potentially infectious materials may be a source of occupational exposure to the employee. Examples include, but are not limited to, hospital and clinic patients; clients in institutions for the developmentally disabled; trauma victims; clients of drug and alcohol treatment facilities; residents of hospices and nursing homes; human remains; and individuals who donate or sell blood or blood components.

Sterilize means the use of a physical or chemical procedure to destroy all microbial life including highly resistant bacterial endospores.

Universal Precautions is an approach to infection control. According to the concept of Universal Precautions, all human blood and certain human body fluids are treated as if known to be infectious for HIV, HBV, and other bloodborne pathogens.

Work Practice Controls means controls that reduce the likelihood of exposure by altering the manner in which a task is performed (e.g., prohibiting recapping of needles by a two-handed technique).

(c) *Exposure control*—(1) *Exposure Control Plan.* (i) Each employer having an employee(s) with occupational exposure as defined by paragraph (b) of this section shall establish a written Exposure Control Plan designed to eliminate or minimize employee exposure.

(ii) The Exposure Control Plan shall contain at least the following elements:

(A) The exposure determination required by paragraph(c)(2),

(B) The schedule and method of implementation for paragraphs (d) Methods of Compliance, (e) HIV and HBV Research Laboratories and Production Facilities, (f) Hepatitis B Vaccination and Post-Exposure Evaluation and Follow-up, (g) Communication of Hazards to Employees, and (h) Recordkeeping, of this standard, and

(C) The procedure for the evaluation of circumstances surrounding exposure incidents as required by paragraph (f)(3)(i) of this standard.

(iii) Each employer shall ensure that a copy of the Exposure Control Plan is accessible to employees in accordance with 29 CFR 1910.20(e).

(iv) The Exposure Control Plan shall be reviewed and updated at least annually and whenever necessary to reflect new or modified tasks and procedures which affect occupational exposure and to reflect new or revised employee positions with occupational exposure.

(v) The Exposure Control Plan shall be made available to the Assistant Secretary and the Director upon request for examination and copying.

(2) *Exposure determination.* (i) Each employer who has an employee(s) with occupational exposure as defined by paragraph (b) of this section shall prepare an exposure determination. This exposure determination shall contain the following:

(A) A list of all job classifications in which all employees in those job classifications have occupational exposure;

(B) A list of job classifications in which some employees have occupational exposure, and

(C) A list of all tasks and procedures or groups of closely related task and procedures in which occupational exposure occurs and that are performed by employees in job classifications listed in accordance with the provisions of paragraph (c)(2)(i)(B) of this standard.

(ii) This exposure determination shall be made without regard to the use of personal protective equipment.

(d) *Methods of compliance*—(1) *General*—Universal precautions shall be observed to prevent contact with blood or other potentially infectious materials. Under circumstances in which differentiation between body fluid types is difficult or impossible, all body fluids shall be considered potentially infectious materials.

(2) *Engineering and work practice controls.* (i) Engineering and work practice controls shall be used to eliminate or minimize employee exposure. Where occupational exposure remains after institution of these controls, personal protective equipment shall also be used.

(ii) Engineering controls shall be examined and maintained or replaced on a regular schedule to ensure their effectiveness.

(iii) Employers shall provide handwashing facilities which are readily accessible to employees.

(iv) When provision of handwashing facilities is not feasible, the employer shall provide either an appropriate antiseptic hand cleanser in conjunction with clean cloth/paper towels or antiseptic towelettes. When antiseptic hand cleansers or towelettes are used, hands shall be washed with soap and running water as soon as feasible.

(v) Employers shall ensure that employees wash their hands immediately or as soon as feasible after removal of gloves or other personal protective equipment.

(vi) Employers shall ensure that employees wash hands and any other skin with soap and water, or flush mucous membranes with water immediately or as soon as feasible following contact of such body areas with blood or other potentially infectious materials.

(vii) Contaminated needles and other contaminated sharps shall not be bent, recapped, or removed except as noted in paragraphs (d)(2)(vii)(A) and (d)(2)(vii)(B) below. Shearing or breaking of contaminated needles is prohibited.

(A) Contaminated needles and other contaminated sharps shall not be recapped or removed unless the employer can demonstrate that no alternative is feasible or that such action is required by a specific medical procedure.

(B) Such recapping or needle removal must be accomplished through the use of a mechanical device or a one-handed technique.

(viii) Immediately or as soon as possible after use, contaminated reusable sharps shall be placed in appropriate containers until properly reprocessed. These containers shall be:

(A) Puncture resistant;

(B) Labeled or color-coded in accordance with this standard;

(C) Leakproof on the sides and bottom; and

(D) In accordance with the requirements set forth in paragraph (d)(4)(ii)(E) for reusable sharps.

(ix) Eating, drinking, smoking, applying cosmetics or lip balm, and handling contact lenses are prohibited in work areas where there is a reasonable likelihood of occupational exposure.

(x) Food and drink shall not be kept in refrigerators, freezers, shelves, cabinets or on countertops or benchtops where blood or other potentially infectious materials are present.

(xi) All procedures involving blood or other potentially infectious materials shall be performed in such a manner as to minimize splashing, spraying, spattering, and generation of droplets of these substances.

(xii) Mouth pipetting/suctioning of blood or other potentially infectious materials is prohibited.

(xiii) Specimens of blood or other potentially infectious materials shall be placed in a container which prevents leakage during collection, handling, processing, storage, transport, or shipping.

(A) The container for storage, transport, or shipping shall be labeled or color-coded according to paragraph (g)(1)(i) and closed prior to being stored, transported, or shipped. When a facility utilizes Universal Precautions in the handling of all specimens, the labeling/color-coding of specimens is not necessary provided containers are recognizable as containing specimens. This exemption only applies while such specimens/containers remain within the facility. Labeling or color-coding in accordance with paragraph (g)(1)(i) is required when such specimens/containers leave the facility.

(B) If outside contamination of the primary container occurs, the primary container shall be placed within a second container which prevents leakage during handling, processing, storage, transport, or shipping and is labeled or color-coded according to the requirements of this standard.

(C) If the specimen could puncture the primary container, the primary container shall be placed within a secondary container which is puncture-resistant in addition to the above characteristics.

(xiv) Equipment which may become contaminated with blood or other potentially infectious materials shall be examined prior to servicing or shipping and shall be decontaminated as necessary, unless the employer can demonstrate that decontamination of such equipment or portions of such equipment is not feasible.

(A) A readily observable label in accordance with paragraph (g)(1)(i)(H) shall be attached to the equipment stating which portions remain contaminated.

(B) The employer shall ensure that this information is conveyed to all affected employees, the servicing representative, and/or the manufacturer, as appropriate, prior to handling, servicing, or shipping so that appropriate precautions will be taken.

(3) Personal protective equipment—(i) Provision. When there is occupational exposure, the employer shall provide, at no cost to the employee, appropriate personal protective equipment such as, but not limited to, gloves, gowns, laboratory coats, face shields or masks and eye protection, and mouthpieces, resuscitation bags, pocket masks, or other ventilation devices. Personal protective equipment will be considered "appropriate" only if it does not permit blood or other potentially infectious materials to pass through to or reach the employee's work clothes, street clothes, undergarments, skin, eyes, mouth, or other mucous membranes under normal conditions of use and for the duration of time which the protective equipment will be used.

(ii) Use. The employer shall ensure that the employee uses appropriate personal protective equipment unless the employer shows that the employee temporarily and briefly declined to use personal protective equipment when, under rare and extraordinary circumstances, it was the employee's professional judgment that in the specific instance its use would have prevented the delivery of health care or public safety services or would have posed an increased hazard to the safety of the worker or co-worker. When the employee makes this judgement, the circumstances shall be investigated and documented in order to determine whether changes can be instituted to prevent such occurences in the future.

(iii) Accessibility. The employer shall ensure that appropriate personal protective equipment in the appropriate sizes is readily accessible at the worksite or is issued to employees. Hypoallergenic gloves, glove liners, powderless gloves, or other similar alternatives shall be readily accessible to those employees who are allergic to the gloves normally provided.

(iv) Cleaning, Laundering, and Disposal. The employer shall clean, launder, and dispose of personal protective equipment required by paragraphs (d) and (e) of this standard, at no cost to the employee.

(v) Repair and Replacement. The employer shall repair or replace personal protective equipment as needed to maintain its effectiveness, at no cost to the employee.

(vi) If a garment(s) is penetrated by blood or other potentially infectious materials, the garment(s) shall be removed immediately or as soon as feasible.

(vii) All personal protective equipment shall be removed prior to leaving the work area.

(viii) When personal protective equipment is removed it shall be placed in an appropriately designated area or container for storage, washing, decontamination or disposal.

(ix) Gloves. Gloves shall be worn when it can be reasonably anticipated that the employee may have hand contact with blood, other potentially infectious materials, mucous membranes, and non-intact skin; when performing vascular access procedures except as specified in paragraph (d)(3)(ix)(D); and when handling or touching contaminated items or surfaces.

(A) Disposable (single use) gloves such as surgical or examination gloves, shall be replaced as soon as practical when contaminated or as soon as feasible if they are torn, punctured, or when their ability to function as a barrier is compromised.

(B) Disposable (single use) gloves shall not be washed or decontaminated for re-use.

(C) Utility gloves may be decontaminated for re-use if the integrity of the glove is not compromised. However, they must be discarded if they are cracked, peeling, torn, punctured, or exhibit other signs of deterioration or when their ability to function as a barrier is compromised.

(D) If an employer in a volunteer blood donation center judges that routine gloving for all phlebotomies is not necessary then the employer shall:

(*1*) Periodically reevaluate this policy;

(*2*) Make gloves available to all employees who wish to use them for phlebotomy;

(*3*) Not discourage the use of gloves for phlebotomy; and

(*4*) Require that gloves be used for

phlebotomy in the following circumstances:

(*i*) When the employee has cuts, scratches, or other breaks in his or her skin;

(*ii*) When the employee judges that hand contamination with blood may occur, for example, when performing phlebotomy on an uncooperative source individual; and

(*iii*) When the employee is receiving training in phlebotomy.

(x) Masks, Eye Protection, and Face Shields. Masks in combination with eye protection devices, such as goggles or glasses with solid side shields, or chin-length face shields, shall be worn whenever splashes, spray, spatter, or droplets of blood or other potentially infectious materials may be generated and eye, nose, or mouth contamination can be reasonably anticipated.

(xi) Gowns, Aprons, and Other Protective Body Clothing. Appropriate protective clothing such as, but not limited to, gowns, aprons, lab coats, clinic jackets, or similar outer garments shall be worn in occupational exposure situations. The type and characteristics will depend upon the task and degree of exposure anticipated.

(xii) Surgical caps or hoods and/or shoe covers or boots shall be worn in instances when gross contamination can reasonably be anticipated (e.g., autopsies, orthopaedic surgery).

(4) *Housekeeping.* (i) General. Employers shall ensure that the worksite is maintained in a clean and sanitary condition. The employer shall determine and implement an appropriate written schedule for cleaning and method of decontamination based upon the location within the facility, type of surface to be cleaned, type of soil present, and tasks or procedures being performed in the area.

(ii) All equipment and environmental and working surfaces shall be cleaned and decontaminated after contact with blood or other potentially infectious materials.

(A) Contaminated work surfaces shall be decontaminated with an appropriate disinfectant after completion of procedures; immediately or as soon as feasible when surfaces are overtly contaminated or after any spill of blood or other potentially infectious materials; and at the end of the work shift if the surface may have become contaminated since the last cleaning.

(B) Protective coverings, such as plastic wrap, aluminum foil, or imperviously-backed absorbent paper used to cover equipment and environmental surfaces, shall be removed and replaced as soon as feasible when they become overtly contaminated or at the end of the workshift if they may have become contaminated during the shift.

(C) All bins, pails, cans, and similar receptacles intended for reuse which have a reasonable likelihood for becoming contaminated with blood or other potentially infectious materials shall be inspected and decontaminated on a regularly scheduled basis and cleaned and decontaminated immediately or as soon as feasible upon visible contamination.

(D) Broken glassware which may be contaminated shall not be picked up directly with the hands. It shall be cleaned up using mechanical means, such as a brush and dust pan, tongs, or forceps.

(E) Reusable sharps that are contaminated with blood or other potentially infectious materials shall not be stored or processed in a manner that requires employees to reach by hand into the containers where these sharps have been placed.

(iii) Regulated Waste.

(A) Contaminated Sharps Discarding and Containment. (*1*) Contaminated sharps shall be discarded immediately or as soon as feasible in containers that are:

(*i*) Closable;

(*ii*) Puncture resistant;

(*iii*) Leakproof on sides and bottom; and

(*iv*) Labeled or color-coded in accordance with paragraph (g)(1)(i) of this standard.

(*2*) During use, containers for contaminated sharps shall be:

(*i*) Easily accessible to personnel and located as close as is feasible to the immediate area where sharps are used or can be reasonably anticipated to be found (e.g., laundries);

(*ii*) Maintained upright throughout use; and

(*iii*) Replaced routinely and not be allowed to overfill.

(*3*) When moving containers of contaminated sharps from the area of use, the containers shall be:

(*i*) Closed immediately prior to removal or replacement to prevent spillage or protrusion of contents during handling, storage, transport, or shipping;

(*ii*) Placed in a secondary container if leakage is possible. The second container shall be:

(*A*) Closable;

(*B*) Constructed to contain all contents and prevent leakage during handling, storage, transport, or shipping; and

(*C*) Labeled or color-coded according to paragraph (g)(1)(i) of this standard.

(*4*) Reusable containers shall not be opened, emptied, or cleaned manually or in any other manner which would expose employees to the risk of percutaneous injury.

(B) Other Regulated Waste Containment.(*1*) Regulated waste shall be placed in containers which are:

(*i*) Closable;

(*ii*) Constructed to contain all contents and prevent leakage of fluids during handling, storage, transport or shipping;

(*iii*) Labeled or color-coded in accordance with paragraph (g)(1)(i) this standard; and

(*iv*) Closed prior to removal to prevent spillage or protrusion of contents during handling, storage, transport, or shipping.

(*2*) If outside contamination of the regulated waste container occurs, it shall be placed in a second container. The second container shall be:

(*i*) Closable;

(*ii*) Constructed to contain all contents and prevent leakage of fluids during handling, storage, transport or shipping;

(*iii*) Labeled or color-coded in accordance with paragraph (g)(1)(i) of this standard; and

(*iv*) Closed prior to removal to prevent spillage or protrusion of contents during handling, storage, transport, or shipping.

(C) Disposal of all regulated waste shall be in accordance with applicable regulations of the United States, States and Territories, and political subdivisions of States and Territories.

(iv) Laundry.

(A) Contaminated laundry shall be handled as little as possible with a minimum of agitation. (1) Contaminated laundry shall be bagged or containerized at the location where it was used and shall not be sorted or rinsed in the location of use.

(2) Contaminated laundry shall be placed and transported in bags or containers labeled or color-coded in accordance with paragraph (g)(1)(i) of this standard. When a facility utilizes Universal Precautions in the handling of all soiled laundry, alternative labeling or color-coding is sufficient if it permits all employees to recognize the containers as requiring compliance with Universal Precautions.

(3) Whenever contaminated laundry is wet and presents a reasonable likelihood of soak-through of or leakage from the bag or container, the laundry shall be placed and transported in bags or containers which prevent soak-through and/or leakage of fluids to the exterior.

(B) The employer shall ensure that employees who have contact with contaminated laundry wear protective gloves and other appropriate personal protective equipment.

(C) When a facility ships contaminated laundry off-site to a second facility which does not utilize Universal Precautions in the handling of all laundry, the facility generating the contaminated laundry must place such laundry in bags or containers which are labeled or color-coded in accordance with paragraph (g)(1)(i).

(e) *HIV and HBV Research Laboratories and Production Facilities.* (1) This paragraph applies to research laboratories and production facilities engaged in the culture, production, concentration, experimentation, and manipulation of HIV and HBV. It does not apply to clinical or diagnostic laboratories engaged solely in the analysis of blood, tissues, or organs. These requirements apply in addition to the other requirements of the standard.

(2) Research laboratories and production facilities shall meet the following criteria:

(i) Standard microbiological practices. All regulated waste shall either be incinerated or decontaminated by a method such as autoclaving known to effectively destroy bloodborne pathogens.

(ii) Special practices.

(A) Laboratory doors shall be kept closed when work involving HIV or HBV is in progress.

(B) Contaminated materials that are to be decontaminated at a site away from the work area shall be placed in a durable, leakproof, labeled or color-coded container that is closed before being removed from the work area.

(C) Access to the work area shall be limited to authorized persons. Written policies and procedures shall be established whereby only persons who have been advised of the potential biohazard, who meet any specific entry requirements, and who comply with all entry and exit procedures shall be allowed to enter the work areas and animal rooms.

(D) When other potentially infectious materials or infected animals are present in the work area or containment module, a hazard warning sign incorporating the universal biohazard symbol shall be posted on all access doors. The hazard warning sign shall comply with paragraph (g)(1)(ii) of this standard.

(E) All activities involving other potentially infectious materials shall be conducted in biological safety cabinets or other physical-containment devices within the containment module. No work with these other potentially infectious materials shall be conducted on the open bench.

(F) Laboratory coats, gowns, smocks, uniforms, or other appropriate protective clothing shall be used in the work area and animal rooms. Protective clothing shall not be worn outside of the work area and shall be decontaminated before being laundered.

(G) Special care shall be taken to avoid skin contact with other potentially infectious materials. Gloves shall be worn when handling infected animals and when making hand contact with other potentially infectious materials is unavoidable.

(H) Before disposal all waste from work areas and from animal rooms shall either be incinerated or decontaminated by a method such as autoclaving known to effectively destroy bloodborne pathogens.

(I) Vacuum lines shall be protected with liquid disinfectant traps and high-efficiency particulate air (HEPA) filters or filters of equivalent or superior efficiency and which are checked routinely and maintained or replaced as necessary.

(J) Hypodermic needles and syringes shall be used only for parenteral injection and aspiration of fluids from laboratory animals and diaphragm bottles. Only needle-locking syringes or disposable syringe-needle units (i.e., the needle is integral to the syringe) shall be used for the injection or aspiration of other potentially infectious materials. Extreme caution shall be used when handling needles and syringes. A needle shall not be bent, sheared, replaced in the sheath or guard, or removed from the syringe following use. The needle and syringe shall be promptly placed in a puncture-resistant container and autoclaved or decontaminated before reuse or disposal.

(K) All spills shall be immediately contained and cleaned up by appropriate professional staff or others properly trained and equipped to work with potentially concentrated infectious materials.

(L) A spill or accident that results in an exposure incident shall be immediately reported to the laboratory director or other responsible person.

(M) A biosafety manual shall be prepared or adopted and periodically reviewed and updated at least annually or more often if necessary. Personnel shall be advised of potential hazards, shall be required to read instructions on practices and procedures, and shall be required to follow them.

(iii) Containment equipment. (A) Certified biological safety cabinets (Class I, II, or III) or other appropriate combinations of personal protection or physical containment devices, such as special protective clothing, respirators, centrifuge safety cups, sealed centrifuge rotors, and containment caging for animals, shall be used for all activities with other potentially infectious materials that pose a threat of exposure to droplets, splashes, spills, or aerosols.

(B) Biological safety cabinets shall be certified when installed, whenever they are moved and at least annually.

(3) HIV and HBV research laboratories shall meet the following criteria:

(i) Each laboratory shall contain a facility for hand washing and an eye wash facility which is readily available within the work area.

(ii) An autoclave for decontamination of regulated waste shall be available.

(4) HIV and HBV production facilities shall meet the following criteria:

(i) The work areas shall be separated from areas that are open to unrestricted traffic flow within the building. Passage through two sets of doors shall be the basic requirement for entry into the work area from access corridors or other contiguous areas. Physical separation of the high-containment work area from access corridors or other areas or activities may also be provided by a double-doored clothes-change room (showers may be included), airlock, or other access facility that requires passing through two sets of doors before entering the work area.

(ii) The surfaces of doors, walls, floors and ceilings in the work area shall be water resistant so that they can be easily cleaned. Penetrations in these surfaces shall be sealed or capable of being sealed to facilitate decontamination.

(iii) Each work area shall contain a sink for washing hands and a readily available eye wash facility. The sink shall be foot, elbow, or automatically operated and shall be located near the exit door of the work area.

(iv) Access doors to the work area or containment module shall be self-closing.

(v) An autoclave for decontamination of regulated waste shall be available within or as near as possible to the work area.

(vi) A ducted exhaust-air ventilation system shall be provided. This system shall create directional airflow that draws air into the work area through the entry area. The exhaust air shall not be recirculated to any other area of the building, shall be discharged to the outside, and shall be dispersed away from occupied areas and air intakes. The proper direction of the airflow shall be verified (i.e., into the work area).

(5) *Training Requirements.* Additional training requirements for employees in HIV and HBV research laboratories and HIV and HBV production facilities are specified in paragraph (g)(2)(ix).

(f) *Hepatitis B vaccination and post-exposure evaluation and follow-up—(1) General.* (i) The employer shall make available the hepatitis B vaccine and vaccination series to all employees who have occupational exposure, and post-exposure evaluation and follow-up to all employees who have had an exposure incident.

(ii) The employer shall ensure that all medical evaluations and procedures including the hepatitis B vaccine and vaccination series and post-exposure evaluation and follow-up, including prophylaxis, are:

(A) Made available at no cost to the employee:

(B) Made available to the employee at a reasonable time and place;

(C) Performed by or under the supervision of a licensed physician or by or under the supervision of another licensed healthcare professional; and

(D) Provided according to recommendations of the U.S. Public Health Service current at the time these evaluations and procedures take place, except as specified by this paragraph (f).

(iii) The employer shall ensure that all laboratory tests are conducted by an accredited laboratory at no cost to the employee.

(2) *Hepatitis B Vaccination.* (i) Hepatitis B vaccination shall be made available after the employee has received the training required in paragraph (g)(2)(vii)(I) and within 10 working days of initial assignment to all employees who have occupational exposure unless the employee has previously received the complete hepatitis B vaccination series, antibody testing has revealed that the employee is immune, or the vaccine is contraindicated for medical reasons.

(ii) The employer shall not make participation in a prescreening program a prerequisite for receiving hepatitis B vaccination.

(iii) If the employee initially declines hepatitis B vaccination but at a later date while still covered under the standard decides to accept the vaccination, the employer shall make available hepatitis B vaccination at that time.

(iv) The employer shall assure that employees who decline to accept hepatitis B vaccination offered by the employer sign the statement in appendix A.

(v) If a routine booster dose(s) of hepatitis B vaccine is recommended by the U.S. Public Health Service at a future date, such booster dose(s) shall be made available in accordance with section (f)(1)(ii).

(3) *Post-exposure Evaluation and Follow-up.* Following a report of an exposure incident, the employer shall make immediately available to the exposed employee a confidential medical evaluation and follow-up, including at least the following elements:

(i) Documentation of the route(s) of exposure, and the circumstances under which the exposure incident occurred;

(ii) Identification and documentation of the source individual, unless the employer can establish that identification is infeasible or prohibited by state or local law;

(A) The source individual's blood shall be tested as soon as feasible and after consent is obtained in order to determine HBV and HIV infectivity. If consent is not obtained, the employer shall establish that legally required consent cannot be obtained. When the source individual's consent is not required by law, the source individual's blood, if available, shall be tested and the results documented.

(B) When the source individual is already known to be infected with HBV or HIV, testing for the source individual's known HBV or HIV status need not be repeated.

(C) Results of the source individual's testing shall be made available to the exposed employee, and the employee shall be informed of applicable laws and regulations concerning disclosure of the identity and infectious status of the source individual.

(iii) Collection and testing of blood for HBV and HIV serological status;

(A) The exposed employee's blood shall be collected as soon as feasible and tested after consent is obtained.

(B) If the employee consents to baseline blood collection, but does not give consent at that time for HIV serologic testing, the sample shall be preserved for at least 90 days. If, within 90 days of the exposure incident, the employee elects to have the baseline sample tested, such testing shall be done as soon as feasible.

(iv) Post-exposure prophylaxis, when medically indicated, as recommended by the U.S. Public Health Service;

(v) Counseling; and

(vi) Evaluation of reported illnesses.

(4) *Information Provided to the Healthcare Professional.* (i) The employer shall ensure that the healthcare professional responsible for the employee's Hepatitis B vaccination is provided a copy of this regulation.

(ii) The employer shall ensure that the healthcare professional evaluating an employee after an exposure incident is provided the following information:

(A) A copy of this regulation;

(B) A description of the exposed employee's duties as they relate to the exposure incident;

(C) Documentation of the route(s) of exposure and circumstances under which exposure occurred;

(D) Results of the source individual's blood testing, if available; and

(E) All medical records relevant to the appropriate treatment of the employee including vaccination status which are the employer's responsibility to maintain.

(5) *Healthcare Professional's Written Opinion.* The employer shall obtain and provide the employee with a copy of the evaluating healthcare professional's written opinion within 15 days of the completion of the evaluation.

(i) The healthcare professional's written opinion for Hepatitis B vaccination shall be limited to whether Hepatitis B vaccination is indicated for an employee, and if the employee has received such vaccination.

(ii) The healthcare professional's written opinion for post-exposure evaluation and follow-up shall be limited to the following information:

(A) That the employee has been informed of the results of the evaluation; and

(B) That the employee has been told about any medical conditions resulting from exposure to blood or other potentially infectious materials which require further evaluation or treatment.

(iii) All other findings or diagnoses shall remain confidential and shall not be included in the written report.

(6) *Medical recordkeeping.* Medical

records required by this standard shall be maintained in accordance with paragraph (h)(1) of this section.

(g) *Communication of hazards to employees*— (1) *Labels and signs.* (i) Labels. (A) Warning labels shall be affixed to containers of regulated waste, refrigerators and freezers containing blood or other potentially infectious material; and other containers used to store, transport or ship blood or other potentially infectious materials, except as provided in paragraph (g)(1)(i)(E), (F) and (G).

(B) Labels required by this section shall include the following legend:

BIOHAZARD

BIOHAZARD

(C) These labels shall be fluorescent orange or orange-red or predominantly so, with lettering or symbols in a contrasting color.

(D) Labels required by affixed as close as feasible to the container by string, wire, adhesive, or other method that prevents their loss or unintentional removal.

(E) Red bags or red containers may be substituted for labels.

(F) Containers of blood, blood components, or blood products that are labeled as to their contents and have been released for transfusion or other clinical use are exempted from the labeling requirements of paragraph (g).

(G) Individual containers of blood or other potentially infectious materials that are placed in a labeled container during storage, transport, shipment or disposal are exempted from the labeling requirement.

(H) Labels required for contaminated equipment shall be in accordance with this paragraph and shall also state which portions of the equipment remain contaminated.

(I) Regulated waste that has been decontaminated need not be labeled or color-coded.

(ii) Signs. (A) The employer shall post signs at the entrance to work areas specified in paragraph (e), HIV and HBV Research Laboratory and Production Facilities, which shall bear the following legend:

BIOHAZARD

BIOHAZARD

(Name of the Infectious Agent)
(Special requirements for entering the area)
(Name, telephone number of the laboratory director or other responsible person.)

(B) These signs shall be fluorescent orange-red or predominantly so, with lettering or symbols in a contrasting color.

(2) *Information and Training.* (i) Employers shall ensure that all employees with occupational exposure participate in a training program which must be provided at no cost to the employee and during working hours.

(ii) Training shall be provided as follows:

(A) At the time of initial assignment to tasks where occupational exposure may take place;

(B) Within 90 days after the effective date of the standard; and

(C) At least annually thereafter.

(iii) For employees who have received training on bloodborne pathogens in the year preceding the effective date of the standard, only training with respect to the provisions of the standard which were not included need be provided.

(iv) Annual training for all employees shall be provided within one year of their previous training.

(v) Employers shall provide additional training when changes such as modification of tasks or procedures or institution of new tasks or procedures affect the employee's occupational exposure. The additional training may be limited to addressing the new exposures created.

(vi) Material appropriate in content and vocabulary to educational level, literacy, and language of employees shall be used.

(vii) The training program shall contain at a minimum the following elements:

(A) Anaccessible copy of the regulatory text of this standard and an explanation of its contents;

(B) A general explanation of the epidemiology and symptoms of bloodborne diseases;

(C) An explanation of the modes of transmission of bloodborne pathogens;

(D) An explanation of the employer's exposure control plan and the means by which the employee can obtain a copy of the written plan;

(E) An explanation of the appropriate methods for recognizing tasks and other activities that may involve exposure to blood and other potentially infectious materials;

(F) An explanation of the use and limitations of methods that will prevent or reduce exposure including appropriate engineering controls, work practices, and personal protective equipment;

(G) Information on the types, proper use, location, removal, handling, decontamination and disposal of personal protective equipment;

(H) An explanation of the basis for selection of personal protective equipment;

(I) Information on the hepatitis B vaccine, including information on its efficacy, safety, method of administration, the benefits of being vaccinated, and that the vaccine and vaccination will be offered free of charge;

(J) Information on the appropriate actions to take and persons to contact in an emergency involving blood or other potentially infectious materials;

(K) An explanation of the procedure to follow if an exposure incident occurs, including the method of reporting the incident and the medical follow-up that will be made available;

(L) Information on the post-exposure evaluation and follow-up that the employer is required to provide for the employee following an exposure incident;

(M) An explanation of the signs and labels and/or color coding required by paragraph (g)(1); and

(N) An opportunity for interactive questions and answers with the person conducting the training session.

(viii) The person conducting the training shall be knowledgeable in the subject matter covered by the elements contained in the training program as it relates to the workplace that the training will address.

(ix) Additional Initial Training for Employees in HIV and HBV Laboratories and Production Facilities. Employees in HIV or HBV research laboratories and HIV or HBV production facilities shall receive the following initial training in addition to the above training requirements.

(A) The employer shall assure that employees demonstrate proficiency in standard microbiological practices and techniques and in the practices and operations specific to the facility before being allowed to work with HIV or HBV.

(B) The employer shall assure that employees have prior experience in the handling of human pathogens or tissue cultures before working with HIV or HBV.

(C) The employer shall provide a training program to employees who have no prior experience in handling human pathogens. Initial work activities shall not include the handling of infectious agents. A progression of work activities shall be assigned as techniques are learned and proficiency is developed. The employer shall assure that employees participate in work activities involving infectious agents only after proficiency has been demonstrated.

(h) *Recordkeeping—(1) Medical Records.* (i) The employer shall establish and maintain an accurate record for each employee with occupational exposure, in accordance with 29 CFR 1910.20.

(ii) This record shall include:

(A) The name and social security number of the employee;

(B) A copy of the employee's hepatitis B vaccination status including the dates of all the hepatitis B vaccinations and any medical records relative to the employee's ability to receive vaccination as required by paragraph (f)(2);

(C) A copy of all results of examinations, medical testing, and follow-up procedures as required by paragraph (f)(3);

(D) The employer's copy of the healthcare professional's written opinion as required by paragraph (f)(5); and

(E) A copy of the information provided to the healthcare professional as required by paragraphs (f)(4)(ii)(B)(C) and (D).

(iii) Confidentiality. The employer shall ensure that employee medical records required by paragraph (h)(1) are·

(A) Kept confidential; and

(B) Are not disclosed or reported without the employee's express written consent to any person within or outside the workplace except as required by this section or as may be required by law.

(iv) The employer shall maintain the records required by paragraph (h) for at least the duration of employment plus 30 years in accordance with 29 CFR 1910.20.

(2) *Training Records. (i) Training records shall include the following information:*

(A) The dates of the training sessions;

(B) The contents or a summary of the training sessions;

(C) The names and qualifications of persons conducting the training; and

(D) The names and job titles of all persons attending the training sessions.

(ii) Training records shall be maintained for 3 years from the date on which the training occurred.

(3) *Availability.* (i) The employer shall ensure that all records required to be maintained by this section shall be made available upon request to the Assistant Secretary and the Director for examination and copying.

(ii) Employee training records required by this paragraph shall be provided upon request for examination and copying to employees, to employee representatives, to the Director, and to the Assistant Secretary in accordance with 29 CFR 1910.20.

(iii) Employee medical records required by this paragraph shall be provided upon request for examination and copying to the subject employee, to anyone having written consent of the subject employee, to the Director, and to the Assistant Secretary in accordance with 29 CFR 1910.20.

(4) *Transfer of Records.* (i) The employer shall comply with the requirements involving transfer of records set forth in 29 CFR 1910.20(h).

(ii) If the employer ceases to do business and there is no successor employer to receive and retain the records for the prescribed period, the employer shall notify the Director, at least three months prior to their disposal and transmit them to the Director, if required by the Director to do so, within that three month period.

(i) *Dates—(1) Effective Date.* The standard shall become effective on March 6, 1992.

(2) The Exposure Control Plan required by paragraph (c)(2) of this section shall be completed on or before May 5, 1992.

(3) Paragraph (g)(2) Information and Training and (h) Recordkeeping shall take effect on or before June 4, 1992.

(4) Paragraphs (d)(2) Engineering and Work Practice Controls, (d)(3) Personal Protective Equipment, (d)(4) Housekeeping, (e) HIV and HBV Research Laboratories and Production Facilities, (f) Hepatitis B Vaccination and Post-Exposure Evaluation and Follow-up, and (g) (1) Labels and Signs, shall take effect July 6, 1992.

Appendix A to Section 1910.1030—Hepatitis B Vaccine Declination (Mandatory)

I understand that due to my occupational exposure to blood or other potentially infectious materials I may be at risk of acquiring hepatitis B virus (HBV) infection. I have been given the opportunity to be vaccinated with hepatitis B vaccine, at no charge to myself. However, I decline hepatitis B vaccination at this time. I understand that by declining this vaccine, I continue to be at risk of acquiring hepatitis B, a serious disease. If in the future I continue to have occupational exposure to blood or other potentially infectious materials and I want to be vaccinated with hepatitis B vaccine, I can receive the vaccination series at no charge to me.

APPENDIX E

Guidelines for Preventing the Transmission of Tuberculosis in Health-Care Settings, with Special Focus on HIV-Related Issues

Samuel W. Dooley, Jr., M.D.
Kenneth G. Castro, M.D.
Mary D. Hutton, B.S.N., M.P.H.
Robert J. Mullan, M.D.
Jacquelyn A. Polder, B.S.N., M.P.H.
Dixie E. Snider, Jr., M.D., M.P.H.

Summary

The transmission of tuberculosis is a recognized risk in health-care settings. Several recent outbreaks of tuberculosis in health-care settings, including outbreaks involving multidrug-resistant strains of Mycobacterium tuberculosis, *have heightened concern about nosocomial transmission. In addition, increases in tuberculosis cases in many areas are related to the high risk of tuberculosis among persons infected with the human immunodeficiency virus (HIV). Transmission of tuberculosis to persons with HIV infection is of particular concern*

This document was prepared in consultation with experts in tuberculosis, acquired immunodeficiency syndrome, infection-control and hospital epidemiology, microbiology, ventilation and industrial hygiene, respiratory therapy, nursing, and emergency medical services.

SOURCE: The contents of Appendix E are reproduced from *Morbidity and Mortality Weekly Report*, Vol. 39, No. RR-17, December 7, 1990, pp. 1-29. Published by Centers for Disease Control, U.S. Department of Health and Human Services, Atlanta, GA.

because they are at high risk of developing active tuberculosis if infected. Health-care workers should be particularly alert to the need for preventing tuberculosis transmission in settings in which persons with HIV infection receive care, especially settings in which cough-inducing procedures (e.g., sputum induction and aerosolized pentamidine [AP] treatments) are being performed.

Transmission is most likely to occur from patients with unrecognized pulmonary or laryngeal tuberculosis who are not on effective antituberculosis therapy and have not been placed in tuberculosis (acid-fast bacilli [AFB]) isolation. Health-care facilities in which persons at high risk for tuberculosis work or receive care should periodically review their tuberculosis policies and procedures, and determine the actions necessary to minimize the risk of tuberculosis transmission in their particular settings.

The prevention of tuberculosis transmission in health-care settings requires that all of the following basic approaches be used: a) prevention of the generation of infectious airborne particles (droplet nuclei) by early identification and treatment of persons with tuberculous infection and active tuberculosis, b) prevention of the spread of infectious droplet nuclei into the general air circulation by applying source-control methods, c) reduction of the number of infectious droplet nuclei in air contaminated with them, and d) surveillance of health-care-facility personnel for tuberculosis and tuberculous infection. Experience has shown that when inadequate attention is given to any of these approaches, the probability of tuberculosis transmission is increased.

Specific actions to reduce the risk of tuberculosis transmission should include a) screening patients for active tuberculosis and tuberculous infection, b) providing rapid diagnostic services, c) prescribing appropriate curative and preventive therapy, d) maintaining physical measures to reduce microbial contamination of the air, e) providing isolation rooms for persons with, or suspected of having, infectious tuberculosis, f) screening health-care-facility personnel for tuberculous infection and tuberculosis, and g) promptly investigating and controlling outbreaks.

Although completely eliminating the risk of tuberculosis transmission in all health-care settings may be impossible, adhering to these guidelines should minimize the risk to persons in these settings.

I. INTRODUCTION

A. Purpose of Document

The purpose of this document is to review the mode and risk of tuberculosis transmission in health-care settings and to make recommendations for reducing the risk of transmission to persons in health-care settings—including workers, patients, volunteers, and visitors. The document may also serve as a useful resource for educating health-care workers about tuberculosis. Several outbreaks of tuberculosis in health-care settings, including outbreaks involving multidrug-resistant strains of *M. tuberculosis*, have been reported to CDC during the past 2 years (*1*; CDC, unpublished data). In addition, CDC has recently received numerous requests for information about reducing tuberculosis transmission in health-care settings. Much of the increased concern is due to the occurrence of tuberculosis among persons infected with HIV (*2*), who are at increased risk of contracting tuberculosis both from reactivation of a latent tuberculous infection (*3*) and from a new infection (*4*).

Therefore, in this document, emphasis is given to the transmission of tuberculosis among persons with HIV infection, although the majority of patients with tuberculosis in most areas of the country do not have HIV infection.

These recommendations consolidate and update previously published CDC recommendations (*5-10*). The recommendations are applicable to all settings in which health care is provided. In this document, the term "tuberculosis," in the absence of modifiers, refers to a clinically apparent active disease process caused by *M. tuberculosis* (or, rarely, *M. Bovis* or *M. africanum*). The terms "health-care-facility personnel" and "health-care-facility workers" refer to all persons working in a health-care setting—including physicians, nurses, aides, and persons not directly involved in patient care (e.g., dietary, housekeeping, maintenance, clerical, and janitorial staff, and volunteers).

B. Epidemiology, Transmission, and Pathogenesis of Tuberculosis

Tuberculosis is not evenly distributed throughout all segments of the population of the United States. Groups known to have a high incidence of tuberculosis include blacks, Asians and Pacific Islanders, American Indians and Alaskan Natives, Hispanics, current or past prison inmates, alcoholics, intravenous (IV) drug users, the elderly, foreign-born persons from areas of the world with a high prevalence of tuberculosis (e.g., Asia, Africa, the Caribbean, and Latin America), and persons living in the same household as members of these groups (*5*).

M. tuberculosis is carried in airborne particles, known as droplet nuclei, that can be generated when persons with pulmonary or laryngeal tuberculosis sneeze, cough, speak, or sing (*11*). The particles are so small (1-5 microns) that normal air currents keep them airborne and can spread them throughout a room or building (*12*). Infection occurs when a susceptible person inhales droplet nuclei containing *M. tuberculosis*, and bacilli become established in the alveoli of the lungs and spread throughout the body. Two to ten weeks after initial human infection with *M. tuberculosis*, the immune response usually limits further multiplication and spread of the tuberculosis bacilli. For a small proportion of newly infected persons (usually <1%), initial infection rapidly progresses to clinical illness. However, for another group (approximately 5%-10%), illness develops after an interval of months, years, or decades, when the bacteria begin to replicate and produce disease (*11*). The risk of progression to active disease is markedly increased for persons with HIV infection (*3*).

The probability that a susceptible person will become infected depends upon the concentration of infectious droplet nuclei in the air. Patient factors that enhance transmission are discussed more fully in section II.B.3. Environmental factors that enhance transmission include a) contact between susceptible persons and an infectious patient in relatively small, enclosed spaces, b) inadequate ventilation that results in insufficient dilution or removal of infectious droplet nuclei, and c) recirculation of air containing infectious droplet nuclei.

Tuberculosis transmission is a recognized risk in health-care settings (*13-21*). The magnitude of the risk varies considerably by type of health-care setting, patient population served, job category, and the area of the facility in which a person works. The risk may be higher in areas where patients with tuberculosis are provided care before diagnosis (e.g., clinic waiting areas and emergency rooms) or where diagnostic or treatment procedures that stimulate patient coughing are performed. Nosocomial transmission of tuberculosis has been associated with close contact with infectious patients, as well as procedures such as bronchoscopy (*16*), endotracheal intubation and suctioning with mechanical ventilation (*17,18*), open abscess irrigation (*19*), and autopsy (*20,21*). Sputum induction and aerosol treatments that induce cough may also increase the potential for tuberculosis transmission (*22*). Health-care workers should be particularly alert to the need for preventing tuberculosis transmission in health-care settings in which persons with HIV infection receive care, especially if cough-inducing procedures such as sputum induction and AP treatments are being performed.

II. General Principles of Tuberculosis Control in Health-Care Settings

A. Approaches to Tuberculosis Control

An effective tuberculosis-control program requires the early identification, isolation, and treatment of persons with active tuberculosis. Health-care facilities in which persons at high risk for tuberculosis work or receive care should periodically review their tuberculosis policies and procedures, and determine the actions necessary to minimize the risk of tuberculosis transmission in their particular settings. The prevention of tuberculosis transmission in health-care settings requires that all of the following basic approaches be used: a) preventing the generation of infectious droplet nuclei, b) preventing the spread of infectious droplet nuclei into the general air circulation, c) reducing the number of infectious droplet nuclei in air contaminated with them, d) following guidelines for cleaning, disinfecting, and sterilizing contaminated items, and e) conducting surveillance for tuberculosis transmission to health-care-facility personnel. Experience has shown that when inadequate attention is given to any of these measures, the probability of tuberculosis transmission is increased.

Specific actions to reduce the risk of tuberculosis transmission should include the following:

- Screening patients for active tuberculosis and tuberculous infection.
- Providing rapid diagnostic services.
- Prescribing appropriate curative and preventive therapy.
- Maintaining physical measures to reduce microbial contamination of the air.
- Providing isolation rooms for persons with, or suspected of having, infectious tuberculosis.
- Screening health-care-facility personnel for tuberculous infection and tuberculosis.
- Promptly investigating and controlling outbreaks.

B. Preventing Generation of Infectious Droplet Nuclei

1. Early identification and treatment of persons with tuberculous infection

Early identification of persons with tuberculous infection and application of preventive therapy are effective in preventing the development of tuberculosis (*5*). Persons at increased risk of tuberculosis (see section I.B.), or for whom the consequences of tuberculosis may be especially severe, should be screened for tuberculous infection to identify those for whom preventive treatment is indicated. The tuberculin skin test is the only method currently available that demonstrates infection with *M. tuberculosis* in the absence of active tuberculosis (*11*).

2. Early identification and treatment of persons with active tuberculosis

An effective means of preventing tuberculosis transmission is preventing the generation of infectious droplet nuclei by persons with infectious tuberculosis. This can be accomplished by early identification, isolation, and treatment of persons with active tuberculosis. Tuberculosis may be more difficult to diagnose among persons with HIV infection; the diagnosis may be overlooked because of an atypical clinical or radiographic presentation and/or the simultaneous occurrence of other pulmonary infections (e.g., *Pneumocystis carinii* pneumonia [PCP]). Among persons with HIV infection, the difficulty in making a diagnosis may be further compounded by impaired responses to tuberculin skin tests (*23,24*), low sensitivity of sputum smears for detecting AFB (*25*), or overgrowth of cultures with *Mycobacterium avium* complex (MAC) among patients with both MAC and *M. tuberculosis* infections (*26*).

A diagnosis of tuberculosis should be considered for any patient with persistent cough or other symptoms compatible with tuberculosis, such as weight loss, anorexia, or fever. Diagnostic measures for identifying tuberculosis should be instituted among such patients. These measures include history, physical examination, tuberculin skin test, chest radiograph, and microscopic examination and culture of sputum or other appropriate specimens (*11,27*). Other diagnostic methods, such as bronchoscopy or biopsy, may be indicated in some cases (*28,29*). The probability of tuberculosis is increased by finding a positive reaction to a tuberculin skin test or a history of a positive skin test, a history of previous tuberculosis, membership in a group at high risk for tuberculosis (see section I.B.), or a history of exposure to tuberculosis. Active tuberculosis is strongly suggested if the diagnostic evaluation reveals AFB in sputum, a chest radiograph is suggestive of tuberculosis, or the person has symptoms highly suggestive of tuberculosis (e.g., productive cough, night sweats, anorexia, and weight loss). Tuberculosis may occur simultaneously with other pulmonary infections, such as PCP.

a. Tuberculin skin test. The Mantoux technique (intradermal injection of 0.1 ml of purified protein derivative [PPD] containing 5 tuberculin units [TU]) should be used as a diagnostic aid to detect tuberculous infection. Although tuberculin skin tests are $<$100% sensitive and specific for detection of infection with *M. tuberculosis*, no better diagnostic method has been devised. Tuberculin skin tests should be interpreted according to current guidelines (*5,11*). For persons with HIV infection, a reaction of $\geq$5 mm is considered positive.

A negative skin test does not rule out tuberculosis disease or infection. Because of the possibility of a false-negative result, *the tuberculin skin test should never be used to exclude the possibility of active tuberculosis among persons for whom the diagnosis is being considered, even if reactions to other skin-test antigens are positive.* Persons with HIV infection are more likely to have false-negative skin tests than are persons without HIV infection (*23,24,30*). The likelihood of a false-negative skin test increases as the stage of HIV infection advances (CDC/Florida Department of Health and Rehabilitative Services/New York City Department of Health, unpublished data). For this reason, a history of a positive tuberculin reaction is meaningful, even if the current skin-test result is negative.

b. Chest radiograph. The radiographic presentation of pulmonary tuberculosis among patients with HIV infection may be unusual (*31*). Typical apical cavitary disease is less common among persons with HIV infection. They may have infiltrates in any lung zone, often associated with mediastinal and/or hilar adenopathy, or they may have a normal chest radiograph.

c. Bacteriology. Smear and culture examination of three to five sputum specimens collected on different days is the main diagnostic procedure for pulmonary tuberculosis (*11*). Sputum smears that fail to demonstrate AFB do not exclude the diagnosis of tuberculosis. Studies indicate that 50%-80% of patients with pulmonary tuberculosis have positive sputum smears. Sputum smears from patients with HIV infection and pulmonary tuberculosis may be less likely to reveal AFB than those from immunocompetent patients, a finding believed to be consistent with the lower frequency of cavitary pulmonary disease observed among HIV-infected persons (*23,25*).

A positive sputum culture, with organisms identified as *M. tuberculosis*, provides a definitive diagnosis of tuberculosis. Conventional laboratory methods may require 4-8 weeks for species identification; however, the use of radiometric culture techniques and genetic probes facilitates more rapid detection and identification of mycobacteria (*32,33*). Mixed mycobacterial infection (either simultaneous or sequential) may occur and may obscure the recognition of *M. tuberculosis* clinically and in the laboratory (*26*). The use of genetic probes for both MAC and *M. tuberculosis* may be useful for identifying mixed mycobacterial infections in clinical specimens.

3. Determining infectiousness of tuberculosis patients

The infectiousness of a person with tuberculosis correlates with the number of organisms that are expelled into the air, which, in turn, correlates with the following factors: a) anatomic site of disease, b) presence of cough or other forceful expirational maneuvers, c) presence of AFB in the sputum smear, d) willingness or ability of the patient to cover his or her mouth when coughing, e) presence of cavitation on chest radiograph, f) length of time the patient has been on adequate chemotherapy, g) duration of symptoms, and h) administration of procedures that can enhance coughing (e.g., sputum induction).

The most infectious persons are those with pulmonary or laryngeal tuberculosis. Those with extrapulmonary tuberculosis are usually not infectious, with the following exceptions: a) nonpulmonary disease located in the respiratory tract or oral cavity, or b) extrapulmonary disease that includes an open abscess or lesion in which the concentration of organisms is high, especially if drainage from the abscess or lesion is extensive (*19*). Although the data are limited, findings suggest that tuberculosis patients with acquired immunodeficiency syndrome (AIDS), if smear positive, have infectiousness similar to that of tuberculosis patients without AIDS (CDC/New York City Department of Health, unpublished data).

Infectiousness is greatest among patients who have a productive cough, pulmonary cavitation on chest radiograph, and AFB on sputum smear (*6*). Infection is more likely to result from exposure to a person who has unsuspected pulmonary tuberculosis and who is not receiving antituberculosis therapy or from a person with diagnosed tuberculosis who is not receiving adequate therapy, because of patient noncompliance or the presence of drug-resistant organisms. Administering effective antituberculosis medications has been shown to be strongly associated with a decrease in infectiousness among persons with tuberculosis (*34*). Effective chemotherapy reduces coughing, the amount of sputum, and the number of organisms in the sputum. However, the length of time a patient must be on effective medication before becoming noninfectious varies (*35*); some patients are never infectious, whereas those with unrecognized or inadequately treated drug-resistant disease may remain infectious for weeks or months. Thus, decisions about terminating isolation precautions should be made on a case-by-case basis.

In general, persons suspected of having active tuberculosis and persons with confirmed tuberculosis should be considered infectious if cough is present, if cough-inducing procedures are performed, or if sputum smears are known to contain AFB, and if these patients are not on chemotherapy, have just started chemotherapy, or have a poor clinical or bacteriologic response to chemotherapy. A person with tuberculosis who has been on adequate chemotherapy for at least 2-3 weeks and has had a definite clinical and bacteriologic response to therapy (reduction in cough, resolution of fever, and progressively decreasing quantity of bacilli on smear) is probably no longer infectious. Most tuberculosis experts agree that noninfectiousness in pulmonary tuberculosis can be established by finding sputum free of bacilli by smear examination on three consecutive days for a patient on effective chemotherapy. Even after isolation precautions have been discontinued, caution should be exercised when a patient with tuberculosis is placed in a room with another patient, especially if the other patient is immunocompromised.

C. Preventing Spread of Infectious Droplet Nuclei via Source-Control Methods

In high-risk settings, certain techniques can be applied to prevent or to reduce the spread of infectious droplet nuclei into the general air circulation. The application of these techniques, which are called source-control methods because they entrap infectious droplet nuclei as they are emitted by the patient, or "source" (*36*), is especially important during performance of medical procedures likely to generate aerosols containing infectious particles.

1. Local exhaust ventilation

Local exhaust ventilation is a source-control technique that removes airborne contaminants at or near their sources (*37*). The use of booths for sputum induction or administration of aerosolized medications (e.g., AP) is an example of local exhaust ventilation for preventing the spread of infectious droplet nuclei generated by these procedures into the general air circulation. Booths used for source control should be equipped with exhaust fans that remove nearly 100% of airborne particles during the time interval between the departure of one patient and the arrival of the next. The time required for removing a given percentage of airborne particles from an enclosed space depends upon the number of air exchanges per hour (Table 1), which is determined by the capacity of the exhaust fan in cubic feet per minute (cfm), the number of cubic feet of air in the room or booth, and the rate at which air is entering the room or booth at the intake source.

The exhaust fan should maintain negative pressure in the booth with respect to adjacent areas, so that air flows into the booth. Maintaining negative pressure in the booth minimizes the possibility that infectious droplet nuclei in the booth will move into adjacent rooms or hallways. Ideally, the air from these booths should be exhausted directly to the outside of the building (away from air-intake vents, people, and animals, in accordance with federal, state, and local regulations concerning environmental discharges). If direct

TABLE 1. Air changes per hour and time in minutes required for removal efficiencies of 90%, 99%, or 99.9% of airborne contaminants*

Air changes per hour	Minutes required for a removal efficiency of		
	90%	99%	99.9%
1	138	276	414
2	69	138	207
3	46	92	138
4	35	69	104
5	28	55	83
6	23	46	69
7	20	39	59
8	17	35	52
9	15	31	46
10	14	28	41
11	13	25	38
12	12	23	35
13	11	21	32
14	10	20	30
15	9	18	28
16	9	17	26
17	8	16	24
18	8	15	23
19	7	15	22
20	7	14	21
25	6	11	17
30	5	9	14
35	4	8	12
40	3	7	10
45	3	6	9
50	3	6	8

*Table prepared according to the formula $t_2 = (\ln (C_2/C_1)/-(Q/V)) \cdot 60$, which is an adaptation of formula for rate of purging airborne contaminants, with $t_1 = 0$ and assuming perfect mixing of the air in the space (*69*). $C_2/C_1 = 1 - (\text{removal efficiency}/100)$.

exhaust to the outside is impossible, the air from the booth could be exhausted through a properly designed, installed, and maintained high-efficiency particulate air (HEPA) filter; however, the efficacy of this method has not been demonstrated in clinical settings (see section II.D.2.a.).

2. Other source-control methods

A simple but important source-control technique is for infectious patients to cover all coughs and sneezes with a tissue, thus containing most liquid drops and droplets before evaporation can occur (*38*). A patient's use of a properly fitted surgical mask or disposable, valveless particulate respirator (PR) (see section II.D.2.c.) also may reduce the spread of infectious particles. However, since the device would need to be worn constantly for the protection of others, it would be practical in only very limited circumstances (e.g., when a patient is being transported within a medical facility or between facilities).

D. Reducing Microbial Contamination of Air

Once infectious droplet nuclei have been released into room air, they should be eliminated or reduced in number by ventilation, which may be supplemented by additional measures (e.g., trapping organisms by high-efficiency filtration or killing organisms with germicidal ultraviolet [UV] irradiation [100-290 nanometers]). Health-care-facility workers may also reduce the risk of inhaling contaminated air by using PRs.

Although for the past 2-3 decades ventilation and, to a lesser extent, UV lamps and face masks have been used in health-care settings to prevent tuberculosis transmission, few published data exist on which to evaluate their effectiveness and liabilities or to draw conclusions about the role each method should play. From a theoretical standpoint, none of the four methods (ventilation, UV irradiation, high-efficiency filtration, and face masks) appears to be ideal. None of the methods used alone or in combination can completely eliminate the risk of tuberculosis transmission; however, when used with the other infection-control measures outlined in this document, they can substantially reduce the risk.

1. General ventilation

Ventilation standards for indoor air quality have been published by the American Society of Heating, Refrigerating, and Air Conditioning Engineers, Inc. (ASHRAE) (*39*). Specific recommendations for health-care facilities have been published by ASHRAE (*40*) and by the Federal Health Resources and Services Administration (*41*). Meeting these standards should reduce the probability of tuberculosis transmission in clinical settings; however, some highly infectious patients may transmit infection even if these ventilation standards are met.

a. Dilution and removal of airborne contaminants. Appropriate ventilation maintains air quality by two processes—dilution and removal of airborne contaminants (*42*). Dilution reduces the concentration of contaminants in a room by introducing air that does not contain those contaminants into the room. Air is then removed from the room by exhaust directly to the outside or by recirculation into the general ventilation system of the building. Continuously recirculating air in a room or in a building may result in the accumulation or concentration of infectious droplet nuclei. Air that is likely contaminated with infectious droplet nuclei should be exhausted to the outside, away from intake vents, people, and animals, in accordance with federal, state, and local regulations for environmental discharges.

b. Air mixing. Proper ventilation requires that within-room mixing of air (ventilation efficiency) must be adequate (*42*). Air mixing is enhanced by locating air-supply outlets at ceiling level and exhaust inlets near the floor, thus providing downward movement of clean air through the breathing zone to the floor area for exhaust.

c. Direction of air flow. For control of tuberculosis transmission, the direction of air flow is as important as dilution. The direction of air flow is determined by the differences in air pressure between adjacent areas, with air flowing from higher pressure areas to lower pressure areas.

In an area occupied by a patient with infectious tuberculosis, air should flow into the potentially contaminated area (the patient's room) from adjacent areas. The patient's room is said to be under lower or negative pressure.

Proper air flow and pressure differentials between areas of a health-care facility are difficult to control because of open doors, movement of patients and staff, temperature, and the effect of vertical openings (e.g., stairwells and elevator shafts) (*40*). Air-pressure differentials can best be maintained in completely closed rooms. An open door between two areas may reduce any existing pressure differential and could reduce or eliminate the desired effect. Therefore, doors should remain closed, and the close fit of all doors and other closures of openings between pressurized areas should be maintained. For critical areas in which the direction of air flow must be maintained while allowing for patient or staff movement between adjacent areas, an appropriately pressurized anteroom may be indicated.

Examples of factors that can change the direction of air flow include the following: a) dust in exhaust fans, filters, or ducts, b) malfunctioning fans, c) adjustments made to the ventilation system elsewhere in the building, or d) or automatic shut down of outside air introduction during cold weather. In areas where the direction of air flow is important, trained personnel should monitor air flow frequently to ensure that appropriate conditions are maintained.

Each area to which an infectious tuberculosis patient might be admitted should be evaluated for its potential for the spread of tuberculosis bacilli. Modifications to the ventilation system, if needed, should be made by a qualified ventilation engineer. Individual evaluations should address factors such as the risk of tuberculosis among the patient population served, special procedures that may be performed, and ability to make the necessary changes.

Too much ventilation in an area can create problems. In addition to incurring additional expense at marginal benefits, occupants bothered by the drafts may elect to shut down the system entirely. Furthermore, if the concentration of infectious droplet nuclei in an area is high, the levels of ventilation that are practical to achieve may be inadequate to completely remove the contaminants (*43*).

2. Potential supplemental approaches

a. HEPA filtration. For general-use areas (e.g., emergency rooms and waiting areas) of health-care facilities, recirculating the air is an alternative to using large percentages of outside air for general ventilation. If air is recirculated, care must be taken to ensure that infection is not transmitted in the process. Although they can be expensive, HEPA filters, which remove at least 99.97% of particles >0.3 microns in diameter, have been shown to be effective in clearing the air of *Aspergillus* spores, which are in the size range of 1.5-6 microns (*44-46*). The ability of HEPA filters to remove tuberculosis bacilli from the air has not been studied, but tuberculosis-containing droplet nuclei are approximately 1-5 microns in diameter, about the same size as *Aspergillus* spores; therefore, HEPA filters theoretically should remove infectious droplet nuclei. HEPA filters may be used in general-use areas, but should not be used to recirculate air from a tuberculosis isolation room back into the general circulation.

Applications in preventing nosocomial *Aspergillus* infection have included using HEPA filters in centralized air-handling units and using whole-wall HEPA filtration units with laminar air flow in patient rooms. In addition, portable HEPA filtration units, which filter the air in a room rather than filtering incoming air, have been effective in reducing nosocomial *Aspergillus* infections (*45,46*). Such units have been used as an interim solution for retrofitting old areas of hospitals. Although these units should not be substituted for other accepted tuberculosis isolation procedures, they may be useful in general-use areas (e.g., waiting rooms and emergency rooms) where an increased risk of exposure to tuberculosis may exist, but where other methods of air control may be inadequate.

When HEPA filters are to be installed at a facility, qualified personnel must assess and design the air-handling system to assure adequate supply and exhaust capacity. Proper installation, testing, and meticulous maintenance are critical if a HEPA filter system is used (*40*). Improper design, installation, or maintenance could permit infectious particles to circumvent filtration and escape into the ventilation (*42*). The filters should be installed to prevent leakage between filter segments and between the filter bed and its frame. A regular maintenance program is required to monitor HEPA filters for possible leakage and for filter loading. A manometer should be installed in the filter system to provide an accurate means of objectively determining the need for filter replacement. Installation should allow for maintenance without contaminating the delivery system or the area served.

HEPA-filtered, recirculated air should not be used if the contaminants contain carcinogenic agents. Qualified personnel should maintain, decontaminate, and dispose of HEPA filters.

b. Germicidal UV irradiation. The use of germicidal UV lamps (wavelengths 100-290 nm) to prevent tuberculosis transmission in occupied spaces is controversial. UV lamps installed in the exhaust air ducts from the rooms of patients with infectious tuberculosis were shown to prevent infection of guinea pigs, which are highly susceptible to tuberculosis (*34*). On the basis of this finding, other studies (*47-50*), and the experience of tuberculosis clinicians and mycobacteriologists during the past 2-3 decades, CDC has continued to recommend UV lamps (with appropriate safeguards to prevent short-term overexposure) as a supplement to ventilation in settings where the risk of tuberculosis transmission is high (*6,8,11,51-54*). Their efficacy in clinical settings has not been demonstrated under controlled conditions, but there is a theoretical and experiential basis for believing they are effective (*43,55,56*). Thus, individual health-care facilities may need to consider, on a case-by-case basis, using these lamps in settings with a high risk of tuberculosis transmission (see section I.B.). UV lamps are less effective in areas with a relative humidity of $>70\%$ (*57*). The potential for serious adverse effects of short- and long-term exposure to germicidal UV has been identified as a major concern (*58*; NIOSH, unpublished report [Health Hazard Evaluation Report, HETA 90-122-L2073]).

The two most common types of UV installation are wall- or ceiling-mounted room fixtures for disinfecting the air within a room and irradiation units for disinfecting air in supply ducts. Wall- or ceiling-mounted fixtures act by disinfecting upper room air, and their effectiveness depends in part upon the mixing of air in the room. Organisms must be carried by air currents from the lower portion of the room to within the range of the UV radiation from the fixtures. These fixtures are most likely to be effective in locations where ceilings are high, but some protection may be afforded in areas with ceilings as low as 8 feet. To be maximally effective, lamps should be left on day and night (*59*).

Installing UV lamps in ventilation ducts may be beneficial in facilities that recirculate the air. UV exposure of air in ducts can be direct and more intense than that provided by room fixtures and may be effective in disinfecting exhaust air. Duct installations provide no protection against tuberculosis transmission to any person who is in the room with an infectious patient. As with HEPA filters, UV installations in ducts may be used in general-use areas but should not be used to recirculate air from a tuberculosis isolation room back into the general circulation.

The main concern about UV lamps is safety. Short-term overexposure to UV irradiation can cause keratoconjunctivitis and erythema of the skin (*60*). However, with proper installation and maintenance, the risk of short-term overexposure is low. Long-term exposure to UV irradiation is associated with increased risk of basal cell carcinoma of the skin and with cataracts (*58*). To prevent overexposure of health-care-facility personnel and patients, UV lamp configurations should meet applicable safety guidelines (*60*).

When UV lamps are used in air-supply ducts, a warning sign should be placed on doors that permit access to the duct lamps. The sign should indicate that looking at the lamps is a safety hazard. In addition, warning lights outside doors permitting access to duct lamps should indicate whether the lamps are on or off. The duct system should be engineered to prevent UV emissions from the duct from radiating into potentially occupied spaces.

Consultation from a qualified expert should be obtained before and after UV lamps are installed. After installation, the safety and effectiveness of UV irradiation must be checked with a UV meter and fixtures adjusted as necessary. Bulbs should be periodically checked for dust, cleaned as needed, and replaced at the end of the rated life of the bulb. Maintenance personnel should be cautioned that fixtures should be turned off before inspection or servicing. A timing device that turns on a red light at the end of the rated life of the lamp is available to alert maintenance personnel that the lamp needs to be replaced.

c. Disposable PRs for filtration of inhaled air.

1.) For persons exposed to tuberculosis patients. Appropriate masks, when worn by health-care providers or other persons who must share air space with a patient who has infectious tuberculosis, may provide additional protection against tuberculosis transmission. Standard surgical masks may not be effective in preventing inhalation of droplet nuclei (*61*), because some are not designed to provide a tight face seal and to filter out particulates in the droplet nucleus size range (1-5 microns). A better alternative is the disposable PR. PRs were originally developed for industrial use to protect workers. Although the appearance and comfort of PRs may be similar to that of cup-shaped surgical masks, they provide a better facial fit and better filtration capability. However, the efficacy of PRs in protecting susceptible persons from infection with tuberculosis has not been demonstrated.

PRs may be most beneficial in the following situations: a) when appropriate ventilation is not available and the patient's signs and symptoms suggest a high potential for infectiousness, b) when the patient is potentially infectious and is undergoing a procedure that is likely to produce bursts of aerosolized infectious particles or to result in copious coughing or sputum production, regardless of whether appropriate ventilation is in place, and c) when the patient is potentially infectious, has a productive cough, and is unable or unwilling to cover coughs.

Comfort influences the acceptability of PRs. Generally, the more efficient the PRs, the greater is the work of breathing through them and the greater the perceived discomfort. A proper fit is vital to protect against inhaling droplet nuclei. When gaps are present, air will preferentially flow through the gaps, allowing the PR to function more like a funnel than a filter, thus providing virtually no protection (*61*).

2.) ***For tuberculosis patients.*** Masks or PRs worn by patients with suspected or confirmed tuberculosis may be useful in selected circumstances (see section II.C.2.). PRs used by patients should be valveless. Some PRs have valves to release expired air, and these would not be appropriate for patients to use.

E. Decontamination: Cleaning, Disinfecting, and Sterilizing

Guidelines for cleaning, disinfecting, and sterilizing equipment have been published (*10,62,63*). The rationale for cleaning, disinfecting, or sterilizing patient-care equipment can be understood more readily if medical devices, equipment, and surgical materials are divided into three general categories (critical items, semi-critical items, and noncritical items) based on the potential risk of infection involved in their use.

Critical items are instruments such as needles, surgical instruments, cardiac catheters, or implants that are introduced directly into the bloodstream or into other normally sterile areas of the body. These items should be sterile at the time of use.

Semi-critical items are items such as noninvasive flexible and rigid fiberoptic endoscopes or bronchoscopes, endotracheal tubes, or anesthesia breathing circuits that may come in contact with mucous membranes but do not ordinarily penetrate body surfaces. Although sterilization is preferred for these instruments, a high-level disinfection procedure that destroys vegetative microorganisms, most fungal spores, tubercle bacilli, and small, nonlipid viruses may be used. Meticulous physical cleaning before sterilization or high-level disinfection is essential.

Noncritical items are those that either do not ordinarily touch the patient or touch only intact skin. Such items include crutches, bedboards, blood pressure cuffs, and various other medical accessories. These items do not transmit tuberculous infection. Consequently, washing with a detergent is usually sufficient.

Facility policies should identify whether cleaning, disinfecting, or sterilizing an item is indicated to decrease the risk of infection. Procedures for each item depend on its intended use. Generally, critical items should be sterilized, semi-critical items should be sterilized or cleaned with high-level disinfectants, and noncritical items need only be cleaned with detergents or low-level disinfectants. Decisions about decontamination processes should be based on the intended use of the item and not on the diagnosis of the patient for whom the item was used. Selection of chemical disinfectants depends on the intended use, the level of disinfection required, and the structure and material of the item to be disinfected.

Although microorganisms are normally found on walls, floors, and other surfaces, these environmental surfaces are rarely associated with transmission of infections to patients or health-care-facility personnel. This is particularly true with organisms such as tubercle bacilli, which generally require inhalation by the host for infection to occur. Therefore, extraordinary attempts to disinfect or sterilize environmental surfaces are rarely indicated. However, routine cleaning (which can be achieved with a hospital-grade, Environmental Protection Agency-approved germicide/disinfectant) is recommended (*63*). The same routine daily cleaning procedures used in other hospital or facility rooms should be used to clean rooms of patients who are on AFB isolation precautions.

F. Conducting Surveillance for Tuberculosis Transmission to Health-Care-Facility Personnel

A tuberculosis screening and prevention program for health-care-facility personnel should be established for protecting both health-care-facility personnel and patients. Personnel with tuberculous infection without evidence of current (active) disease should be identified, because preventive treatment with isoniazid may be indicated (*5*). In addition, the screening program will enable public health personnel to evaluate the effectiveness of current infection-control practices. Recommendations for screening and surveillance are detailed in section III.A.7.

III. Recommendations

The following recommendations are divided into two categories: a) general recommendations applicable to all health-care settings, including special precautions for cough-inducing procedures, and b) recommendations for selected specific health-care settings. Facilities should adapt these recommendations as appropriate for individual circumstances.

A. Recommendations Applicable to All Health-Care Settings

1. Early identification and preventive treatment of persons who have tuberculous infection and are at high risk for active tuberculosis

- Persons belonging to groups at risk for tuberculosis (see section I.B.) should be screened with a Mantoux tuberculin skin test. Those with positive skin tests should be evaluated for preventive therapy according to current guidelines (*5*).
- All persons with HIV infection or with risk factors for HIV infection should be given a Mantoux tuberculin skin test. Those with positive skin tests or histories of positive skin tests, for whom diagnostic evaluation for active tuberculosis is negative, should be evaluated for preventive therapy according to current guidelines (*5*).

2. Early identification and treatment of persons with active tuberculosis

- Vigorous efforts should be made to identify patients with active tuberculosis in a timely manner and to place them on appropriate therapy (see section II.B.2.). Pulmonary tuberculosis should always be included in the differential diagnosis of persons with pulmonary signs or symptoms, and appropriate diagnostic measures should be instituted.
- For patients with pulmonary signs or symptoms that are initially ascribed to other etiologies, evaluation for co-existing tuberculosis should be repeated if the patient does not respond to appropriate therapy for the presumed etiology of the pulmonary abnormalities (see section II.B.2.).
- In health-care facilities, isolation precautions should be applied for patients who are suspected or confirmed to have active tuberculosis and who may be infectious (see sections II.B.3 and III.B.1.a.).
- Procedure-specific precautions should be applied for cough-inducing or aerosol-generating procedures (see section III.A.5.).

- Patients with suspected or confirmed tuberculosis should be reported to the appropriate health department so that standard procedures for identifying and evaluating tuberculosis contacts can be initiated.

3. Ventilation

- Staff of inpatient facilities should either include an engineer or other professional with expertise in ventilation or industrial hygiene, or the facility should have this expertise available from a consultant. These persons should work closely with the infection-control committee in the control of airborne infections.
- Ventilation for health-care facilities should be developed and maintained in consultation with experts in ventilation engineering who also have hospital ventilation experience. Facility design should meet local and state requirements. Specific recommendations for health-care facilities have been published by ASHRAE and HRSA (*40,41*) (see section II.D.).
- The direction of air flow in health-care facilities should be set up and maintained so that air flows from clean areas to less-clean areas. In areas of a facility in which tuberculosis transmission is a potential problem, direction of air flow should be monitored frequently. Periodic checks with smoke tubes or smoke sticks provide a sensitive indication of air-flow direction (see section II.D.1.c.).
- Facilities serving populations with a high prevalence of tuberculosis may need to enhance ventilation or use supplemental approaches in areas of the facility where patients with tuberculosis are likely to be found (e.g., waiting areas, emergency rooms, radiology suites, or treatment rooms) or where skin tests of personnel demonstrate an increased risk of tuberculosis transmission (see section II.D.2.).

4. Potential supplemental environmental approaches

a. High-efficiency filtration

- If air from potentially contaminated general-use areas (e.g., emergency rooms or clinic waiting areas) cannot be exhausted directly to the outside, HEPA filters with test efficiencies of ⩾99.97% may be useful for removing infectious organisms from air before recirculation in a room or before return to common supply ducts. If HEPA filters are used, they must be designed, installed, maintained, and disposed of in accordance with all applicable regulations and manufacturers' recommendations (see section II.D.2.a.). HEPA filters should not be used to recirculate air from a tuberculosis isolation room back into the general circulation.

b. Germicidal UV irradiation

- For settings in which the risk of tuberculosis transmission is high (see section I.B.), UV lamps have been used to supplement ventilation (see section II.D.2.b.). The decision to use UV lamps should be made on a case-by-case basis. If UV lamps are used, applicable safety guidelines should be followed (see section II.D.2.b.). UV lamps are not recommended for use in small rooms or booths where nebulizing devices will be used. UV units installed in ducts should not be used to recirculate air from a tuberculosis isolation room back into the general circulation.

c. Disposable PRs for filtration of inhaled air

- PRs (see section II.D.2.c.) should be provided by health-care facilities and worn by persons in the same room with a patient whose signs and symptoms suggest a high potential for infectiousness and by those performing procedures that are likely to produce bursts of droplet nuclei, such as bronchoscopy, endotracheal suctioning, and administration of AP.

- Wearers should be adequately trained in the use and disposal of PRs and should carefully follow manufacturers' instructions. Ideally, a respirator program consistent with the guidelines found in Department of Health and Human Services (DHHS), National Institute for Occupational Safety and Health (NIOSH), Publication No. 87-116, *Guide to Industrial Respiratory Protection* (*64*) and the requirements of the Occupational Safety and Health Administration (OSHA) General Industry Occupational Safety and Health Standards (29 Code of Federal Regulations Part 1910.134) should be implemented. Such a program includes training, fit testing, care and maintenance, and medical monitoring.

5. Procedure-specific precautions

a. Diagnostic sputum induction

- Sputum induction performed on patients who may have tuberculosis should be carried out in an individual room or booth with negative pressure relative to adjacent rooms and hallways, ideally with room or booth air exhausted directly to the outside and away from all windows and air intake ducts (see section II.C.1.). Patients should remain in the booth or treatment room (or go outside, weather permitting) and not return to common waiting areas until coughing has subsided. Time should be allowed between patients so that any droplet nuclei that have been introduced into the air can be removed. This time will vary according to the efficiency of the ventilation or filtration used (Table 1, page 26). Health-care-facility personnel collecting induced sputum should wear PRs if it is necessary for them to be in the room with the patient during the procedure (see section II.D.2.c.).

b. Administration of AP

- All patients should be screened for active tuberculosis before AP therapy is initiated. Screening should include medical history, tuberculin skin test, and a baseline chest radiograph (see section II.B.2).

- Before each subsequent AP treatment, patients should be evaluated for symptoms highly suggestive of tuberculosis, such as the development of a productive cough or cough and fever. If such symptoms are elicited, a diagnostic evaluation should be initiated.

- If active tuberculosis is found or suspected, the patient should be placed on antituberculosis chemotherapy. AP treatments should be administered to patients who may have active tuberculosis *only* in a room or booth as described for sputum induction.

- Ideally, AP treatments for all patients should be administered in an individual room or booth as described for sputum induction (see sections II.C.1. and III.A.5.a.). Adequate time should be allowed between patients for removal of residual pentamidine and any infectious organisms from the air when treatment rooms or booths are to be reused (Table 1, page 26).
- Workers administering AP should wear PRs whenever they must be in the room or booth during administration of AP to patients who have, or are at high risk of having, tuberculosis (see section II.D.2.c.).
- After they have received AP, patients should not return to common waiting areas until coughing subsides.

c. Bronchoscopy

- Bronchoscopy should be performed in rooms that have adequate ventilation, good distribution of air flow, and air exhausted directly to the outside—in accordance with federal, state, and local regulations for environmental discharges—or recirculated through HEPA filters. Ideally, bronchoscopy should be performed in rooms with negative pressure relative to adjacent areas. If bronchoscopy must be performed in positive-pressure rooms (such as operating rooms), the risk of infectious tuberculosis should be ruled out beforehand.
- Additional protection may be afforded by local exhaust ventilation employed near the patient's head to exhaust most organisms near their source (see section II.C.1.) or by the use of UV lamps in treatment areas where bronchoscopies are performed (see section II.D.2.b.).
- Persons who must be in the room with the patient during bronchoscopy should wear PRs (see section II.D.2.c.).

d. Endotracheal intubation/suctioning

- Rooms occupied by intubated patients who may have active tuberculosis should be provided with ventilation as described for patient isolation rooms (see section III.B.1.a.). Persons performing endotracheal suctioning on patients who have suspected or confirmed active tuberculosis should wear PRs.

e. Other procedures

- Other aerosol treatments, cough-inducing procedures, or aerosol-generating procedures should be administered as described for AP administration (see section II.C.1.).

6. Decontamination: cleaning, disinfecting, and sterilizing

- Decisions about decontamination processes should be based on the intended use of the item and not on the diagnosis of the patient for whom the item was used (see section II.E.).
- Generally, critical items should be sterilized, semi-critical items should be sterilized or cleaned with high-level disinfectants, and noncritical items need only be cleaned with detergents or low-level disinfectants. Meticulous physical cleaning before sterilization or a high level of disinfection is essential (see section II.E.).

- The same routine, daily cleaning procedures used in other hospital or facility rooms should be used to clean rooms of patients who are on AFB isolation precautions (see section II.E.).

7. Conducting surveillance for tuberculosis transmission

a. Surveillance and reporting

- Health-care facilities providing care to patients at risk for tuberculosis should maintain active surveillance for tuberculosis among patients and health-care-facility personnel and for skin-test conversions among health-care-facility personnel. When tuberculosis is suspected or diagnosed, public health authorities should be notified so that appropriate contact investigation can be performed. Data on the occurrence of tuberculosis and skin-test conversions among patients and health-care-facility personnel should be collected and analyzed to estimate the risk of tuberculosis transmission in the facility and to evaluate the effectiveness of infection-control and screening practices.
- At the time of employment, all health-care facility personnel, including those with a history of Bacillus of Calmette and Guerin (BCG) vaccination, should receive a Mantoux tuberculin skin test unless a previously positive reaction can be documented or completion of adequate preventive therapy or adequate therapy for active disease can be documented.
- Initial and follow-up tuberculin skin tests should be administered and interpreted according to current guidelines (*5,11*).
- Health-care-facility personnel with a documented history of a positive tuberculin test, or adequate treatment for disease or preventive therapy for infection, should be exempt from further screening unless they develop symptoms suggestive of tuberculosis.
- Periodic retesting of PPD-negative health-care workers should be conducted to identify persons whose skin tests convert to positive (*11*). In general, the frequency of repeat testing should be based on the risk of developing new infection. Health-care-facility workers who may be frequently exposed to patients with tuberculosis or who are involved with potentially high-risk procedures (e.g., bronchoscopy, sputum induction, or aerosol treatments given to patients who may have tuberculosis) should be retested at least every 6 months. Health-care-facility personnel in other areas should be retested annually. Data on skin-test conversions should be periodically reviewed so that the risk of acquiring new infection may be estimated for each area of the facility. On the basis of this analysis, the frequency of retesting may be altered accordingly.

b. Evaluation of health-care-facility personnel after unprotected exposure to tuberculosis

- In addition to periodic screening, health-care-facility personnel and patients should be evaluated if they have been exposed to a potentially infectious tuberculosis patient for whom the infection-control procedures outlined in this document have not been taken. Unless a negative skin test has been documented within the preceding 3 months, each exposed health-care-facility worker (except those already known to be positive reactors) should receive a Mantoux tuberculin skin test as soon as possible after exposure and should be managed in the same way as other contacts (*5*). If the initial skin test is negative, the test should be repeated 12 weeks after the exposure ended. Exposed persons with skin-test reactions ≥5 mm or with symptoms suggestive of tuberculosis should receive chest radiographs. Persons with previously known positive skin-test reactions who have been exposed to an infectious patient do not require a repeat skin test or a chest radiograph unless they have symptoms suggestive of tuberculosis.

c. ***Evaluation and management of health-care-facility personnel with positive skin tests or symptoms that may be due to tuberculosis***

- Health-care-facility personnel with positive tuberculin skin tests or with skin-test conversions on repeat testing or after exposure should be clinically evaluated for active tuberculosis (*11*). Persons with symptoms suggestive of tuberculosis should be evaluated regardless of skin-test results. If tuberculosis is diagnosed, appropriate therapy should be instituted according to published guidelines (*65*). Personnel diagnosed with active tuberculosis should be offered counseling and HIV-antibody testing (*27*).

- Health-care-facility personnel who have positive tuberculin skin tests or skin-test conversions but do not have clinical tuberculosis should be evaluated for preventive therapy according to published guidelines (*5,65*). Personnel with positive skin tests should be evaluated for risk of HIV infection. If HIV infection is considered a possibility, counseling and HIV-antibody testing should be strongly encouraged (*27*).

- All persons with a history of tuberculosis or positive tuberculin tests are at risk for contracting tuberculosis in the future. These persons should be reminded periodically that they should promptly report any pulmonary symptoms. If symptoms of tuberculosis should develop, the person should be evaluated immediately.

d. ***Routine and follow-up chest radiographs***

- Routine chest films are not required for asymptomatic, tuberculin-negative health-care-facility personnel. After the initial chest radiograph is taken, personnel with positive skin-test reactions do not need repeat chest radiographs unless symptoms develop that may be due to tuberculosis (*66*).

e. ***Work restrictions***

- Health-care-facility personnel with current pulmonary or laryngeal tuberculosis pose a risk to patients and other personnel while they are infectious; therefore, stringent work restrictions for these persons are necessary. They should be excluded from work until adequate treatment is instituted, cough is resolved, and sputum is free of bacilli on three consecutive smears. Health-care-facility personnel with current tuberculosis at sites other than the lung or larynx usually do not need to be excluded from work if concurrent pulmonary tuberculosis has been ruled out. Personnel who discontinue treatment before the recommended course of therapy has been completed should not be allowed to work until treatment is resumed, an adequate response to therapy is documented, and they have negative sputum spears on three consecutive days.

- Health-care-facility personnel who are otherwise healthy and receiving preventive treatment for tuberculous infection should be allowed to continue usual work activities.
- Health-care facility personnel who cannot take or do not accept or complete a full course of preventive therapy should have their work situations evaluated to determine whether reassignment is indicated. Work restrictions may not be necessary for otherwise healthy persons who do not accept or complete preventive therapy. These persons should be counseled about the risk of contracting disease and should be instructed to seek evaluation promptly if symptoms develop that may be due to tuberculosis, especially if they have contact with high-risk patients (i.e., patients at high risk for severe consequences if they become infected).

f. Consultation

- Consultation on tuberculosis surveillance, screening, and other methods to reduce tuberculosis transmission should be available from state health department tuberculosis-control programs. Facilities are encouraged to use the services of health departments in planning and implementing their surveillance and screening programs.

B. Precautions for Specific Settings

1. Hospitals and other inpatient facilities

a. Tuberculosis (AFB) isolation precautions

- In hospitals and other inpatient facilities, any patient suspected or known to have infectious tuberculosis should be placed in AFB isolation in a private room.
- ASHRAE (*40*) and HRSA (*41*) have published recommendations for ventilation in AFB isolation rooms. These recommendations specify that rooms should have at least six total air changes per hour, including at least two outside air changes per hour, with sufficient within-room air distribution to dilute or remove tuberculosis bacilli from locations where health-care-facility personnel or visitors are likely to be exposed.
- The direction of air flow should be set up and maintained so that air flows into the room from the hallway (negative pressure) to minimize possible spread of tuberculosis bacilli into the general health-care setting.
- The direction of air flow should be monitored while the room is being used for AFB isolation. The use of flutter strips provides a means of constantly observing the direction of air flow. Smoke tubes or smoke sticks are also a quick, simple means of determining the direction of air flow.
- Air from the room should be exhausted directly to the outside of the building and away from intake vents, people, and animals, in accordance with federal, state, and local regulations concerning environmental discharges. Germicidal UV lamps may be considered as a supplement to ventilation to further decrease the number of infectious droplet nuclei in the air (see sections II.D.2.b. and III.A.4.b.).

- Isolation-room doors must be kept closed to maintain control over the direction of air flow.
- Optionally, a separate anteroom may serve as an airlock to minimize the potential for droplet nuclei to spread from the patient's area to adjacent areas. To work effectively, the anteroom must have directional airflow.
- Persons who enter a room in which AFB isolation precautions are in place should wear PRs (see section II.D.2.c.).
- The patient should remain in the isolation room with the door closed and should be instructed to cover nose and mouth with a tissue during coughing and sneezing. If the patient must leave the room (e.g., for a medical procedure that cannot be done at the bedside) while potentially infectious, s/he should wear a properly fitted surgical mask or valveless PR (see section II.C.2.).
- AFB isolation precautions may be discontinued and the patient placed in a private room when s/he is improving clinically, cough has substantially decreased, and the number of organisms on sequential sputum smears is decreasing. Usually, this occurs within 2-3 weeks after tuberculosis medications are begun. Failure to take medications as prescribed and the presence of drug-resistant disease are the two most common reasons for a patient's remaining infectious. When a patient is likely to be infected with drug-resistant organisms, AFB precautions should be applied until the patient is improving and the sputum smear is negative for AFB. Placing a tuberculosis patient in a room with other patients is not advisable, especially immunosuppressed patients, until the sputum smear is free of bacilli on three consecutive days (see section II.B.3.).

b. Transport, radiology, and treatment rooms

- When a patient who may have infectious tuberculosis must be transported outside the AFB isolation room, s/he should wear a properly fitted surgical mask or valveless PR (see section II.C.2.)..
- Ideally, an area in the treatment or radiology department should be specially ventilated for AFB isolation patients. If this is not possible, the patient should be returned to the isolation room as soon as is practical.
- Health-care-facility workers performing procedures on patients with potentially infectious tuberculosis should wear a PR, especially if the procedure itself induces cough (see section II.D.2.c.).
- Treatment rooms in which patients who have undiagnosed pulmonary disease and who are at high risk for active tuberculosis are evaluated should meet the ventilation standards for AFB isolation rooms. ASHRAE recommends that treatment rooms have at least six air changes per hour (*40*).
- Treatment rooms in which cough-inducing procedures are performed should meet the specifications outlined under procedure-specific precautions.

c. Intensive-care units (ICUs)

- ASHRAE recommends that ventilation in ICUs should provide at least six total air changes per hour, including at least two outside air changes per hour (*40*). If air is recirculated in the ICU, it should be passed through properly designed, installed, and maintained HEPA filters before being recirculated.

- Installation of UV lamps might be considered in ICUs in which there is a high risk of tuberculosis transmission (see section I.B.).
- Any ICU patient who may have infectious tuberculosis should be placed in a private room in which ventilation meets the recommendations for AFB isolation.
- Endotracheal suctioning of patients who may have infectious tuberculosis should be carried out as described under procedure-specific precautions (see section III.A.5.d.)
- ICU patients with undiagnosed pulmonary symptoms who may have infectious tuberculosis should have respiratory secretions submitted for AFB smear and culture (see section II.B.2.).

d. Emergency rooms

- Ventilation in emergency rooms, including waiting areas, should be designed and maintained to reduce the risk of tuberculosis transmission, (*39-41*). ASHRAE recommends that emergency room waiting areas have at least 10 air changes per hour (*40*).
- In facilities serving populations with a high incidence of tuberculosis (see section I.B.), germicidal UV lamps and/or HEPA filters in the emergency room may provide additional benefit when used to supplement ventilation (see section II.D.2.).

e. Laboratories

- Laboratories should adhere to previously published recommendations concerning control of tuberculosis transmission (*67*).

f. Autopsy rooms

- ASHRAE recommends that autopsy rooms have ventilation that provides at least 12 total air changes per hour (*40*). In addition, these rooms should have good distribution of air flow in the room, negative pressure with respect to adjacent areas, and room air exhausted directly to the outside of the building. PRs should be worn by personnel performing procedures that may aerosolize infectious particles (e.g., sawing, irrigating).

g. Hospices

- All tuberculosis-control recommendations for inpatient facilities apply to hospices.

h. Nursing homes

- Published recommendations for prevention and control of tuberculosis in nursing homes should be followed (*68*).

i. Correctional facilities

- Published recommendations for prevention and control of tuberculosis in correctional facilities should be followed (*54*). Prison medical facilities should follow the recommendations outlined in this document.

2. Ambulatory-care facilities

- Health-care employers in outpatient settings should be aware of the risk of tuberculosis among their patient population. They should be especially aware of the increased risk among persons who have both HIV infection and tuberculous infection, and they should develop infection-control policies accordingly.
- Persons who have HIV infection or who are otherwise at risk for contracting tuberculosis should receive a tuberculin skin test, and the results should be noted in the patient's medical record. Tuberculosis diagnostic procedures should be initiated if signs and symptoms of tuberculosis develop (see section II.B.2.).
- Ambulatory patients who have pulmonary symptoms of uncertain etiology should be instructed to cover their mouths and noses when coughing or sneezing; they should spend a minimum of time in common waiting areas (see section II.C.2.).
- Personnel who are the first point of contact in facilities serving patients at risk for tuberculosis should be trained to recognize, and bring to the attention of the appropriate person, any patients with symptoms suggestive of tuberculosis (see section II.B.2.), such as a productive cough of >3 weeks' duration, especially when accompanied by other tuberculosis symptoms, such as weight loss, fever, fatigue, and anorexia.
- Ventilation in clinics serving patients who are at high risk for tuberculosis (see section I.B.) should be designed and maintained to reduce the risk of tuberculosis transmission (*39-41*) (see section II.D.). This is particularly important if immunosuppressed patients are treated in the same or a nearby area. In some settings, (see section I.B.), enhanced ventilation or air-disinfection techniques (e.g., HEPA filters or germicidal UV lamps, see sections II.D.2.a. and II.D.2.b.) may be appropriate for common areas such as waiting rooms. Air from clinics serving patients at high risk for tuberculosis should not be recirculated unless it is first passed through an effective high-efficiency filtration system.
- In outpatient settings where cough-inducing procedures are carried out, procedure-specific AFB precautions should be implemented (see sections II.C. and III.A.5.).

3. Emergency medical services

- When emergency-medical-response personnel or others must transport patients with confirmed or suspected active tuberculosis, a mask or valveless PR should be fitted on the patient. If this is not possible, the worker should wear a PR (see sections II.C.2. and II.D.2.c.). If feasible, the rear windows of the vehicle should be kept open and the heating and air conditioning system set on a nonrecirculating cycle.
- Emergency-response personnel should be routinely screened for tuberculosis at regular intervals. They should also be included in the follow-up of contacts of a patient with infectious tuberculosis (see section III.A.7.).

4. Home-health services

- For persons visiting the home of patients with suspected or confirmed infectious tuberculosis, precautions may be necessary to prevent exposure to air containing droplet nuclei until infectiousness has been eliminated by chemotherapy. These precautions include instructing patients to cover coughs and sneezes. The worker should wear a PR when entering the home or the patient's room.
- Respiratory precautions in the home may be discontinued when the patient is improving clinically, cough has decreased, and the number of organisms in the sputum smear is decreasing. Usually this occurs within 2-3 weeks after tuberculosis medications are begun. Failure to take medications as prescribed and the presence of drug-resistant disease are the two most common reasons for a patient's failure to improve clinically. Home health-care personnel can assist in preventing tuberculosis transmission by educating the patient about the importance of taking medications as prescribed (unless adverse effects are seen).
- If immunocompromised persons or young children live in the home with a patient who has infectious pulmonary or laryngeal tuberculosis, temporary relocation should be considered until the patient has negative sputum smears.
- If cough-inducing procedures (such as AP) are performed in the home of a patient who may have infectious tuberculosis, they should be administered in a well-ventilated area away from other household members. Persons who perform these procedures should wear PRs while performing them.
- Home health-care workers should be included in an employer-sponsored tuberculosis screening and prevention program (see section III.A.7.).
- Early identification and treatment of persons with tuberculosis is important. Home health-care personnel and patients who are at risk for contracting active tuberculosis should be reminded periodically of the importance of having pulmonary symptoms evaluated.
- Close contacts of any patient with active infectious tuberculosis should be evaluated for tuberculous infection and managed according to CDC and American Thoracic Society guidelines (*5,65*).

IV. Research Needs

Additional research is needed regarding the airborne transmission of tuberculosis including the following: a) better quantitating the risk of tuberculosis transmission in a variety of health-care settings, b) assessing the acceptability, efficacy, adverse impact, and cost-effectiveness of currently available methods for preventing transmission, and c) developing better methods for preventing transmission. These needs also extend to other infections transmitted by the airborne route. Currently, large numbers of immunosuppressed persons, including patients infected with HIV, are being brought together in health-care settings in which procedures are used that induce the generation of droplet nuclei. Research is needed to fill many of the gaps in current knowledge and to lead to new and better guidelines for protecting patients and personnel in these settings.

V. Glossary of Abbreviations

AFB Acid-fast bacilli—organisms that retain certain stains, even after being washed with acid alcohol. Most are mycobacteria. When seen on a stained smear of sputum or other clinical specimen, a diagnosis of tuberculosis should be considered.

AIDS Acquired immunodeficiency syndrome—an advanced stage of disease caused by infection with the human immunodeficiency virus (HIV). A patient with AIDS is especially susceptible to other infections.

AP Aerosolized pentamidine—drug treatment given to patients with HIV infection to treat or to prevent *Pneumocystis carinii* pneumonia. The drug is put into solution, the solution is aerosolized, and the patient inhales the aerosol.

ASHRAE American Society of Heating, Refrigerating, and Air Conditioning Engineers, Inc.

HEPA High-efficiency particulate air filter.

HIV Human immunodeficiency virus—the virus that causes AIDS.

HRSA Health Resources and Services Administration.

PCP *Pneumocystis carinii* pneumonia—this organism does not cause disease among persons with a normal immune system.

PR A disposable, particulate respirator (respiratory protective device [face mask]) that is designed to filter out particles 1-5 microns in diameter.

Tuberculous infection A condition in which tuberculosis organisms (*M. tuberculosis, M. bovis,* or *M. africanum*) are present in the body, but no active disease is evident.

Tuberculosis transmission Spread of tuberculosis organisms from one person to another, usually through the air.

UV Ultraviolet.

References
1. CDC. Nosocomial transmission of multidrug-resistant tuberculosis to health care workers and HIV-infected patients in an urban hospital—Florida. *MMWR* 1990;39:718-22.
2. Pitchenik AR, Fertel D, Bloch AB. Mycobacterial disease: epidemiology, diagnosis, treatment, and prevention. Clin Chest Med 1988;9:425-41.
3. Selwyn PA, Hartel D, Lewis VA, et al. A prospective study of the risk of tuberculosis among intravenous drug users with human immunodeficiency virus infection. N Engl J Med 1989;320:545-50.
4. Di Perri G, Cruciani M, Danzi MC, et al. Nosocomial epidemic of active tuberculosis among HIV-infected patients. Lancet 1989;23/30:1502-04.
5. CDC. Screening for tuberculosis and tuberculous infection in high-risk populations, and The use of preventive therapy for tuberculous infection in the United States: recommendations of the Advisory Committee for Elimination of Tuberculosis. MMWR 1990;39(no. RR-8).
6. American Thoracic Society, CDC. Control of tuberculosis. Am Rev Respir Dis 1983;128: 336-42.

7. American Thoracic Society, Ad Hoc Committee of the Scientific Assembly on Tuberculosis. Screening for pulmonary tuberculosis in institutions. Am Rev Respir Dis 1977;115:901-6.
8. CDC. Guidelines for prevention of TB transmission in hospitals. Atlanta, Georgia: US Department of Health and Human Services, Public Health Service, 1982; DHHS publication no.(CDC)82-8371.
9. Williams WW. Guideline for infection control in hospital personnel. Infect Control 1983; 4(suppl):326-49.
10. Garner JS, Simmons BP. Guideline for isolation precautions in hospitals. Infect Control 1983;4(suppl):245-325.
11. American Thoracic Society, CDC. Diagnostic standards and classification of tuberculosis. Am Rev Respir Dis 1990;142:725-35.).
12. Wells WF. Aerodynamics of droplet nuclei. In: Airborne contagion and air hygiene. Cambridge: Harvard University Press, 1955:13-9.
13. Barrett-Connor E. The epidemiology of tuberculosis in physicians. JAMA 1979;241:33-8.
14. Brennen C, Muder RR, Muraca PW. Occult endemic tuberculosis in a chronic care facility. Infect Control Hosp Epidemiol 1988;9:548-52.
15. Goldman KP. Tuberculosis in hospital doctors. Tubercle 1988;69:237-40.
16. Catanzaro A. Nosocomial tuberculosis. Am Rev Respir Dis 1982;125:559-62.
17. Ehrenkranz NJ, Kicklighter JL. Tuberculosis outbreak in a general hospital: evidence of airborne spread of infection. Ann Intern Med 1972;77:377-82.
18. Haley CE, McDonald RC, Rossi L, et al. Tuberculosis epidemic among hospital personnel. Infect Control Hosp Epidemiol 1989;10:204-10.
19. Hutton MD, Stead WW, Cauthen GM, et al. Nosocomial transmission of tuberculosis associated with a draining tuberculous abscess. J Infect Dis 1990;161:286-95.
20. Kantor HS, Poblete R, Pusateri SL. Nosocomial transmission of tuberculosis from unsuspected disease. Am J Med 1988;84:833-8.
21. Lundgren R, Norrman E, Asberg I. Tuberculous infection transmitted at autopsy. Tubercle 1987;68:147-50.
22. CDC. Mycobacterium tuberculosis transmission in a health clinic—Florida, 1988. MMWR 1989;38:256-64.
23. Pitchenik AE, Cole C, Russell BW, et al. Tuberculosis, atypical mycobacteriosis, and the acquired immunodeficiency syndrome among Haitian and non-Haitian patients in South Florida. Ann Intern Med 1984;101:641-5.
24. Maayan S, Wormser GP, Hewlett D, et al. Acquired immunodeficiency syndrome (AIDS) in an economically disadvantaged population. Arch Intern Med 1985;145:1607-12.
25. Klein NC, Duncanson FP, Lenox TH III, et al. Use of mycobacterial smears in the diagnosis of pulmonary tuberculosis in AIDS/ARC patients. Chest 1989;95:1190-2.
26. Burnens AP, Vurma-Rapp U. Mixed mycobacterial cultures-occurrence in the clinical laboratory. Zbl Bakt 1989; 271:85-90.
27. CDC. Tuberculosis and human immunodeficiency virus infection: recommendations of the Advisory Committee for the Elimination of Tuberculosis (ACET). MMWR 1989;38:236-8,243-50.
28. Willcox PA, Benator SR, Potgieter PD. Use of flexible fiberoptic bronchoscope in diagnosis of sputum-negative pulmonary tuberculosis. Thorax 1982;37:598-601.
29. Willcox PA, Potgieter PD, Bateman ED, Benator SR. Rapid diagnosis of sputum-negative miliary tuberculosis using the flexible fiberoptic bronchoscope. Thorax 1986;41:681-4.
30. Canessa PA, Fasano L, Lavecchia MA, Torraca A, Schiattone ML. Tuberculin skin test in asymptomatic HIV seropositive carriers [Letter]. Chest 1989;96:1215-6.
31. Pitchenik AE, Rubinson HA. The radiographic appearance of tuberculosis in patients with the acquired immune deficiency syndrome (AIDS) and pre-AIDS. Am Rev Respir Dis 1985; 131:393-6.
32. Kiehn TE, Cammarata R. Laboratory diagnosis of mycobacterial infection in patients with acquired immunodeficiency syndrome. J Clin Microbiol 1986;24:708-11.
33. Crawford JT, Eisenach KD, Bates JH. Diagnosis of tuberculosis: present and future. Semin Resp Infect 1989;4:171-81.
34. Riley RL, Mills CC, O'Grady F, Sultan LU, Wittstadt F, Shivpuri DN. Infectiousness of air from a tuberculosis ward. Amer Rev Respir Dis 1962;85:511-25.
35. Noble RC. Infectiousness of pulmonary tuberculosis after starting chemotherapy: review of the available data on an unresolved question. Am J Infect Control 1981;9:6-10.
36. Woods JE. Cost avoidance and productivity in owning and operating buildings [state of the art review]. Occup Med 1989;4:753-70.
37. American Conference of Governmental Industrial Hygienists. Industrial ventilation: a manual of recommended practice. Lansing, Michigan:ACGIH, 1988.
38. Riley RL. Airborne infection. Am J Med 1974;57:466-75.

39. American Society of Heating, Refrigerating, and Air Conditioning Engineers. Ventilation for acceptable indoor air quality. Atlanta, Georgia: ASHRAE, Inc., 1989 Standard 62-1989.
40. American Society of Heating, Refrigerating, and Air Conditioning Engineers. 1987 ASHRAE handbook: heating, ventilating, and air-conditioning systems and applications. Atlanta, Georgia: American Society of Heating, Refrigerating, and Air Conditioning Engineers, Inc., 1987:23.1-23.12.
41. Health Resources and Services Administration. Guidelines for construction and equipment of hospital and medical facilities. Rockville, Maryland.: US Department of Health & Human Services, Public Health Service, 1984;PHS publication no.(HRSA)84-14500.
42. Woods JE, Rask DR. Heating, ventilation, air-conditioning systems: the engineering approach to methods of control. In: Kundsin RB, ed. Architectural design and indoor microbial pollution. New York: Oxford University Press, 1988:123-53.
43. Riley RL, Nardell EA. Clearing the air: the theory and application of UV air disinfection. Am Rev Respir Dis 1989;139:1286-94.
44. Sherertz RJ, Belani A, Kramer BS, et al. Impact of air filtration on nosocomial *Aspergillus* infections. Am J Med 1987;83:709-18.
45. Rhame FS, Streifel AJ, Kersey JH, McGlave PB. Extrinsic risk factors for pneumonia in the patient at high risk of infection. Am J Med 1984;76:42-52.
46. Opal SM, Asp AA, Cannady PB, Morse PL, Burton LJ, Hammer PG. Efficacy of infection control measures during a nosocomial outbreak of disseminated *Aspergillus* associated with hospital construction. J Infect Dis 1986;153:63-47.
47. Collins FM. Relative susceptibility of acid-fast and non-acid-fast bacteria to ultraviolet light. Appl Microbiol 1971;21:411-13.
48. David HL, Jones WD Jr, Newman CM. Ultraviolet light inactivation and photoreactivation in the mycobacteria. Infect Immun 1971;4:318-19.
49. David HL. Response of mycobacteria to ultraviolet light radiation. Am Rev Respir Dis 1973;108:1175-85.
50. Riley RL, Knight M, Middlebrook G. Ultraviolet susceptibility of BCG and virulent tubercle bacilli. Am Rev Respir Dis 1976;113:413-18.
51. National Tuberculosis and Respiratory Disease Association. Guidelines for the general hospital in the admission and care of tuberculous patients. Am Rev Respir Dis 1969; 99:631-3.
52. CDC. Notes on air hygiene: summary of conference on air disinfection. Arch Environ Health 1971;22:473-4.
53. Schieffelbein CW Jr, Snider DE Jr. Tuberculosis control among homeless populations. Arch Intern Med 1988;148:1843-6.
54. CDC. Prevention and control of tuberculosis in correctional institutions: recommendations of the Advisory Committee for the Elimination of Tuberculosis. *MMWR* 1989;38:313-20,325.
55. Stead WW. Clearing the air: the theory and application of ultraviolet air disinfection [Letter]. Am Rev Respir Dis 1989;140:1832.
56. Macher JM. Ultraviolet radiation and ventilation to help control tuberculosis transmission: guidelines prepared for California Indoor Air Quality Program. Berkeley, CA: Air and Industrial Hygiene Laboratory, 1989.
57. Riley RL, Kaufman JE. Effect of relative humidity on the inactivation of airborne *Serratia marcescens* by ultraviolet radiation. Appl Microbiol 1972;23:1113-20.
58. The biological effects of ultraviolet radiation (with emphasis on the skin). In: Urbach F, ed. Proceedings of the 1st International Conference Sponsored Jointly by the Skin and Cancer Hospital, Temple University Health Sciences Center and the International Society of Biometeorology. Oxford, England: Pergamon Press, 1969.
59. Riley RL. Ultraviolet air disinfection for control of respiratory contagion. In: Kundsin RB, ed. Architectural design and indoor microbial pollution. New York: Oxford University Press, 1988:175-97.
60. National Institute for Occupational Safety and Health. Criteria for a recommended standard . . . occupational exposure to ultraviolet radiation. Washington, DC: National Institute for Occupational Safety and Health, 1972;publication no.(HSM)73-110009.
61. Pippin DJ, Verderame RA, Weber KK. Efficacy of face masks in preventing inhalation of airborne contaminants. J Oral Maxillofac Surg 1987;45:319-23.
62. Rutala WA. APIC guidelines for selection and use of disinfectants. Am J Infect Control 1990;18:99-117.
63. Garner JS, Favero MS. Guideline for handwashing and hospital environmental control. Atlanta, Georgia: US Department of Health and Human Services, Public Health Service, CDC, 1985.

64. NIOSH. Guide to industrial respiratory protection. Cincinnati, Ohio: US Department of Health and Human Services, Public Health Service, Centers for Disease Control, National Institute for Occupational Safety and Health. 1987;DHHS (NIOSH) publication no.87-116.
65. American Thoracic Society, CDC. Treatment of tuberculosis and tuberculosis infection in adults and children, 1986. Am Rev Respir Dis 1986;134:355-63.
66. Barrett-Connor E. The periodic chest roentgenogram for the control of tuberculosis in health care personnel. Am Rev Respir Dis 1980;122:153-5.
67. Strong BE, Kubica GP. Isolation and identification of *Mycobacterium tuberculosis*. Atlanta, Georgia: US Department of Health and Human Services, Public Health Service, CDC, 1981; HHS publication no.(CDC)81-8390.
68. CDC. Prevention and control of tuberculosis in facilities providing long-term care to the elderly. MMWR 1990;39(No. RR-10).
69. Mutchler JE. Principles of ventilation. In: National Institute for Occupational Safety and Health. The industrial environment—its evaluation and control. Washington, DC: National Institute for Occupational Safety and Health, 1973.

INDEX

INDEX

A

B

C

D

E

G

H

I

J

K

L

N

O

P

Q

R

S

T

U